Wideband Wave-Propagating Components for Wireless RF Communications

Edited by Mohammad Alibakhshikenari

Published in London, United Kingdom

Wideband Wave-Propagating Components for Wireless RF Communications
http://dx.doi.org/10.5772/intechopen.1004668
Edited by Mohammad Alibakhshikenari

Contributors
Abdelmounaim Belbachir Kchairi, Ali Zidour, Antonio Pino, Asma Bagheri, Bal Virdee, Borja Imaz-Lueje, Daniel Martinez-de-Rioja, Darío Herraiz, David Herraiz, Eduardo Martinez-de-Rioja, Ernesto Limiti, Faouzi Rahmani, Fengyun Zhang, Goksel Turan, Hari Kishore Kakarla, Héctor Esteban, Ignacio Garcia Zuazola, Jia Li, José Antonio Encinar, Mahdi NoroozOliaei, Manuel Arrebola, Marcos R. Pino, Masoumeh Forgonchi Nojavan, Mohammad Alibakhshikenari, Mohammad Naser-Moghadasi, Mohammad Soruri, Moustapha El Bakkali, Naima Amar Touhami, Nihade Taher, Peiman Parand, Raja Kumari Chilukuri, Seyed Ramin Emadian, Subbarao Kakarla, Sunil Kumar, Tajeddin Elhamadi, Takfarinas Saber, Vicente E. Boria, Yolanda Rodriguez-Vaqueiro, Ángel Belenguer

First published in London, United Kingdom, 2025 by IntechOpen
IntechOpen is the global imprint of INTECHOPEN LIMITED, registered in England and Wales, registration number: 11086078, 167-169 Great Portland Street, London, W1W 5PF, United Kingdom

For EU product safety concerns: IN TECH d.o.o., Prolaz Marije Krucifikse Kozulić 3, 51000 Rijeka, Croatia, info@intechopen.com or visit our website at intechopen.com.

British Library Cataloguing-in-Publication Data
A catalogue record for this book is available from the British Library

Wideband Wave-Propagating Components for Wireless RF Communications
Edited by Mohammad Alibakhshikenari
p. cm.
Print ISBN 978-1-83634-134-5
Online ISBN 978-1-83634-133-8
eBook (PDF) ISBN 978-1-83634-135-2

If disposing of this product, please recycle the paper responsibly.

Meet the editor

Dr. Mohammad Alibakhshikenari received a Ph.D. degree with Excellent Cum Laude and a European label in electronics engineering from the University of Rome "Tor Vergata" (UNITOV) in Italy in 2020. In November 2019, he won a two-year post-doctoral research grant awarded by the UNITOV. Between July 2021 and August 2024, he was with the Universidad Carlos III de Madrid (uc3m), Spain, as a Principal Investigator of the CONEX-Plus and Marie Skłodowska-Curie Actions. In September 2024, he joined UNITOV as a senior researcher. His main research interests are electromagnetic systems, antennas and wave-propagations, metamaterials and metasurfaces, sensors, synthetic aperture radars (SAR), wireless communications, substrate integrated waveguides (SIWs), microwave components, millimeter-waves and terahertz integrated circuits, gap waveguide technology, beamforming matrix.

Contents

Preface

Wireless RF communications are rapidly growing because personal and business human activities are significantly dependent on these networks. One of their main challenges is narrow operational bandwidth, which restricts their performances over a broad range of applications. Hence, wideband wave-propagating components are crucial in building high-performance wireless RF communication networks. Some of these fundamental components are antennas, waveguides, sensors, and radar structures, which are considered the main parts of any RF communication devices, such as mobile phones, laptops, satellite networks, automotive radars, transportation systems, imaging and sensing elements, etc. Therefore, implementing high-performance wave-propagating components is crucial to establishing a human society with modern RF communications facilities.

This book presents comprehensive studies to design and implement various types of wideband wave-propagating components, including advanced antennas, sensors, waveguides, and radar structures with the ability to operate through different frequency bands from microwave to millimeter-wave regions for various wireless RF communication applications. Diverse techniques to overcome possible challenges along with feasible solutions to realize such components have been discovered, studied and applied, which makes the proposed book a suitable reference for scholars and even beginners in this field of research.

Mohammad Alibakhshikenari
Electronics Engineering Department,
University of Rome "Tor Vergata",
Rome, Italy

Chapter 1

Advanced Wideband Antenna Arrays for 5G Millimeter-Wave Spectrum at K- and Ka-Bands

Mohammad Alibakhshikenari, Peiman Parand, Bal Virdee, Ignacio Garcia Zuazola, Sunil Kumar, Ali Zidour, Mohammad Soruri, Takfarinas Saber, Mohammad Naser-Moghadasi and Ernesto Limiti

Abstract

Millimeter-wave (mm-Wave) wireless communication systems play a central role in meeting the demands of next-generation wireless technologies such as 5G. This chapter presents the design and analysis of three advanced antenna arrays optimized for mm-Wave 5G wireless networks operating over K-band (18–27 GHz) and Ka-band (27–40 GHz). These structures feature radiation patches suspended above a common substrate, excited using three methods to achieve optimal performance. The first method utilizes a 50 Ω open-ended microstrip-line balun to slot-line transition for effective energy coupling and excitation. The second method employs a 50Ω-microstrip feed network, ensuring consistent power distribution across array elements. Lastly, metallic-rods passing through the substrate provide direct excitation, facilitating robust mechanical and electrical integration. To address challenges, innovative techniques were implemented. Vertical metallic-vias suppress unwanted surface currents, while the spacing between radiating elements was optimized as multiples of half-wavelength to minimize interference and maximize array performance. Additionally, antenna elements were enclosed with metallic-walls to further enhance isolation and ensure predictable performance. Further strategies such as bowtie-like dipoles were incorporated to provide wider radiation apertures and better impedance matching. Power dividers were used to split input signals into multiple equal-phase outputs, boosting overall gain. Furthermore, the patches were embedded with slots to exhibit metasurface characteristics, enhancing bandwidth and radiation performance while maintaining compactness. The design methodologies and optimizations result in antenna arrays with superior performance metrics. The detailed analysis, innovative techniques, and demonstrated performance of the proposed antenna arrays establish them as strong candidates for practical deployment in advanced 5G mm-Wave networks and beyond.

Keywords: antenna array, millimeter-wave (mm-wave) spectrum, metallic-vias, open-ended microstrip-line balun, bowtie-like dipoles, slot-line transition, metallic-walls, power dividers, metasurface characteristics, 5G communications

1. Introduction

The global rollout of 5G wireless communication systems is rapidly progressing, driven by the growing demand for faster, more reliable, and highly connected communication networks [1]. Unlike previous generations of cellular technology, 5G introduces transformative benefits, such as significantly faster internet speeds and the capacity to connect thousands of devices simultaneously [2–4]. These advancements facilitate revolutionary concepts like the Internet of Things (IoT) and the Internet of Everything (IoE), enabling the development of smart services capable of providing real-time information [5, 6]. These innovations are set to redefine industries, enhance automation, and create opportunities for intelligent and connected environments.

In the initial phase of deployment, 5G systems operate in frequency bands between 700 MHz and 3.8 GHz, which were traditionally used for earlier generations of mobile communication technologies. While these frequency bands provide foundational support for the network, the growing demand for bandwidth—driven by the increasing number of connected devices and rising data consumption—necessitates access to higher frequencies [7]. This demand has led to the exploration of the millimeter-wave (mm-Wave) spectrum, operating between 26 and 42 GHz, which offers significantly larger bandwidth and the potential to support high-capacity networks [8–10].

The mm-Wave spectrum, despite its potential, comes with significant challenges. Transmission in this band is affected by high propagation losses due to atmospheric absorption, free-space path loss, dielectric losses, radiation inefficiencies, and metallic losses in transmission lines [11]. To overcome these limitations, the focus of research has shifted to advanced antenna array technologies that can mitigate these losses while leveraging the wide bandwidth available in the mm-Wave spectrum [12].

The mm-Wave band has garnered widespread attention as a critical enabler for next-generation mobile communication standards beyond 5G [13]. Networks powered by mm-Wave technology are expected to deliver multi-gigabit data rates, exceptional capacity, and ultra-reliable mobile broadband speeds, extending even to suburban and rural areas. These capabilities will bridge the digital divide and support a wide array of applications, including telemedicine, autonomous transportation, smart cities, and more [14]. However, millimeter-wave signals have a limited ability to penetrate solid objects, such as buildings, making them suitable primarily for short-range applications in high-density environments like sports arenas, shopping malls, convention centers, and urban streetscapes where data congestion is prevalent [15].

The rising data traffic associated with modern applications exceeds the capabilities of current 5G systems operating at lower frequencies. This has necessitated a shift to mm-Wave frequencies, which provide access to much larger bandwidths capable of accommodating intense data demands [16]. The sub-20 GHz frequency bands are already heavily congested, leaving higher-frequency bands as the only viable solution. Prominent frequency ranges for 5G mm-Wave communication include 24–28, 37–40, and 64–71 GHz, which provide significant bandwidth for high-speed data transmission [17]. However, utilizing these bands introduces new challenges, including increased free-space path loss and hardware impairments [18, 19].

To address the challenges of path loss and attenuation in mm-Wave communications, high-gain beamforming antenna arrays with wide scanning angles are essential [20]. These arrays enhance beam coverage and improve the efficiency of signal propagation over mm-Wave channels. However, this improvement comes at a cost. Increasing the number of antennas in an array to achieve higher gain also increases the complexity and cost of associated electronics, making mm-Wave antenna systems significantly more expensive than their counterparts operating at lower frequencies [21]. Beam-steering antenna arrays, integral components of massive multiple-input multiple-output (MIMO) systems, are indispensable for expanding system capacity and extending coverage in 5G networks [22]. These arrays provide the ability to steer signals in multiple directions simultaneously, enabling higher signal-to-noise ratios (SNR) and improved channel capacity without requiring additional spectrum resources [23].

Despite these advancements, designing compact and efficient antenna systems remains a significant challenge. The size of antennas is directly related to the operating wavelength [24], and reducing the spacing between adjacent radiating elements in antenna arrays is necessary to achieve compactness [25]. However, this reduction in spacing can lead to increased electromagnetic coupling between radiators, which negatively affects performance [26]. These effects include degradation of terminal impedance, reduced impedance bandwidth, and compromised radiation characteristics [27, 28]. This, in turn, impacts key performance metrics such as output signal-to-interference noise ratio, radiation efficiency, and radar cross-section [29, 30].

The primary challenge in designing mm-Wave and sub-terahertz antenna arrays lies in balancing the need for reduced physical size with the requirement to suppress mutual coupling between neighboring radiators while simultaneously enhancing radiation properties [31]. Achieving this balance requires innovative design approaches that address these competing demands. Recent research has explored several methods to mitigate mutual coupling and enhance array performance, such as optimizing element placement, introducing isolation structures, and leveraging novel materials and geometries. Our research team has investigated and reported various approaches to tackle these challenges, many of which will be discussed in this chapter to provide insights into the design and optimization of mm-Wave antenna arrays [2–4].

2. Millimeter-wave antennas for 5G communications systems

In this section, a comprehensive study on the design process of millimeter-wave (mm-Wave) antennas with high-gain performances for 5G applications has been provided. Three published articles that are operate over various 5G spectrums are discussed, and their capability for diverse applications, such as 5G mobile terminals, 5G cellular communications, and 5G on-demand wireless networks, is elaborated in depth [2–4]. Besides the operating working frequencies and radiation properties, interesting techniques have been applied to keep the dimensions of the antennas as small as possible and enhance the isolation between the radiation elements, which are the crucial factors for the proposed 5G applications. Such methods have been also described in detail.

2.1 Wideband endfire antenna array for 5G mm-wave mobile terminals

According to the Third-Generation Partnership Project (3GPP) technical specification (TS 38.101-2 release 16.7.0), the 5G New Radio (NR) sub-millimeter-wave (sub-mm-Wave) spectrum allocations, classified as frequency range 2 (5G NR FR2), have been clearly defined to support the requirements of next-generation wireless communication [23]. Within this context, the broadband frequency ranges from 24.25 to 29.5 GHz, covering a span of 19.5%, and is identified as a key 5G mm-Wave band.

This frequency range is crucial for achieving the high data rates, low latency, and large capacity demanded by modern applications. Consequently, there is an urgent need for wideband antenna designs that not only provide high gain but also exhibit symmetric radiation patterns necessary for advanced beam scanning applications.

Meeting these design requirements poses significant challenges for both academic researchers and industry practitioners. Wideband antennas for 5G mm-Wave applications must simultaneously deliver excellent performance in terms of bandwidth, radiation efficiency, and beam scanning while adhering to stringent constraints on size, profile, and ground clearance. Compact, low-profile designs are particularly desirable for integration into mobile and compact devices. However, developing solutions that achieve these diverse and often competing objectives remains a highly complex task.

2.1.1 The proposed endfire antenna array

To address these challenges, a compact endfire antenna array is presented, operating in the frequency bands of 26, 28, and 30 GHz, which are essential for 5G NR mobile communications [2]. This design not only achieves superior performance but also overcomes some of the most pressing challenges in mm-Wave antenna engineering. The standout features of the proposed antenna array include:

1. *Compact and low-profile design*: The antenna array is highly compact, measuring only 19 mm in size, with an ultra-low profile of 0.712 mm and a ground clearance of just 1.9 mm. This makes the design ideal for space-constrained applications, including mobile devices and wearable technology.

2. *Wideband operation with high isolation*: The array effectively covers a wide frequency bandwidth, including lower bands such as the n258 band. Notably, it maintains an isolation level below 20 dB even with a 4.7 mm inter-element spacing. This performance is achieved without the need for additional decoupling structures, simplifying the design and reducing overall system complexity.

3. *Broad beam scanning capabilities*: The phased array enables a wide beam scanning range of ±68° at 28 GHz, offering exceptional flexibility and adaptability for dynamic beamforming and steering applications.

2.1.2 Antenna element design

The design of the endfire antenna element was optimized for performance, compactness, and efficiency. As illustrated in **Figure 1**, the antenna employs a multilayer printed circuit board (PCB) configuration, with stacked substrates constructed from Rogers RO4350B (dielectric constant, ϵ_r = 3.66; loss tangent, tanδ = 0.0037) and a prepreg bonding layer made from Rogers RO4450F (ϵ_r = 3.52; tanδ = 0.004). This multilayer approach ensures both structural integrity and enhanced electrical performance.

The antenna's radiating elements consist of two bow tie-like dipoles, printed on the top (M1) and bottom (M3) layers of the PCB. These dipoles are connected to a truncated ground plane that serves as a reflector. The bowtie dipole topology has been specifically modified to minimize the overall size of the radiating structure, making it suitable for applications requiring small clearance.

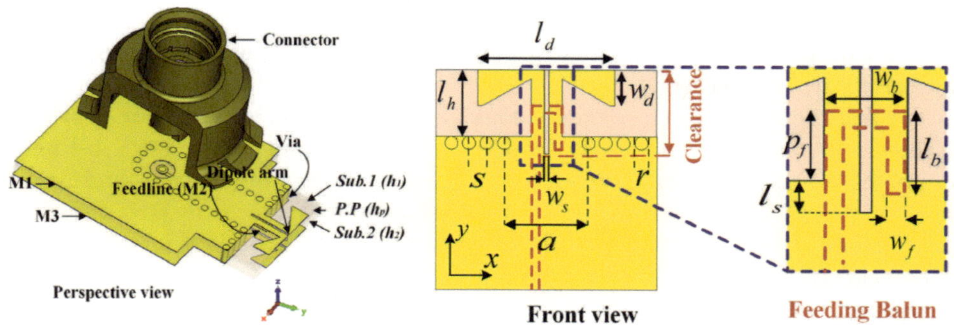

Figure 1.
Configuration of the antenna element [2].

The dipoles are excited *via* a feeding structure etched on the opposite side of the substrate. This feeding mechanism uses a 50 Ω open-end microstrip-line balun (M2) to slot-line transition. The slot-line transition provides efficient energy coupling to the dipoles, ensuring optimal excitation. The stacked substrates not only facilitate the integration of the feeding structure but also improve the bandwidth and radiation characteristics of the antenna.

One of the key challenges in this design is addressing the potential impedance mismatch and radiation efficiency degradation caused by coupling between the dipole and the bottom ground plane (M3). To mitigate this, a clearance is introduced below the dipole, effectively reducing coupling effects. Furthermore, a pair of dipoles is incorporated into both the top and bottom metallic layers (M1 and M3), which enhances impedance matching and improves endfire radiation characteristics.

Metallic vias are strategically introduced between the metallic layers as grounding pins. These vias play a crucial role in suppressing unwanted surface waves, thereby improving impedance bandwidth and overall radiation efficiency. The use of metallic vias also stabilizes the antenna's electrical performance by reducing parasitic effects, making the design robust and reliable for 5G applications.

The proposed endfire antenna array is compact and low-profile configuration making it highly suitable for integration into 5G mobile devices, small cells, and other compact systems. The wideband performance, coupled with excellent beam scanning capabilities, ensures that the antenna can support a diverse range of applications, from high-speed mobile communication to beamforming in dense urban environments. The elimination of additional decoupling structures further simplifies manufacturing and reduces costs, making the design both practical and scalable for widespread deployment.

2.1.3 Antenna array design

To achieve high gain and beamforming capabilities with a wide beam scanning range, the proposed antenna element was extended into a linear 1 × 4 antenna array. This mm-Wave array design features a profile height of less than 0.07λ, making it ultra-low-profile and suitable for compact applications. The spacing between adjacent elements was carefully set at 4.7 mm (0.44λ at 28 GHz), close to the half-wavelength requirement, to ensure adequate port isolation for efficient beam scanning.

The compact radiating aperture results in overall dimensions of just $19 \times 1.9 \times 0.712\ mm^3$, making it ideal for integration into 5G systems where space constraints are critical.

The feeding structure uses MMPX coaxial-type connectors, which are overlaid on the top ground plane. Each array element is excited by an open-end strip-line connected to the inner conductor of the connector through a metal *via*. Surface wave suppression vias were strategically placed around each feeding structure to reduce power loss in the strip-line, thereby improving the overall efficiency of the antenna array.

To validate and verify the proposed design methodology, an antenna array prototype was fabricated and subjected to comprehensive measurements. The S-parameters were evaluated using a Vector Network Analyzer (VNA) model: R&S ZVA40. Photographs of the fabricated prototype and the measurement setup are shown in **Figure 2(a)**. **Figure 2(b)** and **(c)** compare the simulated and measured reflection coefficients (Sii) and mutual coupling values (Sij) between the array elements. The measurements indicate that the array achieves a − 10 dB operating bandwidth spanning from 24 to 34 GHz, with mutual coupling values better than −15 dB, ensuring minimal interference between elements. These results demonstrate excellent agreement between the simulated and measured data, validating the accuracy of the proposed design approach.

Figure 2(d) presents the 2D polar radiation patterns for one of the array elements (port 2) in both the (xoy) plane (E-plane) and the (xoz) plane (H-plane) at 24 GHz, 26 GHz, 28 GHz, and 30 GHz. The measured and simulated results show a strong correlation, particularly in the vertical plane, with stable and symmetric endfire radiation patterns across the operating frequency range. This symmetry ensures consistent performance for beamforming applications.

The realized gain of the antenna array in the $+y$ direction across the operating frequencies is plotted in **Figure 2(e)**. Within the primary frequency band of interest, from 24.25 to 29.5 GHz, the antenna array achieves consistent endfire realized gains. The simulated gain values range from 7.93 to 8.66 dBi, while the measured values are slightly lower, varying between 7.56 and 8.14 dBi. These results highlight the efficiency of the design, as the measured and simulated gains are in close agreement. Moreover, the peak realized gains observed in the simulations are nearly identical to the endfire gains, indicating a minimal loss in the beamforming direction.

While the antenna array performs well within the intended frequency band, a drop in endfire gain is observed in simulations for frequencies above 30 GHz. This reduction in gain is attributed to increased impedance mismatch and radiation inefficiencies at higher frequencies, which is consistent with typical limitations in mm-Wave antenna designs. However, within the designated bands (24.25–29.5 GHz), the antenna array demonstrates stable performance, making it suitable for 5G applications.

The proposed mm-Wave 1×4 antenna array offers an innovative solution for compact, high-performance beamforming in 5G communication systems. Its ultra-low-profile design, compact radiating aperture, and effective integration of surface wave suppression techniques enable wideband operation with excellent radiation and gain characteristics. The fabrication and testing results confirm the reliability and accuracy of the design, with measured parameters closely matching simulations. The stable and symmetric endfire radiation patterns, along with consistent realized gains across the operating band, make this antenna array a strong candidate for advanced mm-Wave applications requiring efficient and compact beam scanning solutions.

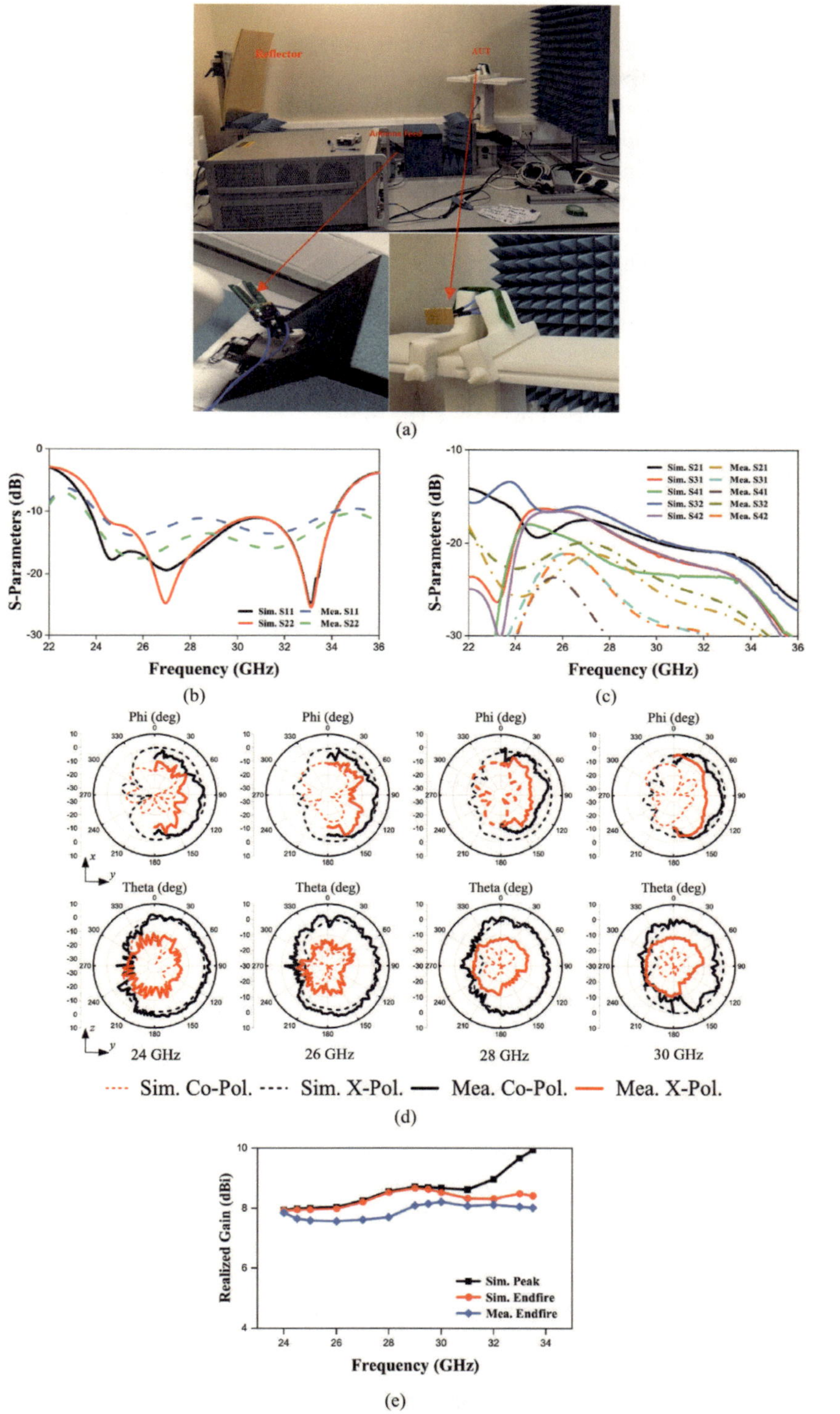

Figure 2.
(a) Photographs of the fabricated antenna array and far-field measurement setup, (b) reflection coefficient response, (c) mutual coupling response, (d) radiation patterns of the array @ 24, 26, 28, and 30 GHz, and (e) realized gain response of the array in the endfire direction [2].

2.1.4 State-of-the-art comparison

To evaluate the proposed design's performance, a comparison with other relevant mm-Wave endfire antenna arrays is presented, as summarized in **Table 1**. This section highlights how the proposed antenna array compares with its peers in terms of key performance metrics, including bandwidth, gain, compactness, and beam scanning range.

The proposed design achieves a bandwidth of 34%, which is comparable to [32] (37%) and [36] (27%) and significantly better than [35] (15%) and [37] (22%). While [34] achieves a higher bandwidth (42%), the proposed design's bandwidth is well-suited for accommodating the entire 5G NR band (24.25–29.5 GHz), which is a critical requirement for modern 5G systems. Additionally, the proposed design maintains excellent performance without requiring additional structural complexity.

In terms of realized gain, the proposed design achieves values ranging from 7.5 to 8.1 dBi. This gain is slightly lower than the higher ranges observed in Ref. [32] (9.1–13.8 dBi), [33] (10–12 dBi), and [36] (≥12 dBi). However, it is comparable to [34] (8 dBi) and [37] (6.6–8.1 dBi). While the gain is somewhat limited, it meets the minimum acceptable levels set by 3GPP regulations and remains practical for 5G applications where compactness is prioritized.

The proposed design achieves a beam scanning range of ±68°, which is broader than [37] (±54°) and comparable to Ref. [36] (±70°). However, Refs. [33, 34] achieve wider scanning ranges of ±75° and ± 90°, respectively. The proposed design balances a good scanning range with structural simplicity and compactness, whereas some designs with larger scanning ranges may require additional complexities or larger inter-element spacing.

The compactness of the proposed design is one of its standout features. With an element width of 0.29λ, a profile of 0.07λ, and a clearance of 0.18λ, it is smaller than most other designs, such as Ref. [32] (0.16λ profile and 0.82λ clearance), [34] (0.08λ profile and 0.73λ clearance), and [35] (0.07λ profile and 0.23λ clearance). Only Ref. [33] achieves a lower profile (0.02λ) and clearance (0.51λ) but at the cost of a larger element width of 0.55λ, which affects overall compactness. The proposed design's low profile and small clearance make it particularly well-suited for integration into compact mobile devices.

The proposed design features an inter-element spacing of 4.7 mm (0.44λ at 28 GHz) and achieves good isolation without requiring additional decoupling structures. This contrasts with some other designs, such as Refs. [32, 34], which rely on more complex methods to manage mutual coupling. This simplicity makes the proposed design more practical for manufacturing and integration.

The proposed antenna array effectively balances bandwidth, gain, beam scanning range, and compactness, making it highly competitive with existing designs. While it does not achieve the highest gain or scanning range, its exceptional compactness, low profile, and small clearance make it particularly suitable for integration into modern mobile devices. Furthermore, its ability to maintain good isolation without requiring additional decoupling structures ensures both simplicity and cost-effectiveness, positioning it as a practical and high-performing solution for 5G NR mm-Wave applications.

2.2 Design of high gain base station antenna array with wide bandwidth for mm-wave cellular communication systems

In recent years, significant research has focused on developing antennas for 5G wireless systems operating in the 37–40 GHz frequency band [6, 13–17]. However,

Ref.	Antenna Type	Array Number	Operating Band (GHz)	BW (%)	Array Gain (dBi)	Scanning (degree)	Element Width (λ)	Profile (λ)	Clearance (λ)
[32]	SIW-fed Metasurface	1 × 4	26.6-38.7	37	9.1-13.8	—	0.16	0.16	0.82
[33]	Dipole	1 × 8	26.5-38.0	36	10-12	±75	0.55	0.02	0.51
[34]	Quasi-Yagi	1 × 4	26-40	42	8	±90	0.55	0.08	0.73
[35]	Quasi-Yagi	1 × 4	25.9-30.2	15	9.0-9.6	—	0.30	0.07	0.23
[36]	Dipole	1 × 8	25-33	27	≥12	±70	0.36	0.15	0.13
[37]	Quasi-Yagi	1 × 4	24-30	22	6.6-8.1	±54	0.32	0.12	0
Authors' Work in [2]	**Bow-Tie Dipole**	**1 × 4**	**24-34**	**34**	**7.5-8.1**	**±68**	**0.29**	**0.07**	**0.18**

Table 1.
Comparison with other relevant endfire mm-Wave antenna arrays [2].

many of these designs cannot be directly adopted for base station applications due to their insufficient gain, which is critical for achieving the required coverage and signal quality in mm-Wave base station deployments.

A novel antenna array suitable for 5G mm-Wave base station applications was investigated and practically tested in Ref. [3]. The design addresses the limitations of previous works by employing a 2 × 2 array configuration for each radiating structure within the antenna array. This approach significantly enhances the overall gain, making the array well-suited for the demanding requirements of mm-Wave base stations.

2.2.1 Single antenna element design

The single element antenna that forms the foundation of the 5G mm-Wave array here is shown in **Figure 3**. The geometry resembles a two-pronged fork, with excitation provided through a microstrip feedline edge-connected to the U-shaped arm of the fork [19, 25]. This unconventional configuration was selected for its superior impedance matching characteristics compared to a standard rectangular patch antenna. Enhanced impedance matching ensures minimal power loss at the feed point, which is critical for high-frequency operations like mm-Wave communication.

While the quasi-fork-like antenna configuration offers improved impedance matching, it does have a smaller effective aperture compared to a conventional rectangular patch antenna operating at the same center frequency. This limitation results in a slightly lower gain for a single element. Despite this, the configuration's simplicity, improved matching, and suitability for integration into compact arrays make it a practical choice for base station designs.

The single antenna element is constructed on a Rogers 5880 dielectric substrate, known for its excellent electrical properties, which make it ideal for high-frequency applications such as mm-Wave communication. The substrate has a relative permittivity of 2.2, a thickness of 0.254 mm, and a low loss tangent of 0.0009, minimizing dielectric losses and ensuring efficient radiation performance. The overall dimensions of the antenna element are 10 × 6 mm^2, which balances compactness with acceptable performance metrics.

2.2.2 Antenna array design

To enhance the performance and scalability of the proposed antenna design, it was implemented into a series of progressively larger antenna arrays, starting with

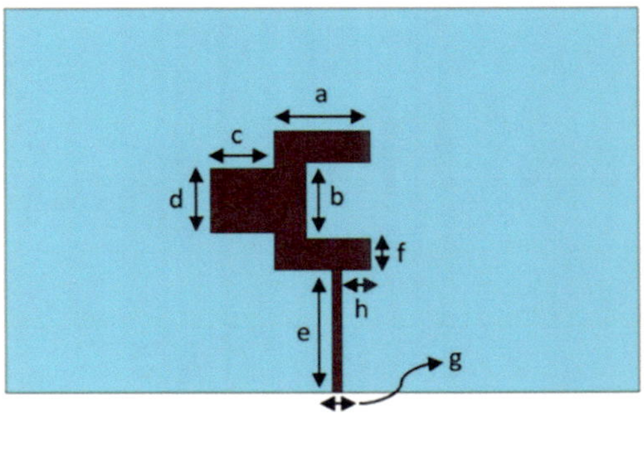

Figure 3.
Geometry of the proposed single antenna element [3].

an 8 × 8 array and extending to 8 × 16 and 8 × 32 arrays. Each iteration of the design maintains careful attention to factors such as spacing, feedline design, and impedance matching to optimize performance for mm-Wave applications.

2.2.2.1 8 × 8 antenna array

The first stage of the array design utilizes 16 individual antennas arranged in an 8 × 8 configuration, as shown in **Figure 4(a)**. These antennas are interconnected using power dividers and excited through a 50-Ω feed network. Several key design considerations were applied to ensure optimal performance:

- The spacing between the radiating elements was set as a multiple of 0.5λ. This minimizes mutual coupling effects, which can degrade array performance by introducing interference between adjacent elements. The precise spacing was optimized through simulation and analysis to achieve the best radiation characteristics.

- The feedline lengths were calculated to ensure phase coherency and power parity at all individual antennas. This ensures uniform excitation of all elements, which is critical for maintaining the gain and beamforming capabilities of the array. Without this optimization, the array would suffer from uneven radiation patterns and reduced efficiency.

- The compact design results in an overall array size of 25.5 × 27.5 mm^2, making it suitable for applications where space is at a premium.

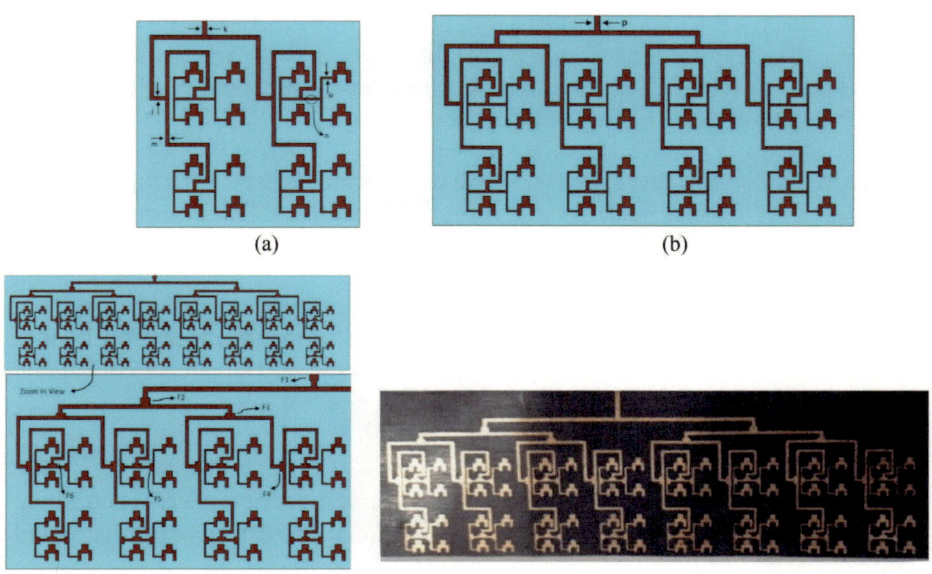

Figure 4.
Geometry of the antenna array (a) 8 × 8 array, (b) 8 × 16 array, (c) simulated layout of 8 × 32 array, and (d) fabricated prototype [3].

2.2.2.2 8 × 16 antenna array

To further boost gain and improve coverage, the 8 × 8 array was expanded into an 8 × 16 configuration, consisting of 32 antenna elements, as illustrated in **Figure 4(b)**. This was achieved by combining two 8 × 8 arrays using a power divider. Key enhancements include:

- A power divider splits the input signal into two equal-phase output signals, which are then applied to each of the two 8 × 8 arrays. This approach ensures consistent performance across the entire 32-element array.
- A quarter-wave transformer is integrated with the power divider to ensure impedance matching at the input. Proper matching minimizes signal reflection and maximizes power transfer to the array elements.
- Like the 8 × 8 array, the feedlines are designed to maintain phase coherency and power equity across all elements of the array. This is critical for achieving uniform radiation and preventing degradation of the array's gain and radiation characteristics.
- The overall size of the 8 × 16 array is 27.5 × 55 mm^2. Despite the increased number of elements, the design maintains a relatively compact form factor.

2.2.2.3 8 × 32 antenna array

Building on the success of the 8 × 16 design, the array was further scaled to an 8 × 32 configuration, consisting of 64 antenna elements, as shown in **Figure 4(c)**. This design maintains the same foundational principles of power division, impedance matching, and feedline optimization:

- The 8 × 32 array was constructed by combining two 8 × 16 arrays using additional power dividers. This modular approach ensures scalability while preserving phase coherency and power balance.
- As with the 8 × 16 array, a quarter-wave transformer is used with the power divider to maintain impedance matching. This ensures efficient signal distribution across all 64 elements.
- Despite the large number of elements, the array achieves a compact overall size of 30 × 110 mm^2. This makes it suitable for use in base stations or other high-performance mm-Wave communication systems.

2.2.2.4 Key design considerations across arrays

1. The element spacing (a multiple of 0.5λ) effectively reduces mutual coupling, which can otherwise lead to performance degradation in terms of gain and radiation patterns.
2. By ensuring phase coherency and power parity across all elements, the design achieves consistent radiation characteristics and prevents erosion of gain, which is critical in beamforming applications.

3. The modular approach to extending the array ensures that performance enhancements, such as increased gain and broader beam coverage, can be achieved without introducing significant complexity.

4. Each configuration maintains a balance between high performance and a compact form factor, ensuring suitability for mm-Wave applications, such as 5G base stations.

2.2.2.5 Simulated and measured performance of the antenna arrays

The simulated and measured performance of the proposed antenna design, including the single element and its array configurations (8 × 8, 8 × 16, and 8 × 32), demonstrates its suitability for 5G mm-Wave applications. The reflection coefficients for each configuration are shown in **Figure 5**, with the single element resonating at 38.5 GHz, the 8 × 8 array at 39.0 GHz, the 8 × 16 array at 39.1 GHz, and the 8 × 32 array at 37.2 GHz. The measured results are in close agreement with the simulations, validating the design accuracy.

The radiation patterns for the single antenna element and the arrays in the φ = 0° and φ = 90° planes are displayed in **Figures 5(c)** and **(d)**. The single antenna element achieves a gain of 7.6 dBi, which is adequate for mobile handset applications but insufficient for 5G base stations that must overcome significant signal attenuation due to path loss, multipath effects, and atmospheric absorption. Its radiation beam is wide,

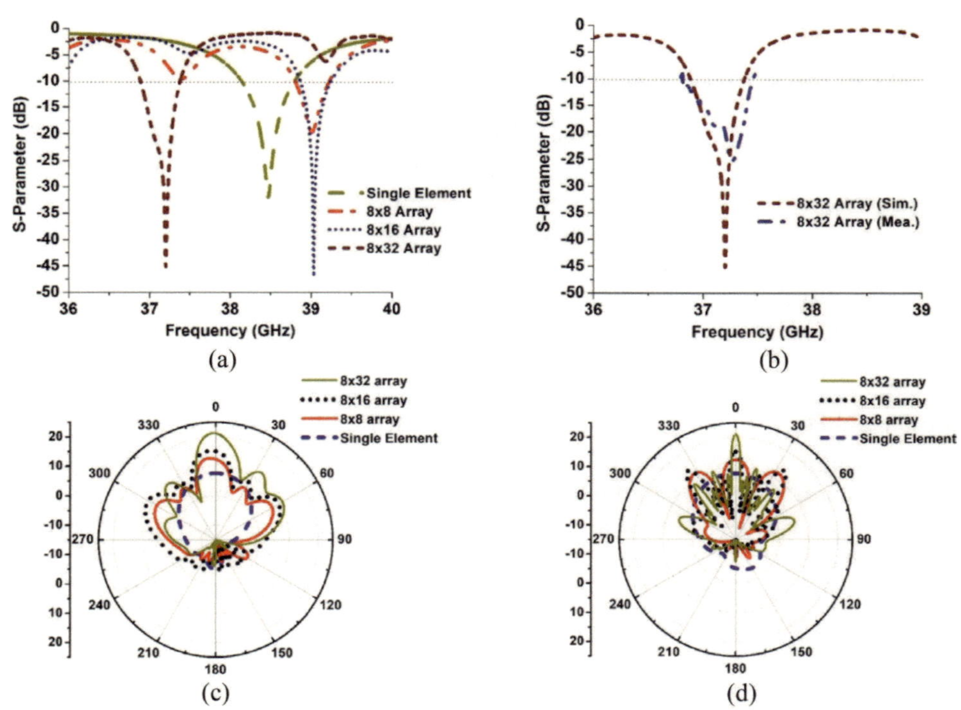

Figure 5.
(a) Simulated reflection coefficient response of the single element and the antenna arrays, that is, 8 × 8, 8 × 16, 8 × 32, and (b) simulated and measured reflection coefficient responses of the 8 × 32 array. Polar radiation patterns of the single antenna element and the three arrays (8 × 8, 8 × 16, and 8 × 32), (c) φ = 0° plane radiation pattern, and (d) φ = 90° plane radiation pattern [3].

with an angular width of 82.5° in the $\varphi = 90°$ plane, making it less directed and less suitable for base station applications.

The 8 × 8 array improves the gain to 13 dBi, but the side lobes in the $\varphi = 0°$ plane are suppressed to less than −6 dB, while in the $\varphi = 90°$ plane they are suppressed to less than −3 dB. These higher side lobe levels can lead to interference, limiting its effectiveness. The array narrows the angular width to 18.4°, providing a more focused beam compared to the single element.

The 8 × 16 array further enhances the gain to 15.3 dBi, which is acceptable for base station use. However, the side lobes remain a concern, with suppression to less than −6 dB in the $\varphi = 0°$ plane and less than −4 dB in the $\varphi = 90°$ plane. The angular width narrows to 7.2° in the $\varphi = 90°$ plane, offering improved directivity compared to the smaller arrays.

The 8 × 32 array achieves a substantial gain of 21.2 dBi, meeting the high-gain requirements of 5G mm-Wave base stations. It also demonstrates significant improvement in side lobe suppression, with side lobes reduced to less than −10 dB in the $\varphi = 0°$ plane and less than −12 dB in the $\varphi = 90°$ plane, making it suitable for practical applications. The beam becomes highly directed, with an angular width of just 4.1° in the $\varphi = 90°$ plane, ensuring efficient energy concentration and minimal interference.

This study verifies the feasibility of the proposed antenna design for large array configurations, particularly for 5G mm-Wave base stations. The results highlight the progressive improvement in gain and beam focusing as the array size increases. However, further work is needed to optimize the side lobe levels, especially for the smaller arrays, and to develop dual-polarized configurations to enhance the array's performance for advanced 5G applications.

2.3 State-of-the-art comparison

Over the years, significant advancements have been made in mm-Wave antenna array designs for 5G communication systems. **Table 2** compares the proposed 8 × 32 antenna array with several recently reported designs, focusing on parameters such as the number of elements, center frequency, gain, substrate permittivity, and dimensions. The analysis highlights the advantages of the proposed design in terms of compactness and performance while noting areas for potential improvement.

The proposed 8 × 32 array achieves a gain of 21.2 dBi at 37.2 GHz, which is competitive with several state-of-the-art designs. For instance, the 8 × 8 array in Refs. [46, 47] achieves gains of 22 dBi at 39 and 28 GHz, respectively, using a stacked multilayer PCB structure and independent control of phase and amplitude at each radiating element. However, the proposed design is more compact, with dimensions of 110 × 30 mm^2 compared to the 100 × 90 mm^2 and 100.6 × 96.2 mm^2 of these arrays.

The 8 × 32 array in Ref. [48] demonstrates a higher gain of 25.7 dBi at 35 GHz but comes with a much larger form factor (212 × 56 mm^2), approximately 3.6 times the size of the proposed design. While Ref. [48] achieves better gain, its significantly larger size limits its suitability for compact applications, where the proposed design offers a more practical solution.

The compact dimensions of the proposed antenna array (110 × 30 mm^2) are a standout feature, making it significantly smaller than most comparable designs. For example: (1) The dual-band 4 × 3 MIMO array in Ref. [38] has a gain of 13 dBi at 38 GHz but has much larger dimensions of 122 × 156 mm^2, (2) The 8 × 8 metallic array in Ref. [45] achieves a high gain of 24 dBi at 26 GHz but lacks detailed size specifications, and metallic structures are generally bulkier, and (3) The aperture-coupled

Refs.	No. of elements	Center freq. (GHz)	Gain (dBi)	Substrate permittivity (ε_r)	Dimensions (mm)
[38]	4 × 3	28	12.6	2.2	122 × 156
[38]	4 × 3	38	13	2.2	122 × 156
[39]	1 × 4	26.87	~5–6	3.35	5 × 5
[39]	1 × 4	38.5	~5–6	3.35	5 × 5
[40]	1 × 2	26.87	~5–6	3.35	5 × 5
[41]	2 × 2	37	12.8	2.2	20 × 8
[42]	2 × 2	29	14	2.2	9.7 × 9.7
[43]	4 × 4	30.1	11.35	2.94	90 × 54
[44]	1 × 8	33.5	14.1	2.2	58 × 5.5
[12]	1 × 8	36	16.9	2.2	50 × 40
[45]	8 × 8	26	24	—	—
[46]	8 × 8	39	22	3.3	100 × 90
[47]	8 × 8	28	22	2.55	100.6 × 96.2
[48]	8 × 32	35	25.7	2.22	212 × 56
Authors' Work in [3]	**8 × 32**	37.2	**21.2**	2.2	**110 × 30**

Table 2.
Comparison of the proposed antenna array with recent works [3].

patch array in Ref. [43] operates at 30.1 GHz with a gain of 11.35 dBi but occupies a large area of 90 × 54 mm^2. The proposed design's compact form factor makes it highly suitable for applications requiring high performance in limited space, such as 5G base stations or other mm-Wave systems in urban or dense environments.

The proposed array demonstrates excellent beamwidth and side lobe suppression. Its angular width of 4.1° and side lobe levels of less than −10 dB (in the $\varphi = 0°$ plane) and less than −12 dB (in the $\varphi = 90°$ plane) represents a substantial improvement in performance. By contrast: (1) The 8 × 8 array in Refs. [46, 47] achieves good gain but does not explicitly address side lobe suppression, a critical factor for reducing interference; and (2) The array in Ref. [48] offers high gain but with limited information on beamwidth and side lobe levels, which are crucial for practical deployment in 5G systems.

The proposed design uses a substrate with a permittivity of 2.2, which is commonly used in high-frequency designs due to its low dielectric loss and high efficiency. Other works, such as Refs. [38, 41, 42], also employ substrates with similar permittivity, demonstrating a standard approach for minimizing loss in mm-Wave designs. However, designs such as Refs. [43, 46] use substrates with higher permittivity (e.g., $\varepsilon r = 2.94$ and 3.3), which can increase the dielectric loss but may help reduce size in specific configurations.

When considering overall performance, the proposed design strikes a balance between gain, compactness, and practical features such as side lobe suppression. While designs like Ref. [48] achieve higher gain, their significantly larger size limits their practicality in compact applications. Similarly, while smaller arrays such as Refs. [38, 41, 43] are compact, they deliver lower gains and do not match the beamwidth or side lobe suppression of the proposed design.

The proposed 8 × 32 antenna array offers a strong balance between gain, compact size, and side lobe suppression. With a gain of 21.2 dBi, an angular width of 4.1°, and superior side lobe performance, it is well-suited for 5G mm-Wave applications. Compared to other designs, it stands out for its compactness, achieving high performance in a much smaller form factor. While higher gain designs exist, such as Ref. [48], their larger size makes them less practical for applications requiring compact integration. This balance of features positions the proposed array as a competitive and versatile solution in the state-of-the-art mm-Wave antenna landscape.

2.4 Broadband 3-D shared aperture high isolation nine-element antenna array for on-demand millimeter-wave 5G applications

A novel 3D shared aperture 3 × 3 antenna array design, introduced in Ref. [4], is proposed for millimeter-wave (mm-Wave) 5G applications operating in the 24–28.4 GHz frequency range. The array employs hexagonal-shaped patches elevated above a common substrate and excited *via* metallic rods passing through the substrate. A key feature of this design is the inclusion of isolation walls that decouple individual antennas, effectively reducing mutual coupling and preserving far-field performance. To enhance gain and radiation efficiency, the patches are engineered with metasurface characteristics by incorporating hexagonal-shaped concentric slots of decreasing diameters. This innovative design enables the array to serve both single-aperture and shared aperture applications by utilizing subarrays to cover specific portions of the bandwidth for different time- or frequency-shared applications.

2.4.1 3D shared aperture nine-element antenna array

The geometry of the proposed 3D shared aperture nine-element antenna array is illustrated in **Figure 6(a)**. The array features hexagonal-shaped patches constructed on a base dielectric substrate made of Rogers RT5880, a material selected for its low loss tangent (tanδ = 0.0009) and relative permittivity (ε_r = 2.2), which ensure high efficiency at mm-Wave frequencies. The dielectric substrate has a thickness of 0.2 mm and is supported by a metallic ground plane of 0.8 mm thickness.

The radiating patches are elevated 3.35 mm above the dielectric substrate using metallic rods with a diameter of 0.4 mm. These rods serve a dual purpose: they physically support the patches and function as the excitation mechanism by connecting to the feed network. The rods protrude through the base substrate and are insulated from the ground plane metallization layer, ensuring that unwanted interactions with the ground plane do not degrade antenna performance. Elevating the patches effectively reduces surface current interactions between adjacent antennas, which could otherwise lead to performance degradation. The overall size of the basic 3 × 3 antenna array is compact, measuring 20 × 20 × 3.35 mm^3.

Mutual coupling between closely spaced radiating elements in an antenna array can distort radiation patterns and degrade overall far-field performance. To address this challenge, isolation walls are introduced around each individual antenna, as shown in **Figure 6(b)**. These walls, made from a perfect electric conductor, effectively suppress near-field radiation emanating from adjacent antennas. This technique ensures high isolation without altering the compact size of the array, maintaining the integrity of its physical footprint while preserving performance.

To further improve the performance of the antenna array, each hexagonal patch is loaded with concentric hexagonal slot rings of decreasing diameters, as depicted

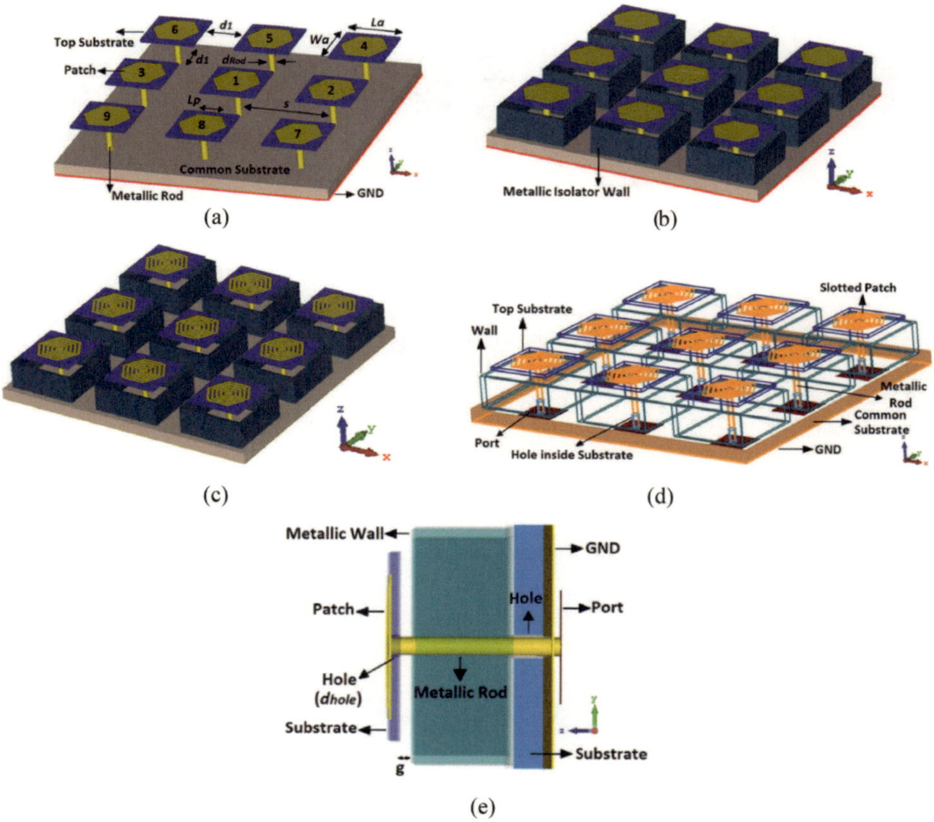

Figure 6.
3-D shared aperture nine-element 3 × 3 array antennas, (a) basic array, (b) antenna array with the isolation walls, (c) array antennas with isolation walls and loaded with metasurface slots, (d) the array's schematic-view, and (e) the single element's side-view [4].

in **Figure 6(c)–(e)**. These slots are of subwavelength circumference, and each slot has a width of 0.08 mm. When exposed to electromagnetic (EM) waves, the slot-loaded patches behave as metasurfaces, as described in Refs. [14–16]. Metasurfaces are engineered surfaces with unique electromagnetic properties that enhance the effective aperture of the antenna, resulting in improved gain and radiation efficiency.

The integration of metasurface characteristics magnifies the aperture of the antenna array without increasing its physical dimensions. This technique leverages the interaction between EM waves and the subwavelength slots, enabling the array to deliver high performance across a broad bandwidth.

2.4.2 Performance improvements with isolation walls and metasurface slots

The results obtained from 3D electromagnetic solvers, namely computer systems technology (CST) Microwave Studio and Ansoft high frequency simulation software (HFSS), as shown in **Figure 7**, demonstrate the significant performance enhancements achieved by incorporating isolation walls and metasurface slots into the antenna array. Notably, these improvements are realized without increasing the physical size of the array, maintaining its compact form factor.

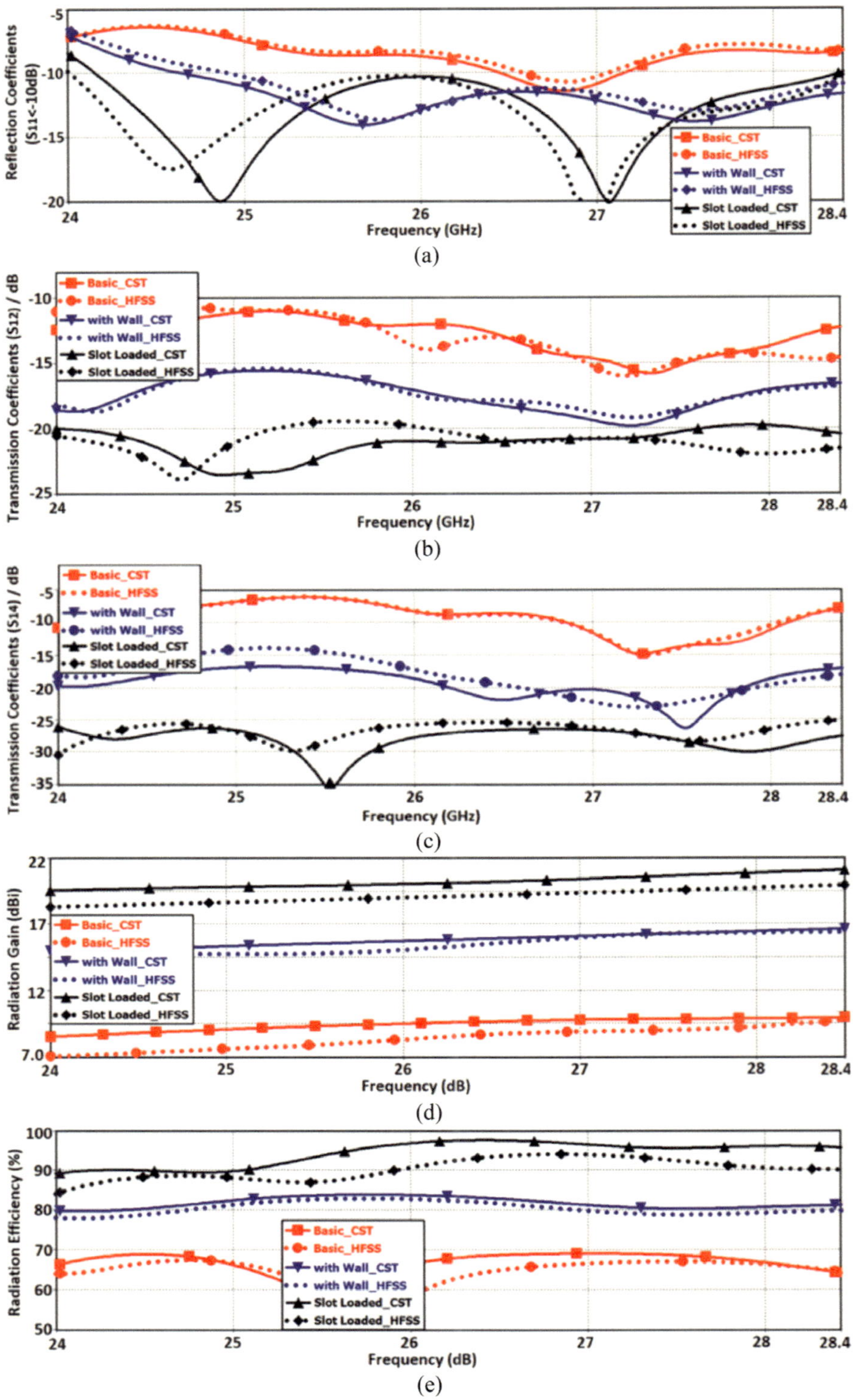

Figure 7.
S-parameter response of the proposed antenna array, (a) reflection coefficient (S_{11}), (b) transmission-coefficient between antennas #1&#2 (S_{12}), (c) transmission-coefficient between antennas #1&#4 (S_{14}), (d) radiation gain, and (e) radiation efficiency [4].

The introduction of isolation walls and metasurface slots substantially increases the array's impedance bandwidth. As shown in **Figure 7(a)**, the bandwidth of the final structure is extended by 4.4 GHz compared to the basic array design. This represents a sixfold increase in bandwidth, making the array suitable for broader operational frequencies within the mm-Wave 5G spectrum. The ability to achieve such a substantial improvement without increasing the physical dimensions of the antenna highlights the effectiveness of these design modifications.

Figures 7(b) and **(c)** illustrate the enhanced isolation between radiating elements in the array due to the incorporation of isolation walls. The symmetric nature of the design allows for the isolation performance to be assessed between representative pairs of elements, such as elements 1 & 2 and elements 1 & 4. The results indicate an isolation improvement of more than 10 dB in both cases compared to the basic model. This significant reduction in mutual coupling ensures better far-field performance and minimizes interference, which is critical for multi-element arrays used in 5G applications.

In addition to the improvements in bandwidth and isolation, the gain and radiation efficiency of the array are also significantly enhanced. **Figures 7(d)** and **(e)** show that the gain of the array increases by an average of 10.5 dBi compared to the basic design. Similarly, radiation efficiency is improved by 25%, which is a notable enhancement for an array operating in the challenging mm-Wave spectrum. These improvements ensure better signal strength and reduced power losses, making the array more effective for practical 5G applications.

The integration of isolation walls and metasurface slots into the 3 × 3 antenna array design results in significant performance enhancements across key metrics, including bandwidth, isolation, gain, and efficiency. These improvements are achieved without compromising the array's compact size, demonstrating the effectiveness of these design modifications. The significant gains in bandwidth (a sixfold increase), isolation (10 dB improvement), and radiation efficiency (25% enhancement) position this array as a highly capable and practical solution for on-demand mm-Wave 5G applications.

2.4.3 State-of-the-art comparison

A comparative analysis of the proposed 3 × 3 antenna array with previously published designs operating within the same frequency band is presented in **Table 3**. The parameters considered for comparison include array dimensions, substrate properties, bandwidth, maximum gain, and maximum efficiency. This analysis highlights the advantages of the proposed design in terms of gain and efficiency while maintaining a compact footprint, making it a viable solution for mm-Wave applications such as 5G wireless communications.

The proposed antenna array has a compact size of 20 × 20 mm^2, which is significantly smaller than the arrays in Ref. [50] (99.2 × 17.45 mm^2, 53] (130 × 90 mm^2). The only comparable designs in terms of size are those in Refs. [49, 51], which also feature compact footprints of 20 × 20 mm^2 and 27 × 25 mm^2, respectively. Despite its small size, the proposed design achieves superior performance in terms of gain and efficiency compared to these designs.

The proposed design utilizes a 0.8 mm Rogers RT5880 substrate, which has excellent electrical properties, including a low loss tangent (tanδ = 0.0009) and a relative permittivity (ε_r = 2.2), making it ideal for high-frequency applications. Other designs, such as Refs. [49, 51], use Rogers substrates (RO4350B and RO4003C) with slightly

Ref.	Size (mm^2)	Substrate	Bandwidth/freq. Range (GHz)	Max. gain (dBi)	Max. eff. (%)
[49]	20 × 20	Two different layers: (i) 1 mm FR4 (ii) 0.254 mm Rogers RO4350B	5.25/24.25–29.5	10	85
[50]	99.2 × 17.45	0.254 mm Rogers RO5880	6.78/24.35–31.13	19.88	86
[51]	27 × 25	0.508 mm Rogers RO4003C	5/24–29	9	—
[52]	130 × 90	0.508 mm Rogers RO4003C	3/28–31	9.6	—
Authors' Work in [4]	**20 × 20**	**0.8 mm Rogers RO5880**	**4.4/24.0–28.4**	**21**	**98**

Table 3.
The proposed antenna array is compared with other arrays reported in the literature [4].

different thicknesses and properties, but they do not achieve the same level of performance as the proposed design.

The bandwidth of the proposed array is 4.4 GHz (24.0–28.4 GHz), which is slightly narrower than that of [49] (5.25 GHz) and [50] (6.78 GHz) but still adequate for most mm-Wave 5G applications. The designs in Refs. [51, 52] have bandwidths of 5 GHz and 3 GHz, respectively, which are narrower than [50] and comparable to the proposed design.

The proposed array achieves a maximum gain of 21 dBi, which is significantly higher than the gains of other designs, including Refs. [49] (10 dBi), [51] (9 dBi), and [52] (9.6 dBi). The closest competitor in terms of gain is Ref. [50], which achieves a gain of 19.88 dBi. However, Ref. [50] has a much larger footprint (99.2 × 17.45 mm^2), making the proposed design superior for applications requiring compact form factors.

The proposed design demonstrates a maximum efficiency of 98%, surpassing the reported efficiencies of other designs. For instance, Refs. [49, 50] achieve efficiencies of 85% and 86%, respectively. The efficiency of Refs. [51, 52] is not reported, but their relatively lower gain suggests that their efficiency is likely inferior to that of the proposed array.

3. Conclusion

This chapter presented a comprehensive design and analysis of three innovative antenna arrays optimized for wideband operation in 5G millimeter-wave spectra. Various advanced techniques have been proposed and implemented to enhance key performance metrics such as frequency bandwidth, isolation, and radiation properties, all while maintaining compact physical dimensions. These techniques include the use of isolation walls, metasurface-inspired slots, and carefully optimized feeding mechanisms, which collectively improve array performance without adding complexity to the overall structure. The simplicity of the feeding mechanisms also ensures the practicality and feasibility of the designs for real-world implementations.

The antenna arrays presented here exhibit several notable advantages, including compact dimensions, wide operational bandwidth, high gain, and excellent efficiency.

Additionally, the arrays demonstrate stable and acceptable radiation patterns, low profiles, and cost-effectiveness, making them ideal candidates for 5G millimeter-wave wireless networks. The straightforward manufacturing process further enhances their appeal for large-scale production, ensuring scalability and reliability for widespread deployment.

Through a detailed comparison with state-of-the-art designs, the antenna arrays described here are shown to achieve a superior balance of compactness, performance, and practicality. With significant bandwidth enhancements—up to six times greater than baseline designs—alongside improvements in isolation (10 dB or more) and radiation efficiency (an increase of up to 25%), these arrays address the key challenges of mm-Wave communication. Notably, the integration of metasurface-inspired elements and isolation walls provides substantial performance gains without increasing the physical size, positioning these arrays at the forefront of compact high-performance antenna solutions.

The exceptional gain levels, such as 21 dBi for the 8 × 32 array configuration, and efficiency up to 98%, surpass the performance of many comparable designs in the literature. Despite their superior performance, the proposed arrays maintain a compact form factor, which is critical for modern 5G applications where space constraints are a key concern. Furthermore, their ability to support shared aperture applications and subarray configurations allows for flexible use in time- or frequency-shared scenarios, catering to the diverse demands of dynamic 5G networks.

The cost-effective and straightforward manufacturing process, leveraging commercially available materials and simple design principles, ensures that these arrays are not only technically advanced but also economically viable. This balance of innovation, efficiency, and scalability makes the proposed antenna arrays strong candidates for widespread adoption in 5G millimeter-wave wireless networks and potentially beyond, in future communication systems.

By addressing the core challenges of mm-Wave antenna design—such as bandwidth, isolation, efficiency, and compactness—the work presented in this chapter provides a foundational approach for developing robust, high-performance solutions for next-generation communication technologies. The designs presented in this chapter exemplify the potential to bridge the gap between cutting-edge research and practical deployment, making a significant contribution to the evolution of wireless networks.

Author details

Mohammad Alibakhshikenari[1*], Peiman Parand[1], Bal Virdee[2], Ignacio Garcia Zuazola[2], Sunil Kumar[2], Ali Zidour[3], Mohammad Soruri[4], Takfarinas Saber[5], Mohammad Naser-Moghadasi[6] and Ernesto Limiti[1]

1 Electronics Engineering Department, University of Rome "Tor Vergata", Rome, Italy

2 Center for Communications Technology, London Metropolitan University, London, United Kingdom

3 Faculty of Technology, Department of Electrical Systems Engineering, University of M'hamed Bougara, Boumerdes, Algeria

4 Faculty of Ferdows Technical, University of Birjand, Birjand, Iran

5 Lero, Science Foundation Ireland Centre for Software, School of Computer Science, University of Galway, Galway, Ireland

6 Computer and Electrical Engineering Department, Science and Research Branch, Islamic Azad University, Tehran, Iran

*Address all correspondence to: alibakhshikenari@ing.uniroma2.it

References

[1] Mendonça S et al. The rise of 5G technologies and systems: A quantitative analysis of knowledge production. Telecommunications Policy. 2022;**46**(4):102327

[2] Zidour A et al. Wideband endfire antenna Array for 5G mmWave mobile terminals. IEEE Access. 2024;**12**:39926-39935

[3] Sehrai DA et al. Design of high gain base station antenna array for mm-wave cellular communication systems. Scientific Reports. 2023;**13**:4907

[4] Alibakhshikenari M et al. Broadband 3-D shared aperture high isolation nine-element antenna Array for on-demand millimeter-wave 5G applications. Optik. 2022;**267**:169708

[5] Rappaport TS, Xing Y, MacCartney GR, Molisch AF, Mellios E, Zhang J. Overview of millimeter wave communications for fifth-generation (5G) wireless networks—With a focus on propagation models. IEEE Transactions on Antennas and Propagation. 2017;**65**(12):6213-6230

[6] Nguyen VL, Lin P, Hwang R. Enhancing misbehavior detection in 5G vehicle-to-vehicle communications. IEEE Transactions on Vehicular Technology. 2020;**69**(9):9417-9430

[7] Storck CR, Duarte-Figueiredo F. A survey of 5G technology evolution, standards, and infrastructure associated with vehicle-to-everything communications by internet of vehicles. IEEE Access. 2020;**8**:117593-117614

[8] Sharaf MH, Zaki AI, Hamad RK, Omar MMM. A novel dual-band (38/60 Ghz) patch antenna for 5G Mobile handsets. Sensors. 2020;**20**:2541

[9] Peng M, Zhao A. High performance 5G millimeter-wave antenna array for 37-40 GHz mobile application. In: International Workshop on Antenna Technology (iWAT). Nanjing, China: IEEE; 2018. pp. 1-4

[10] Park J, Choi D, Hong W. 37-39 GHz vertically-polarized end-fire 5G antenna array featuring electrically small profile. In: 2018 IEEE International Symposium on Antennas and Propagation & USNC/URSI National Radio Science Meeting, Boston, MA, USA, 08-13 July 2018. IEEE; 2018. pp. 637-638

[11] Abbasi MAB, Abbasi QH. Development Challenges of Millimeter-Wave 5G Beamformers. In: Tafazolli R, Wang C-L, Chatzimisios P, editors. Wiley 5G Ref. Wiley; 2020. DOI: 10.1002/9781119471509.w5GRef226

[12] Li A, Luk K-M. Millimeter-wave end-fire magneto-electric dipole antenna and arrays with asymmetrical substrate integrated coaxial line feed. IEEE Open Journal of Antennas and Propagation. 2021;**2**:62-71

[13] Xiao M et al. Millimeter wave communications for future mobile networks. IEEE Journal on Selected Areas in Communications. 2017;**35**(9):1909-1935

[14] Bang J, Chung H, Hong J, Seo H, Choi J, Kim S. Millimeter-wave communications: Recent developments and challenges of hardware and beam management algorithms. IEEE Communications Magazine. 2021;**59**(8):86-92

[15] Ghafoor KZ et al. Millimeter-wave communication for internet of vehicles: Status, challenges, and perspectives.

IEEE Internet of Things Journal. 2020;**7**(9):8525-8546

[16] Hong W et al. The role of millimeter-wave technologies in 5G/6G wireless communications. IEEE Journal of Microwaves. 2021;**1**(1):101-122

[17] Khan Z et al. Performance evaluation of 5G millimeter-wave-based vehicular communication for connected vehicles. IEEE Access. 2022;**10**:31031-31042

[18] Zakeri H et al. Path loss model estimation at indoor environment by using deep neural network and CatBoost for wireless application. IEEE Access. 2024;**12**:159070-159085

[19] Azizpour R et al. Multi-channel radio-over-fiber communication systems through modulation instability phenomenon. IEEE Photonics Journal. 2024;**16**(5):1-13

[20] Karim BA, Ali HK. A novel beamforming technique using mmWave antenna arrays for 5G wireless communication networks. Digital Signal Processing. 2023;**134**:103917

[21] Siddiqui SI et al. A dual-band high-gain beam steering antenna Array for 5G Sub-6 GHz Base station. Scientific Reports. 2024;**14**:26517

[22] El Mrini M, Mchbal A. Naima Amar Touhami, design of a 240 elements antenna system for 5G massive MIMO applications with an inclined beam and two operating modes. Journal of Electromagnetic Waves and Applications. 2024;**38**(16):1796-1810

[23] Wang Y, Xue Q, Hu Z, Liao S. Mixed-modes-enabled element-level beamforming antenna with enhanced isolation for phased array applications. IEEE Transactions on Antennas and Propagation. 2024;**72**(5):4577-4582

[24] Zandamela A et al. Digital pattern synthesis with a compact MIMO antenna of half-wavelength diameter. AEU - International Journal of Electronics and Communications. 2021;**135**:153728

[25] Pan W, Liu Y-H, Zhao Z-Q, Liu Q-H. Sparse antenna Array design methodologies: A review. Journal of Electronic Science and Technology. 2024;**22**(3):100276.3

[26] Abdalmalak KA, Yousfi AE, Vargas DS. Massive MIMO antenna arrays for low-frequency bands. In: Innovation in MIMO Systems [Working Title]. IntechOpen; 17 Dec 2024. DOI: 10.5772/intechopen.1007820

[27] Yousfi AE et al. Miniaturized dual-polarized, high-gain, and wideband dielectric resonator antenna for low band massive MIMO applications. Progress in Electromagnetics Research-PIER. 2024;**179**:101-111

[28] Yousfi AE, Atia Abdalmalak K, Lamkaddem A, Vargas DS. Miniaturized broadband dual-polarized dielectric resonator antenna using characteristic modes. In: 2023 17th European Conference on Antennas and Propagation (EuCAP), Florence, Italy, 26-31 March. IEEE; 2023. pp. 1-4

[29] Althuwayb AA et al. Design technique to mitigate unwanted coupling in densely packed radiating elements of an antenna array for electronic devices and wireless communication systems operating in the millimeter-wave band. AEU - International Journal of Electronics and Communications. 2023;**159**:154464

[30] Zheng Q, Wang J, PourMohammadi P, Pang X. Dual-band metasurface-based closely packed antennas by controlling surface wave propagation. IEEE Antennas

and Wireless Propagation Letters. 2024;**23**(5):1633-1637

[31] Ghosh S, Sen D. An inclusive survey on array antenna design for millimeter-wave communications. IEEE Access. 2019;7:83137-83161

[32] Li T, Chen ZN. Wideband substrate-integrated waveguide-fed endfire metasurface antenna array. IEEE Transactions on Antennas and Propagation. 2018;**66**(12):7032-7040

[33] Ta SX, Choo H, Park I. Broadband printed-dipole antenna and its arrays for 5G applications. IEEE Antennas Wireless Propag. Lett. 2017;**16**:2183-2186

[34] Di Paola C, Zhang S, Zhao K, Ying Z, Bolin T, Pedersen GF. Wideband beam-switchable 28 GHz quasi-yagi array for mobile devices. IEEE Transactions on Antennas and Propagation. 2019;**67**(11):6870-6882

[35] Hwang I-J et al. Quasi Yagi antenna array with modified folded dipole driver for mmWave 5G cellular devices. IEEE Antennas and Wireless Propagation Letters. 2019;**18**:971-975

[36] Syrytsin I, Zhang S, Pedersen GF, Morris AS. Compact quad mode planar phased array with wideband for 5G mobile terminals. IEEE Transactions on Antennas and Propagation. 2018;**66**(9):4648-4657

[37] Liu Y, Zhao C, Yue Z, Ren A, Jia Y. A horizontally polarized endfire antenna with complete ground for 5GmmWave applications. Microwave and Optical Technology Letters. 2020;**62**(12):3936-3944

[38] Elhabbash T, Skaik T. Design of dual-band dual-polarized MIMO antenna for mm-wave 5G base stations with octagonal prism structure. In: IEEE 7th Palestinian International Conference on Electrical and Computer Engineering (PICECE). Gaza, Palestine: IEEE; 2019. pp. 1-6

[39] Siddiqui Z et al. Dual-band dual-polarized antenna for mm-wave 5G base station antenna Array. In: 14th European Conference on Antennas and Propagation (EuCAP). Copenhagen, Denmark: IEEE; 2020. pp. 1-4

[40] Siddiqui Z et al. Dual-polarized filtering antenna for mm-wave 5G base station antenna array. In: 15th European Conference on Antennas and Propagation (EuCAP), Dusseldorf, Germany, 22-26 March 2021. IEEE; 2021. pp. 1-4

[41] Khan J et al. Design of a millimeter-wave MIMO antenna array for 5G communication terminals. Sensors. 2022;**22**:2768

[42] Li K-X, Wu Y-W, Hao Z-C. A 5G millimeter-wave circularly polarized planar antenna array. In: 9th Asia-Pacific Conference on Antennas and Propagation (APCAP), Xiamen, China, 04-07 August. IEEE; 2020. pp. 1-2

[43] Afifi A, Sebak A-R. Wideband 4×4 butler matrix in the printed ridge gap waveguide technology for millimeter-wave applications. IEEE Transactions on Antennas and Propagation. 2020;**68**(11):7670-7675

[44] Yu Y, Jiang ZH, Zhang J-D, Wu W. Broadband millimeter-wave endfire circularly polarized array with a low-profile feeding structure. IEEE Transactions on Antennas and Propagation. 2022;**70**(8):7270-7275

[45] Tan YCM, Guan Hong N. 64-elements mmWave detachable phased array antenna for 5G 26GHz band. In: 2020 IEEE International Symposium

on Antennas and Propagation and North American Radio Science Meeting. Montreal, QC, Canada: IEEE; 2020. pp. 1317-1318

[46] Valkonen R, Doumanis E. Analysis and design of mm-Wave phased array antennas for 5G access. In: IEEE International Symposium on Antennas and Propagation and USNC-URSI Radio Science Meeting, Atlanta, GA, USA, 07-12 July 2019. IEEE; 2019. pp. 611-612

[47] Islam MA, Karmakar NC. An 8×8 mm-wave LP ACMPA array for a long-range mm-wave chipless RFID tag-sensor reader. IEEE Journal of Radio Frequency Identification. 2021;**5**(1):53-63

[48] Verma PK, Kumar R, Singh VK, Kumar A. Dual radiator based low profile fan beam antenna for millimeter wave fencing system. International Journal of Advances in Microwave Technology. 2021;**6**(1):254-259

[49] Lima de Paula I et al. Cost-effective high-performance air-filled SIW antenna array for the global 5G 26 GHz and 28 GHz bands. IEEE Antennas and Wireless Propagation Letters. 2021;**20**(2):194-198

[50] Khalily M et al. Broadband mm-wave microstrip array antenna with improved radiation characteristics for different 5G applications. IEEE Transactions on Antennas and Propagation. 2018;**66**(9):4641-4647

[51] Borhani Kakhki M et al. Millimeter-wave beam-tilting vivaldi antenna with gain enhancement using multilayer FSS. IEEE Antennas and Wireless Propagation Letters. 2018;**17**(12):2279-2283

[52] Mantash M, Kesavan A, Denidni TA. Beam-tilting endfire antenna using a single-layer FSS for 5G communication networks. IEEE Antennas and Wireless Propagation Letters. 2018;**17**(1):29-33

Chapter 2

Advancements in Ultrawideband and Compact Microwave Components Using Ridge Empty Substrate Integrated Waveguides (RESIWs) for Satellite Communications

David Herraiz, Héctor Esteban, Darío Herraiz, Ángel Belenguer and Vicente E. Boria

Abstract

Ridge empty substrate integrated waveguides (RESIWs) present a significant leap in the development of compact and ultrawideband microwave components. RESIWs offer a class of efficient integrated transmission lines by combining empty substrate integrated waveguide (ESIW) technology with conventional ridge waveguides. This integration results in a transmission line that exhibits low losses, high-quality factors, cost-effective solutions compared to conventional waveguides, and high bandwidth capabilities. The chapter describes the technology and applications of RESIW transmission lines, with particular emphasis on their compact and ultrawideband features (where increments of more than 100% in terms of operational bandwidth can be reached compared to ESIW). Several microwave components commonly found in conventional communication systems, such as filters, transitions, and power dividers, will be presented as examples of this type of transmission line, showcasing the main advantages of RESIW technology compared to other substrate-integrated technologies. The compactness, easy-integration, and wideband responses offered by RESIW make it an excellent choice for next-generation ultrawideband microwave circuits, where the size, weight, power, and cost (SWaP-C) are crucial factors.

Keywords: wideband communication devices, ridge empty substrate integrated waveguide, microwave components, broadband applications, ridge waveguides

1. Introduction

In recent years, significant efforts have been devoted to integrating three-dimensional structures into planar technologies, known as substrate integrated circuits (SICs). This necessity is driven by the requirements of next-generation satellite systems, particularly constellations operating in the low Earth orbit (LEO), which demand low cost and reduced weight components [1–3]. Consequently, the payload communication systems of these satellites need to achieve competitive electrical performance, while maintaining high integration and minimizing mass and size [4, 5].

On the one hand, three-dimensional structures, such as waveguides and coaxial lines, offer superior electrical performance, characterized by low losses and high-quality factors. However, these technologies are bulky, expensive, and heavy. On the other hand, planar technologies such as microstrip (MS) and coplanar waveguides (CPWs) offer low cost and lightweight solutions, but compromise on the electrical performance (high losses and lower quality factor), and are susceptible to crosstalk due to their unshielded nature. To bridge the gap between these two contrasting technologies, substrate integrated circuits (SICs) emerged as a promising solution (see **Figure 1**).

The most widely known SIC technological platform is the substrate integrated waveguide (SIW) [6], which integrates a rectangular waveguide into a printed circuit board (PCB), using the copper cladding metallization of the substrate and two rows of metallic vias to confine the electromagnetic wave (see **Figure 1**). SIW offers a higher quality factor than conventional planar technologies while maintaining important features such as low cost, lightweight, and low profile. Additionally, it provides a platform for developing both passive and active microwave components on the same substrate. Several microwave devices and antennas have been developed using this technology [7, 8]. However, compared to the conventional rectangular waveguide, the SIW has a lower quality factor and higher insertion losses due to propagation through the dielectric medium. Finally, to enhance the electrical performance of SIW technology, the empty substrate integrated waveguide (ESIW) [9] and the air-filled substrate integrated waveguide (AFSIW) [10] were introduced. In these solutions, the dielectric through which the electromagnetic wave propagates in SIW is completely (ESIW) or partially (AFSIW) removed, leading to high electrical performance while maintaining other key advantages. This chapter will focus on ESIW, due to its exceptional electrical performance and resemblance to traditional waveguides. A variety of ESIW devices have been designed and manufactured [11–17], showcasing their excellent performance. The ESIW comprises three layers: an empty central layer with metallized outer

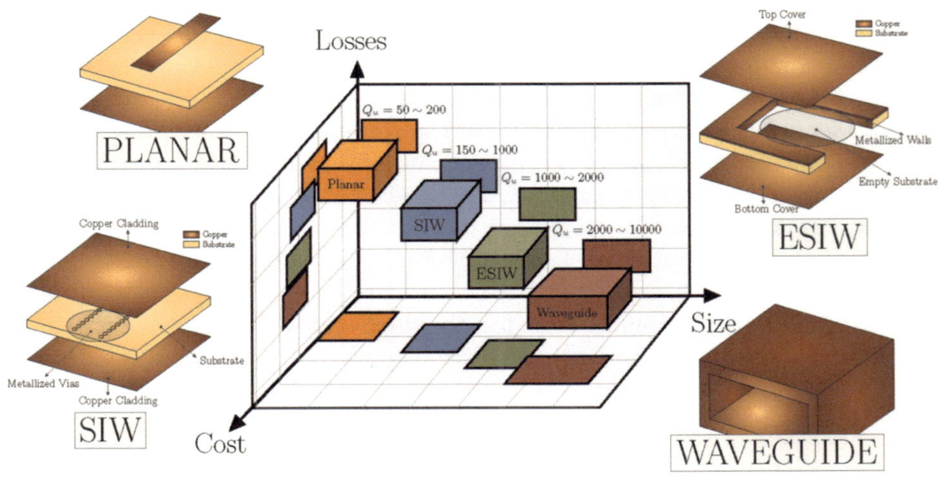

Figure 1.
Comparison of ESIW technology to conventional technologies taking into account the cost, size, and losses. Source: Author's original work.

edges forming the lateral waveguide walls. The top and bottom covers confine the electromagnetic waves, creating a reduced-height waveguide implemented using PCBs. The 3D-view of the ESIW can be seen in **Figure 1**. Both SIW and ESIW exhibit a bandwidth limitation due to single-mode propagation, restricting the maximum fractional bandwidth (FBW) to 66%. Furthermore, in practical implementations, this FBW is constrained due to the waveguide dispersion to a value of 44%.

In order to overcome the bandwidth limitation, single and double ridge empty substrate integrated waveguide (SRESIW and DRESIW) technologies were proposed in [18–20]. These technologies introduce one or more additional layers to the conventional ESIW, incorporating a ridge structure. The ridge enhances the single-mode bandwidth while preserving the competitive advantages of ESIW technology. This chapter explores these technologies (SRESIW and DRESIW), detailing the design of various microwave components and practical manufacturing considerations, as well as highlighting its significant advantages compared to other technologies.

The chapter is structured as follows: In Section 2, the transversal section of the SRESIW and DRESIW is analyzed, demonstrating the potential of this technology to achieve a higher monomode bandwidth. In Section 3, several microwave components commonly found in satellite communication payloads are discussed, followed by an examination of the manufacturing considerations for this technology in Section 4. The chapter concludes with a summary in Section 5, where the conclusions and future prospects of the considered technology are discussed.

2. Single and double ridge empty substrate integrated waveguide (SRESIW/DRESIW)

Single and double ridge empty substrate integrated waveguides (SRESIWs and DRESIWs) are conventional ridge waveguides with one or more ridges created by stacking multiple substrates [21]. **Figure 2** illustrates the 3D view of both structures, highlighting the ridge layers. The presence of the ridge causes a perturbation in the electric field of the fundamental mode, creating a fringing effect, that reduces the cut-off frequency of the fundamental mode (TE_{10}) and increases the cut-off frequency of the first higher order mode (TE_{20}).

As the SRESIW and DRESIW are implemented using several substrates, the total height of the waveguide (b) and the height of the ridge layer (h_r) are fixed parameters that depend on the available commercial substrates. **Figure 3** presents several transversal sections of SRESIW and DRESIW, each comprising a different number of layers. To simplify the structure, all layers are assumed to have the same height (h_s), resulting in $b = h_r + d$ and $h_r = nh_s$, where n represents the number of ridge layers. Therefore, the only design parameters are the ridge width (w_r) and the number of ridge layers (n). The following equations can be used in order to calculate both $f_{c_{10}}$ and $f_{c_{20}}$:

$$\cot\left(\frac{2\pi f_{c_{10}}\left(\frac{a-w_r}{2}\right)}{c}\right) - \frac{b}{d}\tan\left(\frac{\pi w_r f_{c_{10}}}{c}\right) - \frac{B}{Y_{01}} = 0 \tag{1}$$

$$\cot\left(\frac{2\pi f_{c_{20}}\left(\frac{a-w_r}{2}\right)}{c}\right) + \frac{b}{d}\cot\left(\frac{\pi w_r f_{c_{20}}}{c}\right) - \frac{B}{Y_{01}} = 0 \tag{2}$$

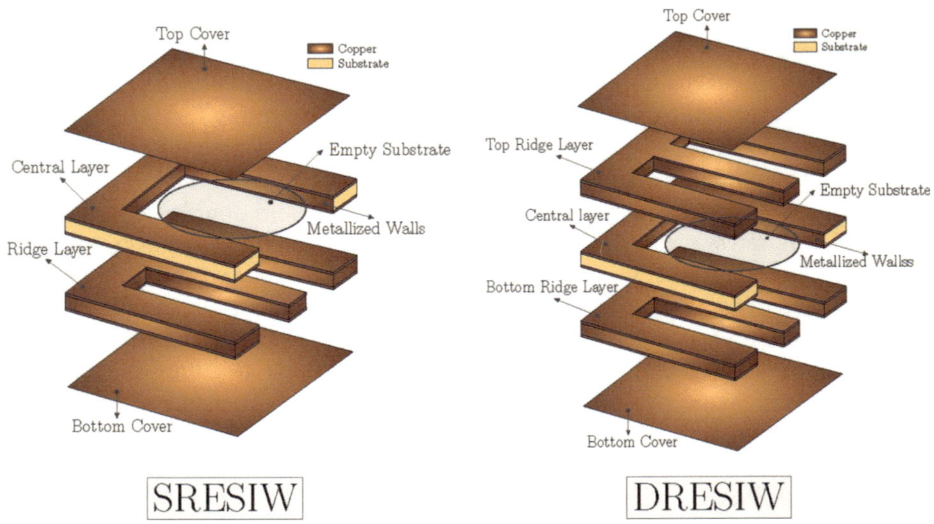

Figure 2.
Three-dimensional view of piled layers of the SRESIW and DRESIW. Sources: D. Herraiz, H. Esteban, J. V. Morro, D. Herraiz, A. Belenguer, and V.E. Boria. [Updated 2022 Aug 16 and 2022 Nov 10]. Available from: [22, 23]. CC-BY-NC-ND 4.0. Used with permission.

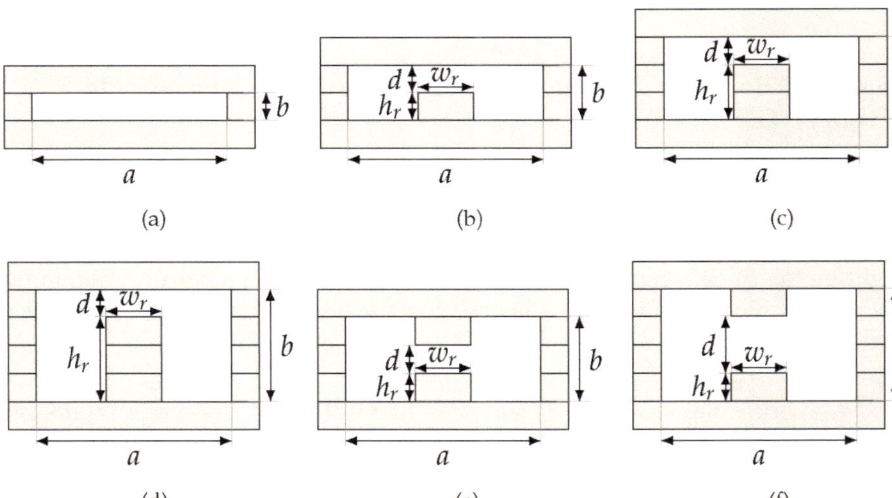

Figure 3.
Single ridge and double ridge ESIW transversal sections, with n additional layers: empty waveguide ($n = 0$ ridge layers) in (a), SRESIW with a single ridge layer ($n = 1$) in (b), SRESIW with two ridge layers ($n = 2$) in (c), SRESIW with three ridge layers ($n = 3$) in (d), DRESIW with two ridge layers ($n = 2$) in (e), and DRESIW with three additional layers ($n = 2$ ridge layers) in (f). Source: D. Herraiz, H. Esteban, D. Herraiz, A. Belenguer, and V.E. Boria. [Updated 2020 Feb 24]. Available from: [19]. CC-BY-NC-ND 4.0. Used with permission.

where c represents the velocity of light and a, b, w_r, and d are geometric parameters which can be seen in **Figure 3**. The term B/Y_{01} denotes the normalized susceptance that characterizes the impact of the transverse discontinuity generated by the ridge structure. This susceptance can be approximated, as explained in [24], for both ridge cases as follows:

$$\frac{B}{Y_{01}} \approx 4\left(\frac{b}{d}\right)\left(\frac{af_c}{c}\right) \ln\csc\left(\frac{\pi d}{2b}\right) \tag{3}$$

$$\frac{B}{Y_{01}} \approx 2\left(\frac{b}{d}\right)\left(\frac{af_c}{c}\right) \ln\csc\left(\frac{\pi d}{2b}\right) \tag{4}$$

where f_c corresponds to $f_{c_{10}}$ and $f_{c_{20}}$ for solving Eqs. (1) and (2), respectively. Eq. (3) is applicable for the SRESIW, whereas for the case of a DRESIW, Eq. (4) is used. The maximum usable bandwidth (BW) can be calculated as follows:

$$BW = f_{c_{20}} - f_{c_{10}} \tag{5}$$

Eqs. (1), (2), and (5) have been solved for various numbers of layers in both SRESIW and DRESIW structures, determining the cut-off frequency for the fundamental mode (f_{c10}), the first higher-order mode (f_{c20}) and the absolute bandwidth (BW), which depend on the ridge width (w_r). For these results, the ROGERS 4003C substrate with a height of $h_s = 0.813$ mm and $a = 15.7988$ mm (the same as WR62) has been used. Several configurations with $n = 0,1,2,$ and 3 have been considered for both SRESIW and DRESIW (see **Figure 3**).

Figure 4 shows the cut-off frequencies of both modes using the equations (marked with solid lines and markers) and compares them with the results using a commercial electromagnetic simulator Computer Simulation Technology (CST) Studio Suite (marked with dashed lines). It can be observed that the equations and CST tool provide similar results. Additionally, it is noted that the maximum absolute bandwidth is obtained when $w_r/a \approx 0.2 - 0.35$ for all configurations compared.

Table 1 compares the geometric structures depicted in **Figure 3** across several indices, including the cut-off frequencies of the fundamental and the first higher-order mode, the absolute and fractional bandwidth, and the optimum value of w_r that provides the highest monomode bandwidth. It can be observed that the SRESIW configuration with $n = 1, 2,$ and 3 layers offers a broader bandwidth compared to the ESIW ($n = 0$). Nevertheless, the bandwidth increment diminishes as additional layers are added. Regarding the DRESIW, it is observed that increasing the number of ridge layers does not yield a substantial improvement in bandwidth relative to other solutions. Therefore, considering these conclusions, a suitable compromise between the

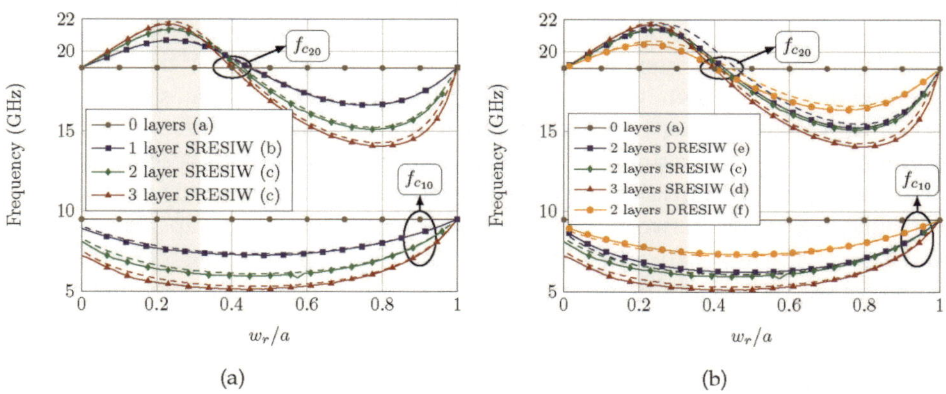

Figure 4.
*Cut-off frequencies for the fundamental mode $f_{c_{10}}$ and higher-order mode $f_{c_{20}}$ for the geometries shown in **Figure 3** [19, 20]. Source: Author's original work.*

n	Type	Figure	f_{10}(GHz)	f_{20}(GHz)	BW(GHz)	FBW(%)	w_r(mm)
0	ESIW	(a)	9.480	18.970	9.490	63.3	—
1	SRESIW	(b)	7.530	20.660	13.310	87.5	3.848
2	SRESIW	(c)	6.227	21.352	15.125	100.8	3.751
3	SRESIW	(d)	5.420	21.692	16.272	100.5	3.660
2	DRESIW	(e)	6.369	21.723	15.354	102.4	4.059
2	DRESIW	(f)	7.537	20.662	13.135	87.6	3.822

Table 1.
*Comparison between the geometries of **Figure 3**.*

bandwidth achieved, and factors such as the number of additional layers, cost, volume and manufacturing complexity, is to use a single layer ($n = 1$) for SRESIW (see **Figure 3b**) and two layers ($n = 2$) for DRESIW (see **Figure 3e**). Throughout the chapter, these selected configurations will be used, along with the aforementioned substrate type, waveguide, and ridge widths.

3. Microwave passive components

Several passive microwave components have been developed using these advanced technologies, indicating a highly promising future. These devices are integrated into satellite communications payloads, performing functions such as transitions to planar lines, filtering, and power division. In this section, we will provide a detailed explanation of these three components.

3.1 Transitions

3.1.1 Microstrip-to-SRESIW

The SRESIW and DRESIW have higher monomode bandwidth; however, these structures require transitions to planar technologies to fully leverage their potential. These transitions must be designed to ensure good matching across as much of the available bandwidth as possible. Therefore, a microstrip-to-SRESIW transition presented in [19] is shown in **Figure 5**. This proposed transition needs to couple the electromagnetic waves between the microstrip line and the SRESIW, which exhibit different characteristic impedances, electric field distributions, and geometric dimensions. Therefore, the transition includes a dielectric iris (denoted as 1) to couple the electromagnetic wave between the two technologies, a microstrip taper that changes the characteristic impedance (denoted as 2), and finally, a widening SRESIW section (denoted as 3), which modifies the characteristic impedance in this section and enhances the matching level. In the last section, the ridge is widened of width w_{tr} and length l_{t2} in order to achieve a good coupling between the fundamental modes.

A design procedure based on the initial expresions for the design parameters, which are summarized in [19], has been proposed for this transition. These parameters ($w_{ti}, w_t, l_t, a_c, b_c, w_{tr}, l_{t2},$ and l_2) are optimized to improve the return losses of the proposed transition in the maximum available bandwidth, leading to the optimum

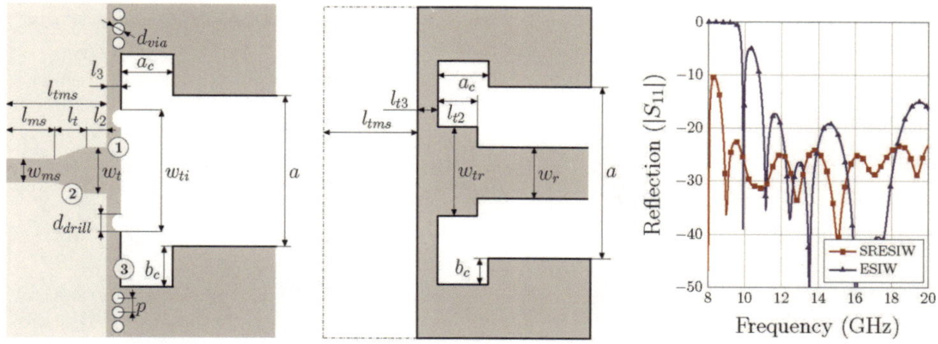

(a) (b) (c)

Figure 5.
Layout of the microstrip to SRESIW transition. (a) Central layer and (b) ridge layer. White represents empty spaces, light gray represents dielectric body, and dark gray represents copper surfaces. (c) Simulated reflection of the back-to-back microstrip to SRESIW transition with the optimum values and comparison with microstrip to ESIW transition without a ridge [25]. Source: Author's original work.

Dimension	w_{ti}	w_t	l_t	a_c	b_c	w_{tr}	l_{t2}	l_2
Value (mm)	10.598	2.360	2.954	10.507	4.566	11.169	4.533	1.181

Table 2.
*Dimensions of the microstrip to SRESIW transition illustrated in **Figure 5**.*

values which are listed in **Table 2**. The optimization has been performed using CST Microwave Studio. **Figure 5c** illustrates a comparison of the reflection coefficient for the final optimized design, as well as that of a microstrip to ESIW transition [25]. Results have been obtained in a back-to-back configuration. It can be observed that the return losses are better than 23 dB within the frequency range of 9 to 20 GHz, corresponding to an absolute and fractional bandwidths of 11 GHz and 76%, respectively. Compared to the 43% fractional bandwidth of the microstrip to ESIW transition [25], the proposed solution exhibits a 90% increase in terms of relative bandwidth, with return losses better than 20 dB.

Figure 6 illustrates the manufactured prototype in a back-to-back configuration, with a comparison of the simulated and measured results. As depicted, measured return losses are better than 11 dB, whereas the insertion losses remain below 1.5 dB in the frequency range of 8 to 20.5 GHz. These results correspond to an absolute bandwidth of 12.5 GHz and a fractional bandwidth of 87%, covering almost the maximum bandwidth achieved using the SRESIW technology.

3.1.2 Microstrip-to-DRESIW

The aforementioned microstrip to SRESIW transition exhibits a non-competitive performance when it is directly translated for the DRESIW structure. Consequently, to achieve higher bandwidth, a new microstrip to DRESIW transition was proposed in [20]. **Figure 7** illustrates the layout of the proposed transition. As observed, the transition comprises four key elements: an iris that couples the electromagnetic wave between both technologies, a microstrip taper that modifies the characteristic impedance, a widening DRESIW section with a fixed ridge width serving the same objective,

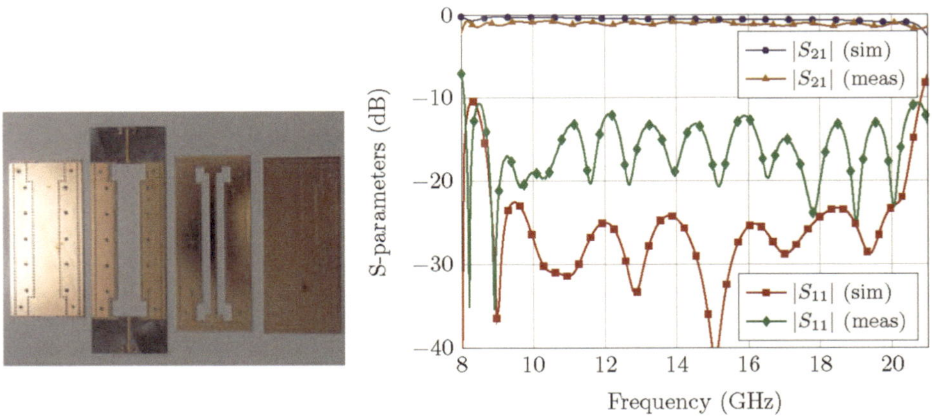

(a) (b)

Figure 6.
Microstrip to SRESIW transiton. (a) Manufactured prototype. (b) Comparison of measurements and simulations. Source: D. Herraiz, H. Esteban, D. Herraiz, A. Belenguer, and V.E. Boria. [Updated 2020 Feb 24]. Available from: [19]. CC-BY-NC-ND 4.0. Used with permission.

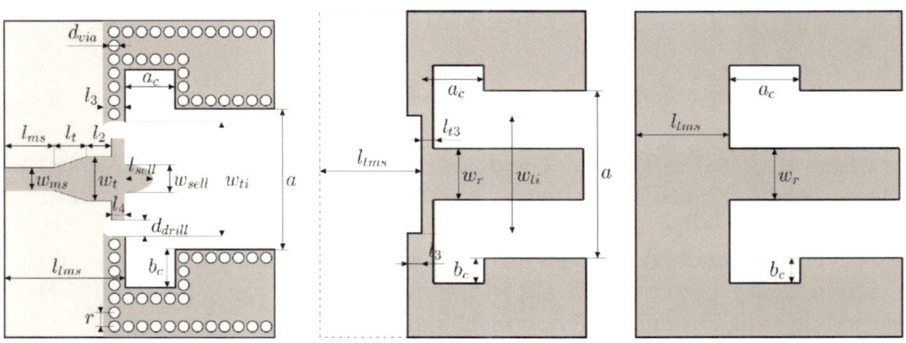

(a) Central Layer (b) Top Ridge Layer (c) Bottom Ridge Layer

Figure 7.
Layout of the microstrip line to DRESIW transition. White represents empty spaces, light gray represents dielectric body, and dark gray represents copper surfaces [20]. Source: D. Herraiz, H. Esteban, D. Herraiz, A. Belenguer, and V.E. Boria. [Updated 2021 Dec 08]. Available from: [20]. CC-BY 4.0.

and a superelliptical dielectric taper inside the DRESIW. Due to the concentrated electric field within the ridges in the DRESIW, the dielectric taper inside this section enhanced the electromagnetic coupling between the two technologies, thus improving the return losses. This dielectric taper provides a gradual transition between the empty section and the dielectric of the microstrip line. The width of the dielectric taper is described by the following equation:

$$\left(\frac{x}{l_{sell}}\right)^p + \left(\frac{y}{\frac{w_{sell}}{2}}\right)^q = 1 \tag{6}$$

where p and q are parameters that independently control the curvature of the superellipse along the x and y axes, respectively, and l_{sell} and $w_{sell}/2$ correspond to the semi-axes of the generalized superellipse. Consequently, it is possible to obtain a taper structure with a sharp nose shape, or a widened ellipse, by adjusting the extreme values of p and q ($p, q = 0$ or $p, q \to \infty$).

The design parameters ($w_{ti}, w_t, l_t, a_c, b_c, w_{sell}, l_{sell}, l_2$, p, and q) are optimized to improve the return losses of the proposed transition in the maximum available bandwidth, leading to the optimum values which are listed in **Table 3**. The same features of the previous transition has been used regarding to the waveguide width and substrate material. The S-parameters of the proposed transition are shown in **Figure 8**. For comparison purposes, results for the microstrip to ESIW and microstrip [25] to SRESIW (see Section 3.1.1) have also been included. These results have been obtained using the simulator CST. It can be observed that, for the DRESIW case, the return losses are better than 26 dB, whereas the insertion losses are lower than 0.7 dB within the frequency range of 7.3 to 20.64 GHz. These results correspond to an absolute bandwidth of 13.3 GHz and a fractional bandwidth of 95.4%, covering nearly the entire monomode bandwidth available in the DRESIW. Compared to the responses obtained for microstrip to ESIW and microstrip to SRESIW cases, the fractional bandwidth achieved for the proposed DRESIW transition represents an increase of 123% and 17% over the fractional bandwidths of ESIW and SRESIW cases, respectively.

Figure 9 shows the manufactured prototype in a back-to-back configuration, and the comparison of the simulated and measured S-parameters results. It can be observed that the measured return and insertion losses exhibit excellent values, reaching 19.7 dB and 1.5 dB, respectively, within the frequency range of 7.2 to 20.7 GHz. This range corresponds to an absolute bandwidth of 13.4 GHz, with a value of 96.4% fractional bandwidth.

Dimension	w_{ti}	w_t	l_t	a_c	b_c	w_{sell}	l_{sell}	l_2	p	q
Value (mm)	10.190	2.656	5.006	11.328	3.121	1.710	4.945	1.002	3.506	2.702

Table 3.
Dimensions of the microstrip to DRESIW transition illustrated in ***Figure 7****.*

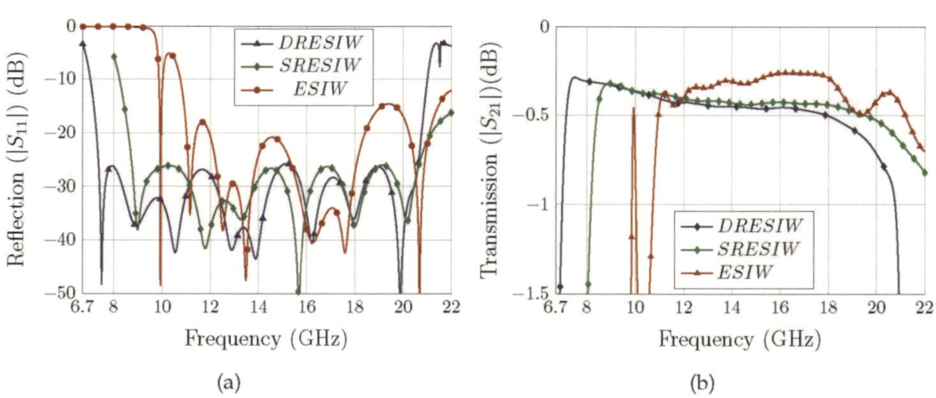

Figure 8.
Simulated results for reflection (a) and transmission (b) of the back-to-back microstrip to DRESIW transition, compared with microstrip to ESIW transition without a ridge [25] and microstrip to SRESIW transition presented in Section 3.1.1. Source: D. Herraiz, H. Esteban, D. Herraiz, A. Belenguer, and V.E. Boria. [Updated 2021 Dec 08]. Available from: [20]. CC-BY 4.0.

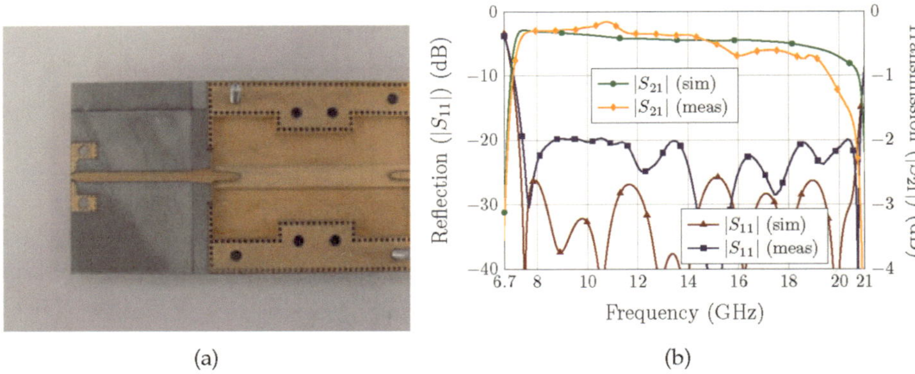

Figure 9.
Microstrip to DRESIW transiton. (a) Manufactured prototype. (b) Comparison of measurements and simulations [20]. Source: D. Herraiz, H. Esteban, D. Herraiz, A. Belenguer, and V.E. Boria. [Updated 2021 Dec 08]. Available from: [20]. CC-BY 4.0.

3.1.3 Comparison

Table 4 presents a detailed comparison of the proposed microstrip-to-SRESIW/ DRESIW transitions with other reported solutions. The proposed microstrip-to-SRESIW transition exhibits a higher bandwidth than the other reported SIW and ESIW solutions. Additionally, the microstrip-to-DRESIW transitions provide a greater bandwidth compared to the other solutions.

3.2 Filters

The SRESIW technology offers the advantage of developing several filter topologies using the same cross-section, thereby achieving more compact solutions. Nonetheless, the conventional inverter structure implemented with an inductive iris cannot be manufactured using SRESIW, due to the minimal gaps between the ridge and the iris waveguide sides. To address this limitation, a novel U-shape configuration for inverters was introduced in [22]. This filter was designed using the traditional theory of cascading inverters and series SRESIW resonators of initial length of $\lambda_g/2$ [28]. The inverters present a U-shaped structure, where the ridge is connected to the waveguide sides in the ridge layer, while the central layer maintains the geometry of a traditional iris, as depicted in **Figure 10**.

Transition	Number of layers	*BW* (GHz)	*FBW* (%)	R_L (dB)	I_L (dB)
MS-SIW [26]	1	17.5–30	52%	15	1.26
MS-ESIW [25]	3	12–18	40%	20	1.2
MS-ESIW [27]	3	12–18	40%	21	1
MS-SRESIW [19] Subsection 3.1.1	4	8–20.5	87.7%	11	1.5
MS-DRESIW [20] Subsection 3.1.2	5	7.2–20.6	96.4%	19.7	1.5

Table 4.
Comparison of the most relevant features of the proposed MS-to-SRESIW and MS-to-DRESIW transition in this chapter, with other solutions proposed in the literature. R_L and I_L refer to the minimum return losses and maximum insertion losses in the bandwidth BW, respectively. FBW refers to fractional bandwidth.

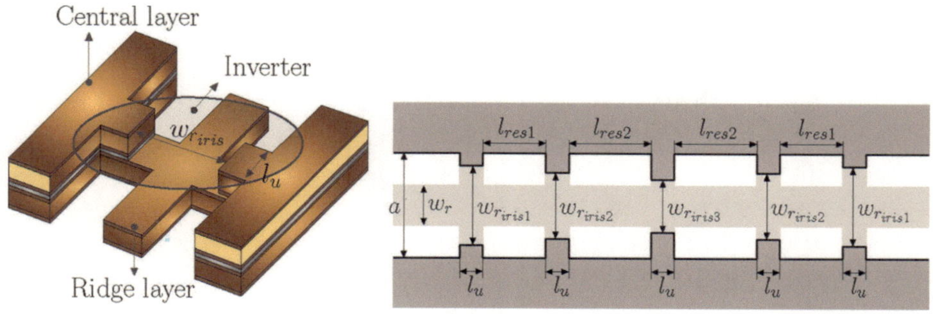

(a) 3D-view of the inverter.

(b) Layout of the filter.

Figure 10.
Geometry of the SRESIW filter. Light gray represents the ridge layer, while dark gray represents the central layer. Source: D. Herraiz, H. Esteban, J. V. Morro, D. Herraiz, A. Belenguer, and V.E. Boria. [Updated 2022 Aug 16]. Available from: [22]. CC-BY-NC-ND 4.0. Used with permission.

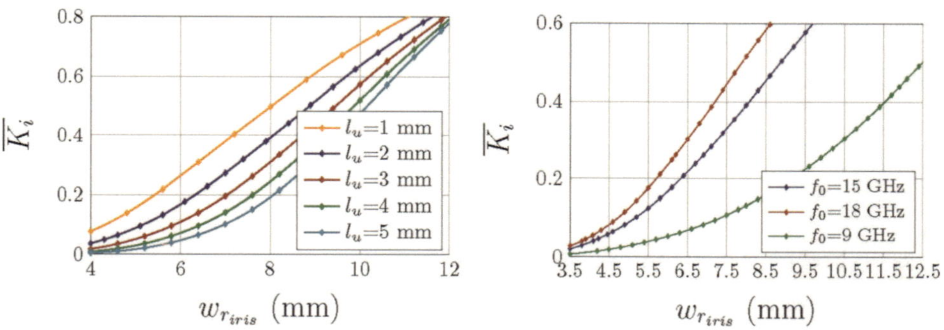

(a)

(b)

Figure 11.
Normalized impedance inverter constant $\overline{K_i}$ versus ridge width $w_{r_{iris}}$: (a) for different values of l_u at frequency of $f_o = 15$ GHz (b) for the three different central frequencies ($f_o = 9, 15,$ and 18 GHz). Source: D. Herraiz, H. Esteban, J. V. Morro, D. Herraiz, A. Belenguer, and V.E. Boria. [Updated 2022 Aug 16]. Available from: [22]. CC-BY-NC-ND 4.0. Used with permission.

The value for the coupling coefficient (K) of this inverter can be controlled by adjusting the iris width ($w_{r_{iris}}$) and the length of this section (l_u) (see **Figure 10a**). A parametric analysis was conducted using the simulator CST as shown in **Figure 11a**, revealing that the $w_{r_{iris}}$ parameter has a more influence on the normalized coupling coefficient $(\overline{K_i})$ than the l_u parameter. Consequently, the l_u parameter will be fixed at $l_u = 2$ mm, to achieve an optimal balance between manufacturing tolerances and the desired coupling coefficient value.

A set of filters will be designed at center frequencies $f_0 = 9, 15$ and 18 GHz, addressing whole frequency range which is not achievable with a conventional ESIW of the same waveguide width. The SRESIW section has been fixed according to the specifications outlined in Section 2. The bandpass filters do present a Chebyshev response of order $N = 4$, FBW of 3%, and return losses better than 20 dB. **Figure 11b** illustrates the normalized inverter constant $(\overline{K_i})$ as a function of iris width ($w_{r_{iris}}$) at various frequencies, which are utilized to design the filters and achieve the desired coupling coefficients $(\overline{K_i})$. Such values are calculated using the conventional microwave filter theory [28–30].

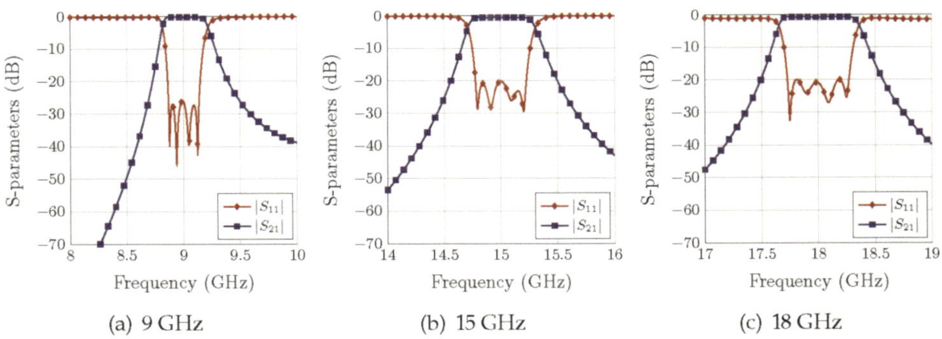

Figure 12.
Simulated results for the three designed SRESIW bandpass filters. Source: Author's original work.

Frequency (GHz)	w_{iris1}	w_{iris2}	w_{iris3}	l_{res1}	l_{res2}
9	11.970	8.7128	8.0161	23.710	26.933
15	6.978	4.526	4.213	9.326	10.399
18	6.126	4.062	3.812	7.334	8.215

Table 5.
Final values (in mm) of the design parameters of the filters.

Using **Figure 11b**, the filter responses were obtained and optimized, considering also the input transitions as described in Section 3.1.1. As illustrated in **Figure 12**, the filter performance aligns well with the specified requirements, and the optimum values of the filter dimensions are summarized in **Table 5**. These results have been obtained using the electromagnetic simulator CST.

The filter operating at the frequency of 15 GHz was manufactured and measured. **Figure 13** depicts the layers of the prototype and compares the simulated and measured S-parameters. The results demonstrate a high degree of correlation between the simulated and measured responses. The measured return losses are better than 18 dB between 14.80 and 15.23 GHz, and the insertion losses are lower than 1.05 dB.

It is important to highlight that, by fixing the waveguide width, various filters can be implemented without altering the geometry of the employed structure. This feature offers versatility and flexibility in design, allowing the adaptation of the structure to several applications without needing to change the waveguide width. This contributes to enhanced efficiency and simplicity in the implementation process of different filter configurations.

3.3 Power divider

Power dividers are critical microwave components in a wide variety of communication system front-ends, particularly in feeding antenna arrays. The DRESIW technology offers the potential to realize a high-performance and ultrawideband power divider, characterized by by low losses and seamless integration within the substrate. The structure of the proposed power divider is illustrated in **Figure 14** [23].

This design features two compensation elements: a post and an impedance matching adapter at the input port. The post is in the central layer of the power

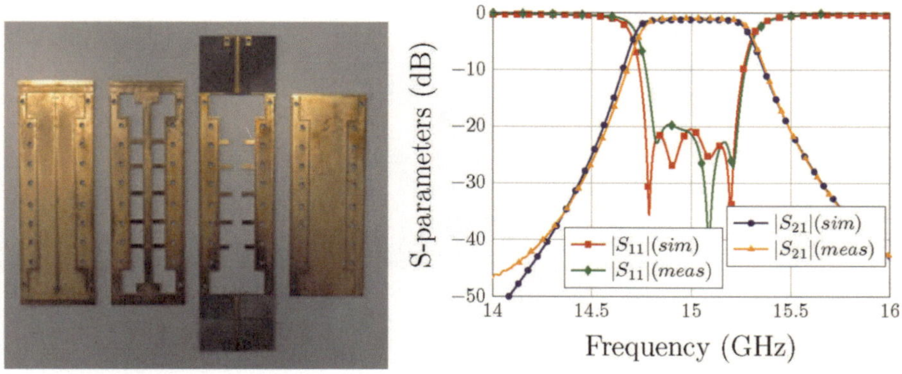

(a) (b)

Figure 13.
Layers of the filter prototype before assembly (a). Comparison between measured and simulated results for the prototype (b). Source: D. Herraiz, H. Esteban, J. V. Morro, D. Herraiz, A. Belenguer, and V.E. Boria. [Updated 2022 Aug 16]. Available from: [22]. CC-BY-NC-ND 4.0. Used with permission.

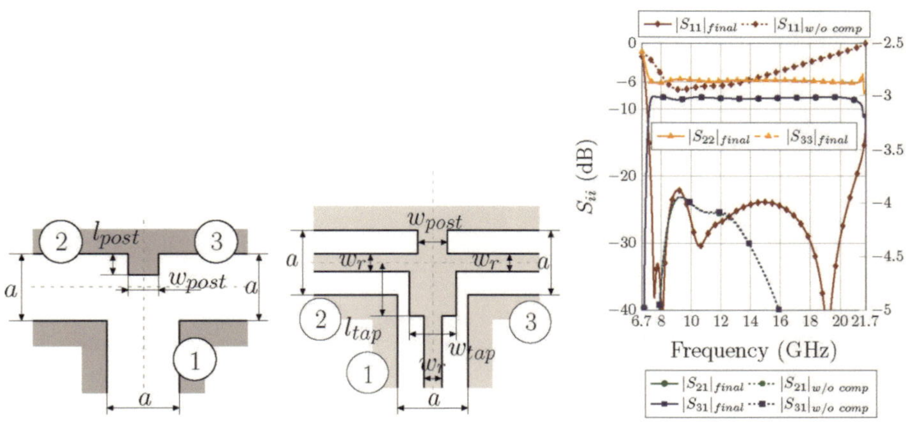

(a) (b) (c)

Figure 14.
Layer layouts in the DRESIW power divider: (a) Central layer. (b) Ridge layer. (c) Comparison of S-parameters with compensations elements and without them (note that results in green for S_{21}, with and without compensation, are, respectively, superimposed to those in blue for S_{31}). Source: Author's original work.

divider, allowing for the control of power division and compensating for the reactive effects of the power divider arms (see **Figure 14**). Defined by the parameters l_{post} and w_{post}, the post is centrally located and connected to the ridge with each ridge layer. The impedance matching adapter is a DRESIW widening section controlled by the parameters w_{tap} and l_{tap} in both ridge layers (see **Figure 14**).

A power divider designed to cover the X- and Ku-bands has been developed using the design procedure outlined in [23], leading to the optimum values $w_{post} = 2.291$ mm, $l_{post} = 5.297$ mm, $w_{tap} = 5.494$ mm, and $l_{tap} = 6.803$ mm. **Figure 14c** illustrates the response of the power divider along with the initial values without any compensation element. It can be clearly seen that the compensation elements are required. It can also be observed that the return losses are better than 22 dB within the frequency range of 7.2 to 21.7 GHz. Moreover, the power division between output ports was

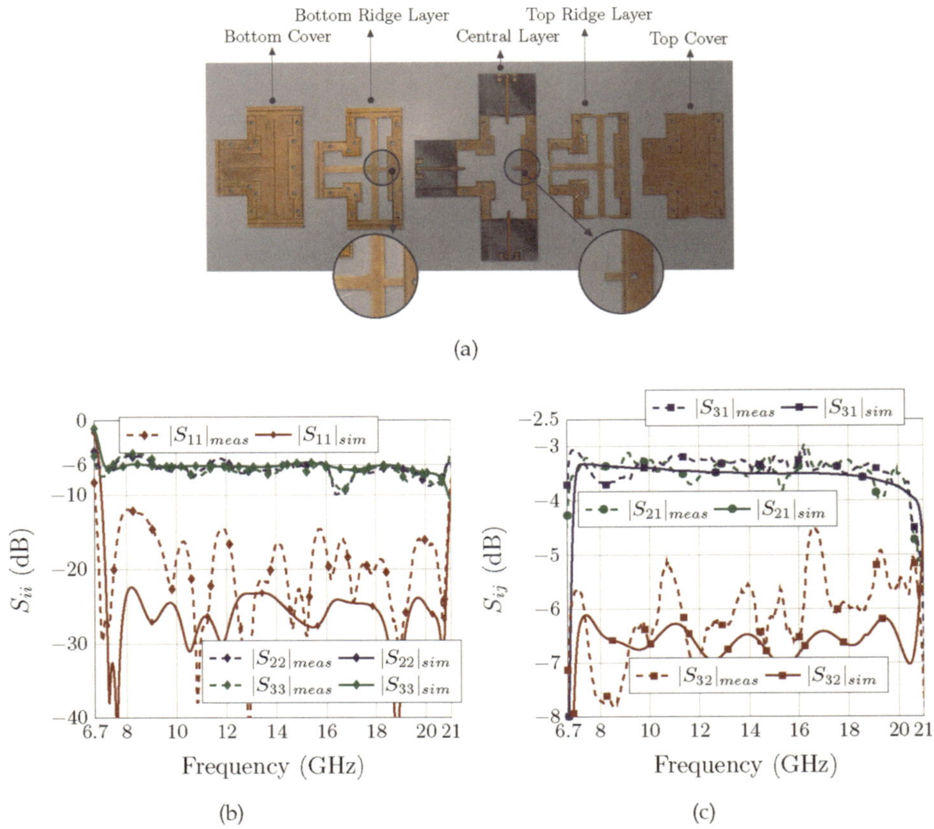

Figure 15.
Layers of the proposed power divider before assembly (a). Comparison between measured and simulated results of the prototype: (b) S_{ii}. (c) S_{ij}. Source: D. Herraiz, H. Esteban, D. Herraiz, A. Belenguer, and V.E. Boria. [Updated 2022 Nov 01]. Available from: [23]. CC-BY-NC-ND 4.0. Used with permission.

effectively achieved, as indicated by the transmission parameters $|S_{31}|$ and $|S_{21}|$, which were both approximately -3 dB.

The power divider has been fabricated using standard manufacturing techniques, as depicted in **Figure 15**. The transition described in Subsection 3.1.2 was employed for measurement purposes, and a comparison between simulated and measured S-parameters is shown in **Figure 15**. Measured return losses are better than 12 dB between 6.9 and 20.4 GHz. The power ratio (i.e., $|S_{21}|$ and $|S_{31}|$) exhibited a maximum

Reference	Technology	*BW* (GHz)	*FBW* (%)	R_L (dB)	P_W (dB)
[31]	Ridge SIW	5–11	75	11	−4.6
[32]	SIW	8–11	31.58	15	−4.2
[33]	ESIW	12–18	40	12	−4
[23]	DRESIW	6.9–20.4	99	12	−3.98

Table 6.
Comparison of the most relevant features of the proposed DRESIW power divider with other solutions proposed in the literature. R_L and P_W refer to the minimum return losses and maximum insertion losses in the bandwidth BW, respectively. FBW refers to fractional bandwidth.

value of −2.97 dB and a minimum value of −3.98 dB, demonstrating promising results. The maximum levels of amplitude and phase imbalance are 0.57 dB and 6.2°, respectively.

Table 6 compares the DRESIW power divider with other reported solutions, demonstrating a higher fractional bandwidth achieved alongside good performance in other indices.

4. Manufacturing process

SRESIW and DRESIW follow the same manufacturing principles as ESIW. To manufacture this type of circuits, the classical planar manufacturing guidelines and techniques for multilayer ESIW circuits can be followed [34]. In some cases, the increased bandwidth can induce to miss-matchings specially at low frequencies. Although the manufacturing process is very similar to the one used with ESIWs, there are three main aspects while manufacturing ridge ESIW circuits that must be considered as follows:

1. Edges of the ridge have to be fully metallized.

2. Consecutive layers must be well welded together, specially the ones which contain the ridge. Otherwise, the performance of the device will degrade.

3. Different press covers for the layers must be used in order to guarantee a good soldering.

4.1 Metallized edges

In order to obtain the metallized edges of the ridge layers, some processes such electrogalvanic plating can be used. **Figure 16a** shows the machine which can be used to realize this process.

A coating of copper, or other metal, is added to the substrate where the layers of the circuits are being manufactured. This copper is deposited in holes and edges. The thickness of the copper cladding can be adjusted directly in the machine configurations, or in older models, it can be estimated mathematically taking into account the temperature, time, and current.

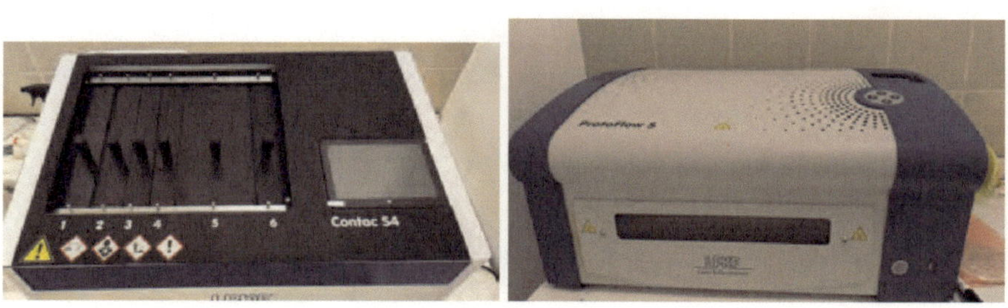

(a) (b)

Figure 16.
Metallizer machine (a) and reflow oven (b). Source: Author's original work.

4.2 Welding of consecutive layers

To guarantee the correct welding of the ridge layer with the consecutive ones, separate welding processes are recommended. By the use of tin pastes with different melting temperatures, these layers can be welded in separated welding stages. The process begins with the highest melting temperature and proceeds to lower melting temperature to prevent the circuit from becoming unsoldered during successive soldering stages.

For a deeper understanding of the soldering process, the previously discussed SRESIW and DRESIW structures will be explained in more detail. In the SRESIW case, the welding process should be done in two stages with the already mentioned two tin pastes. In the first one, the ridge layer and bottom cover (see **Figure 2**) must be welded together;[1] otherwise, small air gaps could be formed and currents will go through them, resulting in incresed insertion losses and reflections. Small overflows of tin in the solder joints between the ridge and the cover layer are a positive indication, as they result from proper welding between the two layers. These minor overflows do not cause significant losses and can be minimized by spreading the tin with a soldering iron (see **Figure 17b**). Then, the second welding stage must include the rest of the layers of the structure (i.e., the central layer and the top cover).

In case of a DRESIW structure, the process is quite similar; in this case, there are two ridge layers: one that connects with the top cover and a second one that connects with the bottom cover. There are three different stages in the soldering processes; in the first stage, top cover and top ridge layer (same ideas as before) are soldered using a tin paste with a high melting temperature; then, in the second stage, bottom cover and bottom ridge layer are soldered using the same tin paste as in the previous stage; and finally, the remaining layers of the DRESIW structure, along with the resulting structures from the previous stages, are soldered using tin paste with a lower melting temperature.

(a) (b)

Figure 17.
Press plates (a) and welding details for the SRESIW filter [22] (b). Source: Author's original work.

---

[1] The number of layers that conform the SRESIW is not relevant; the ridge layer can be compound of several layers. As all these layers have the same structure they are all the same ridge, and consequently they can be welded together.

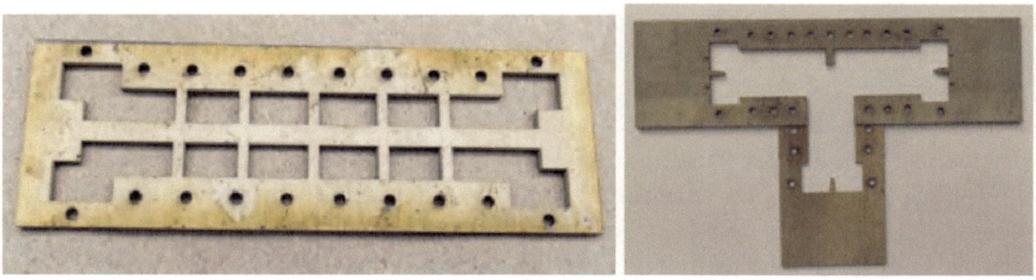

(a) (b)

Figure 18.
Press covers for welding the ridge layers: (a) for an SRESIW filter for the ridge layer and bottom cover [22]; (b) for the central layer of the DRESIW power divider [23]. Source: Author's original work.

4.3 Press covers

The soldering of integrated circuits can be realized in several ways. In the ESIW case, the reduced covers that do not need a reflow oven or pressing machine can be used [35]. However, for the SRESIW and DRESIW, the use of either a reflow oven or a pressing machine is necessary. **Figure 16b** illustrates the reflow oven machine, whereas the pressing system which is combined with the reflow oven is shown in **Figure 17a**. In this process, the stacked and aligned layers of the prototype are heated to high temperatures in the reflow oven machine, causing the solder paste to melt and join the layers together. After cooling, the device is fully soldered.

It is critical, as explained before, that ridge layers weld as well as possible with consecutive layers. A very good system to achieve this, is using pressing covers. In **Figure 18**, an example of this technique can be seen for the SRESIW filter (see Subsection 3.2) and DRESIW power divider (see Subsection 3.3). This system ensures that the required pressure is applied to the circuit, preventing small gaps between consecutive layers and ensuring that the tin paste is evenly distributed across the surface. The pressing covers, made from dielectric substrate remnants or other materials, are positioned above and below the layers during the welding process, to ensure proper pressure is applied in the relevant areas. These covers are removed once the welding is complete. Different press covers are used for various welding stages, depending on the specific pressing requirements.

5. Conclusions

In this chapter, an overview of the single and double ridge empty substrate integrated waveguide technologies has been presented. The motivation for the empty substrate integrated waveguide has been highlighted, emphasizing its low cost, high electrical performance, low mass, and compact profile. The bandwidth limitations of the ESIW have been addressed by the introduction of the single and double ridge empty substrate waveguides (SRESIWs and DRESIWs). Analytical expressions for the increased bandwidth of the SRESIW and DRESIW have been used, and several configurations have been studied with the objective of achieving a balance between manufacturing complexity and bandwidth.

Several microwave components have also been presented. First, transitions from microstrip to SRESIW and DRESIW have been shown and explained, providing an efficient method for seamless integration while maintaining the enhanced bandwidth. Subsequently, a set of bandpass filters in SRESIW has been designed using a novel U-shaped inverter, offering a new compact filter solution. Furthermore, a DRESIW power divider has also been studied, offering a solution with low losses and wideband performance. Finally, practical manufacturing considerations for SRESIW/DRESIW have been discussed, along with their application to some of the previously considered devices.

All these various microwave components underscore the potential of SRESIW and DRESIW in achieving ultrawideband and compact devices, making them promising solutions for future satellite communication systems, where size, weight, cost, and power are crucial parameters.

Acknowledgements

This work was supported by MICIU/AEI/10.13039/501100011033, Spanish Government, under the coordinated Research Project PID2022-136590OB, Subprojects C41 and C44, as well as by MICIU/AEI/10.13039/501100011033, Spanish Government, and by the European Union Next Generation EU/PRTR, under the coordinated Research Project TED2021-129196B, Subprojects C41 and C44. This work has been financed by European Social Fund + (ESF+).

Author details

David Herraiz[1*†], Héctor Esteban[1†], Darío Herraiz[2†], Ángel Belenguer[2†] and Vicente E. Boria[1†]

1 Instituto de Telecomunicaciones y Aplicaciones Multimedia, Universitat Politècnica de València, Valencia, Spain

2 Departamento de Ingeniería Eléctrica, Electrónica, Automática y Comunicaciones, Universidad de Castilla-La Mancha, Escuela Politécnica de Cuenca, Cuenca, Spain

*Address all correspondence to: daherza@teleco.upv.es; david.herraiz@ieee.org

† These authors contributed equally.

References

[1] Al Homssi B, Al-Hourani A, Wang K, Conder P, Kandeepan S, Choi J, et al. Next generation mega satellite networks for access equality: Opportunities, challenges, and performance. IEEE Communications Magazine. 2022;**60**(4): 18-24

[2] Wang Y, Meng X, Luo H, Kong C, Li K-X, Wang X, et al. When leo mega-constellations meet new radio: A holistic performance evaluation framework. IEEE Access. 2023;**11**:146 122-146 135

[3] Chen G, Wu S, Deng Y, Jiao J, Zhang Q. VLEO satellite constellation design for regional aviation and marine coverage. IEEE Transactions on Network Science and Engineering. 2024;**11**(1): 1188-1201

[4] Barnes A, Boetti A, Marchand L, Hopkins J. An overview of microwave component requirements for future space applications. European Gallium Arsenide and Other Semiconductor Application Symposium, GAAS. 2005; **2005**:5-12

[5] Tasić A, Serdijn WA, Larson LE, Setti G. Circuits and Systems for Future Generations of Wireless Communications. Springer; 2009

[6] Deslandes D, Wu K. Integrated microstrip and rectangular waveguide in planar form. IEEE Microwave and Wireless Components Letters. 2001; **11**(2):68-70

[7] Iqbal A, Tiang JJ, Wong SK, Alibakhshikenari M, Falcone F, Limiti E. Miniaturization trends in substrate integrated waveguide (SIW) filters: A review. IEEE Access. 2020;**8**:223 287-223 305

[8] Alibakhshikenari M, Ali EM, Soruri M, Dalarsson M, Naser-Moghadasi M, Virdee BS, et al. A comprehensive survey on antennas on-chip based on metamaterial, metasurface, and substrate integrated waveguide principles for millimeter-waves and terahertz integrated circuits and systems. IEEE Access. 2022;**10**: 3668-3692

[9] Belenguer A, Esteban H, Boria V. Novel empty substrate integrated waveguide for high-performance microwave integrated circuits. IEEE Transactions on Microwave Theory and Techniques. 2014;**62**(4): 832-839

[10] Parment F, Ghiotto A, Vuong T-P, Duchamp J-M, Wu K. Air-filled substrate integrated waveguide for low-loss and high power-handling millimeter-wave substrate integrated circuits. IEEE Transactions on Microwave Theory and Techniques. 2015;**63**(4):1228-1238

[11] Herraiz D, Esteban H, Herraiz D, Belenguer A, Boria VE. A novel in-line ESICL-to-ESIW transition for high-performance empty substrate integrated microwave systems. IEEE Microwave and Wireless Technology Letters. 2024; **34**(4):363-366

[12] Yan N, Ji C, Luo Y, Ma K. Wideband and miniaturized phase-corrected empty substrate-integrated h-plane horn antenna for 5g millimeter waves. IEEE Transactions on Antennas and Propagation. 2023;**71**(9):7638-7643

[13] Qiang J, Xu F, Yang L, Zhan J. Enhanced performance of multilayer bandpass filter using slow-wave empty substrate- integrated waveguide (SW-ESIW). IEEE Microwave and Wireless Components Letters. 2021;**31**(12): 1279-1282

[14] Máximo-Gutiérrez C, Hinojosa J, Melcon AA. Narrowband and wideband bandpass filters based on empty substrate integrated waveguide loaded with dielectric elements. IEEE Access. 2021;**9**:32 094-32 105

[15] Fernandez MD, Herraiz D, Herraiz D, Alomainy A, Belenguer A. Dielectricless floating-patch antenna at 28 ghz fed with an empty substrate integrated waveguide. IEEE Access. 2023;**11**:65 599-65 607

[16] Peng H, Zhao F, Liu Y, Tatu SO, Yang T. Robust microstrip to empty substrate-integrated waveguide transition using tapered artificial dielectric slab matrix. IEEE Microwave and Wireless Components Letters. 2020; **30**(9):849-852

[17] Qi Z, Li X, Xiao J, Zhu H. Low-cost empty substrate integrated waveguide slot arrays for millimeter-wave applications. IEEE Antennas and Wireless Propagation Letters. 2019; **18**(5):1021-1025

[18] Murad NA, Lancaster MJ, Wang Y, Ke M. Micromachined rectangular coaxial line to ridge waveguide transition. In: 2009 IEEE 10th Annual Wireless and Microwave Technology Conference. Clearwater, FL, USA: IEEE; 2009. pp. 1-5

[19] Herraiz D, Esteban H, Martínez JA, Belenguer A, Boria V. Microstrip to ridge empty substrate-integrated waveguide transition for broadband microwave applications. IEEE Microwave and Wireless Components Letters. 2020; **30**(3):257-260

[20] Herraiz D, Esteban H, Herraiz D, Vidal A, Belenguer A, Boria VE. Microstrip to double ridge empty substrate integrated waveguide transitions based on exponential and superelliptical dielectric taper. IEEE Access. 2021;**9**:165 745-165 753

[21] Hopfer S. The design of ridged waveguides. IRE Transactions on Microwave Theory and Techniques. 1955;**3**(5):20-29

[22] Herraiz D, Esteban H, Morro JV, Herraiz D, Belenguer A, Boria V. Bandpass filter in ridge empty substrate integrated waveguide with U-shaped impedance inverters. IEEE Microwave and Wireless Technology Letters. 2023; **33**(1):31-34

[23] Herraiz D, Esteban H, Herraiz D, Belenguer A, Boria VE. Wideband H-plane T-junction power divider with compensation elements using double ridged empty substrate integrated waveguide. IEEE Microwave and Wireless Technology Letters. 2023;**33**(3):255-258

[24] Marcuvitz N. Waveguide handbook. IET. 1951;**21**:249-253

[25] Esteban H, Belenguer A, Sánchez JR, Bachiller C, Boria VE. Improved low reflection transition from microstrip line to empty substrate-integrated waveguide. IEEE Microwave and Wireless Components Letters. 2017; **27**(8):685-687

[26] Taringou F, Dousset D, Bornemann J, Wu K. Substrate-integrated waveguide transitions to planar transmission-line technologies. IEEE/MTT-S International Microwave Symposium Digest. 2012;**2012**:1-3

[27] Liu Z, Xu J, Wang W. Wideband transition from microstrip line-to-empty substrate-integrated waveguide without sharp dielectric taper. IEEE Microwave and Wireless Components Letters. 2019;**29**(1):20-22

[28] Matthaei G. Microwave Filters, Impedance-Matching Networks and Coupling Structures. Artech House Book; 1980. pp. 775-809

[29] Cameron RJ. Advanced filter synthesis. IEEE Microwave Magazine. 2011;**12**(6):42-61

[30] Soto P, Tarín E, Boria VE, Vicente C, Gil J, Gimeno B. Accurate synthesis and design of wideband and inhomogeneous inductive waveguide filters. IEEE Transactions on Microwave Theory and Techniques. 2010;**58**(8):2220-2230

[31] Kazemi R, Fathy AE. Design of single-ridge SIW power dividers with over 75% bandwidth. In: 2014 IEEE MTT-S International Microwave Symposium (IMS2014). Vol. 2014. Tampa, FL, USA: IEEE; 2014. pp. 1-3

[32] Lorente D, Schreiber A, Limbach M, Esteban H, Boria VE. High permittivity CPW-SIW power divider for antenna feed networks in airborne phased arrays applications. In: 2020 17th European Radar Conference (EuRAD). Utrecht, Netherlands: IEEE; 2021. pp. 254-257

[33] Reddy MG, Bharath K, Nandigama SV, Dasari R. Design of half power divider at Ku-band using empty substrate integrated waveguide. In: 2020 IEEE MTT-S Latin America Microwave Conference (LAMC 2020). Cali, Colombia: IEEE; 2021. pp. 1-3

[34] Herraiz D, Martínez L, Ballesteros JA, Fernandez MD, Esteban H, Belenguer Á. Manufacturing methods based on planar circuits. In: Fernandez MD, Ballesteros JA, Esteban H, Belenguer Á, editors. Hybrid Planar. Rijeka: IntechOpen; 2022. [Online]. Available from:. DOI: 10.5772/intechopen.109114

[35] Fernandez MD, Belenguer A, Herraiz D, Martinez L, Ballesteros JA. Simplified high-efficiency soldering method for the fabrication of devices on empty substrate integrated waveguides. IEEE Access. 2024;**12**:61 336-61 342

Chapter 3

High-Gain Antennas for Space Applications Based on Multifaceted and Curved Reflectarrays

Borja Imaz-Lueje, Daniel Martinez-de-Rioja, Eduardo Martinez-de-Rioja, Yolanda Rodriguez-Vaqueiro, Manuel Arrebola, Marcos R. Pino, Antonio Pino and José Antonio Encinar

Abstract

Recent and upcoming space applications demand advanced high-gain antennas onboard the satellites to meet challenging specifications in terms of bandwidth or multibeam operation. Reflectarray antennas are emerging as a potential candidate for these applications due to their planar and lightweight reflecting surface, easily integrated with the satellite platform. In the recent SWOT, MARCO, and ISARA missions from NASA, reflectarrays have been integrated into the spacecraft by dividing the surface into multiple flat panels and deploying them as a single-facet aperture. However, this approach limits the bandwidth of the antenna due to the differential spatial phase delay effect. The use of conformal apertures approaching a curved profile can mitigate this effect, providing broader bandwidths and improved antenna performances. This chapter is dedicated to the comprehensive analysis and design of different non-planar aperture antennas based on reflectarrays, including multifaceted and doubly curved surfaces, engineered to exhibit electrical and mechanical performance suitable for space applications.

Keywords: satellite communications, multifaceted structures, doubly curved surfaces, multibeam antennas, reflectarray antennas

1. Introduction

A reflectarray antenna consists of an array of reflecting elements, typically implemented in printed circuit technology (e.g., microstrip patches, rings, etc.), which is illuminated by a primary feed. **Figure 1** depicts a schematic view of a flat reflectarray antenna in offset-fed configuration, including the reference coordinate system [1]. Each unitary element forming the reflecting surface (i.e., reflectarray cell) must be properly designed to introduce a certain phase-shift in the incident electric field radiated by the feed. Thanks to the phasing capability of the reflecting elements, reflectarray antennas are able to focus high-gain collimated beams [2, 3] or to generate

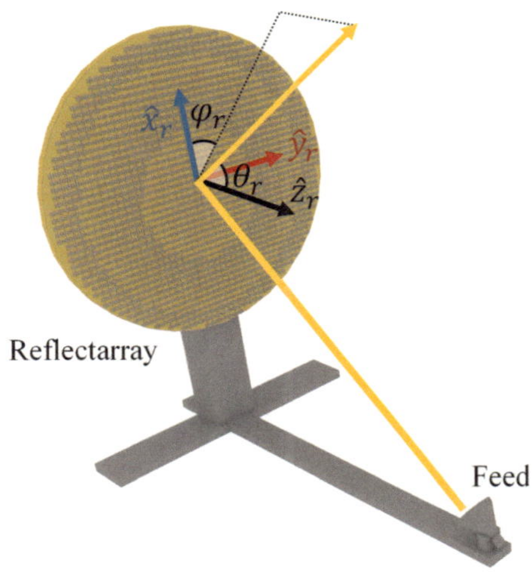

Figure 1.
Schematic representation of a printed reflectarray.

contoured beams with a specific shaping [4, 5]. There are different design techniques to achieve the required phase-shift in each reflectarray cell, such as: use of variable-size elements (the most used), fixed-size elements with variable rotation angles, variable-length delay lines attached to the elements, and so on.

Since reflectarrays are spatially fed antennas (as traditional parabolic reflectors), they do not require expensive feeding or beamforming networks, so they achieve a significant reduction in ohmic losses, fabrication complexity, and cost with respect to phased arrays [1]. Compared to metallic reflectors, reflectarrays offer additional capabilities to operate at multiple frequencies [6, 7] and to generate independent beams in orthogonal polarizations [8, 9] by selecting a suitable configuration for the reflectarray cells. For example, multi-frequency reflectarrays can be realized by stacking several layers of reflecting elements (one layer for each frequency), as in [7], or by combining different types of reflecting elements (with a different resonance frequency) in a single layer, as in [8].

One significant advantage of reflectarray antennas when considered for space applications is that they can be easily integrated with the satellite platform, thanks to their planar and lightweight reflecting surface [10–13]. Typically, reflectarrays are integrated into the spacecraft by dividing the surface into multiple flat panels and deploying them as a single-facet aperture, as in the recent SWOT, MARCO, and ISARA missions from NASA [11, 13]. The feed can be attached to the spacecraft [11, 12] or deployed together with the reflecting panels [13].

The operating bandwidth of reflectarray antennas is limited by two factors: the bandwidth of the reflecting elements and the variation with frequency of the spatial phase delay. Concerning the first factor, several solutions have been shown for bandwidth improvement, such as those involving broadband reflectarray cells based on multi-resonant elements [14] and cells formed by several stacked layers [15]. However, in those configurations with a reduced focal distance-to-diameter ratio (F/D), the use of a planar aperture limits the operating bandwidth of the reflectarray antenna due to the differential spatial phase delay effect [1]. The use of conformal apertures following a parabolic profile can mitigate this issue, leading to a broader bandwidth, as will be explained in the next section.

2. Antenna optics

Reflectarray antennas are typically arranged following a planar single-offset structure based on a flat reflective surface [1], as depicted in **Figure 2**. Here, the reflective surface is in a chordal plane of an equivalent parabolic reflector (gray shape shown in **Figure 2(a)**) and a primary feed, located in the focus of the quadric surface. The elements are expressed in a Global Coordinate System (GCS), and the parameters that fully characterize the antenna optics are the focal distance (F), offset of the surface (H), the subtended angle (θ_0), and the projection aperture (D).

However, the reflectarray planar surface is sensitive to the spatial phase delay effect, as demonstrated in [16] and illustrated in **Figure 2(b)**. Let us consider two elements of the reflectarray surface, whose paths from the focal point are S_1, S_2. The phase difference between both paths can be expressed as [1]:

$$\Delta S = S_1 - S_2 = (N + d)\lambda_0 \tag{1}$$

N and d are a multiple and fraction, respectively, of the operational wavelength (λ_0). If the wavelength varies $\Delta\lambda$, the phase difference defined in Eq. (1) depends on a term $(N + d)\Delta\lambda$ that can be several wavelengths or phase cycles (360°). Since the reflectarray elements are designed to only compensate the phase difference at central frequency, the frequency excursion $\Delta\lambda$ derivates in an error in the re-radiated phase front. The greater the value of N (i.e., the greater the difference between paths), the higher the sensitivity of the planar reflector to frequency deviations. There are several strategies to minimize the spatial phase delay effect, including the use of multi-resonant unit-cell topologies [17, 18] or design methods based on optimization [15, 19]. Another alternative is the use of multifaceted or parabolic reflectarray surfaces, as the ones shown in **Figure 3**. Such structures seek to mimic the parabolic curvature of a classic reflector to reduce the difference between paths and, therefore, the differential spatial phase delay.

One approach is the implementation of a multifaceted reflectarray [20] consisting of multiple panels, assembled edge to edge, that approximates the equivalent parabolic reflector surface (see **Figure 3(a)**), defined according to Eq. (1). These panels are in consecutive or tangent planes of the equivalent reflector surface. In the simplest multifaceted topologies, the panels approximate the paraboloid surface along only one

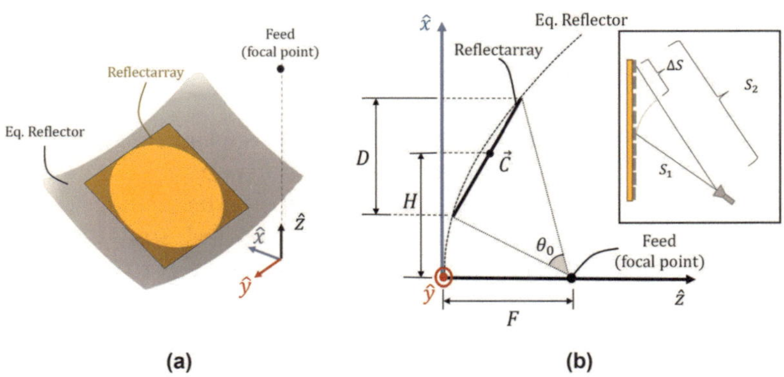

Figure 2.
Typical optical configuration of a flat reflectarray antenna: (a) 3D perspective; (b) cross section (XZ plane) and detail of the differential spatial phase delay.

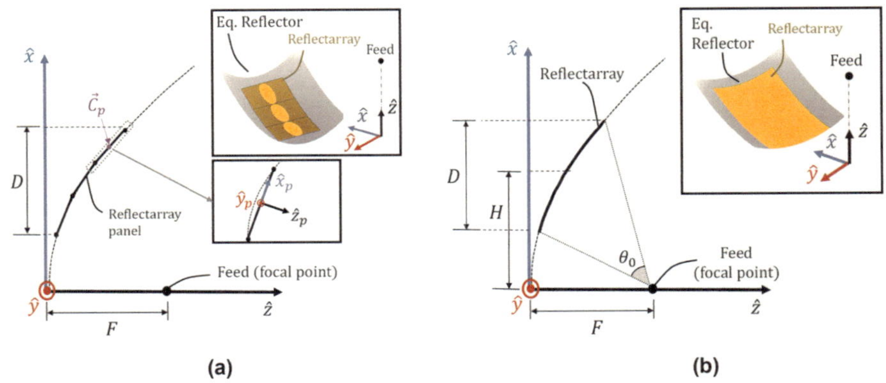

Figure 3.
Antenna optics of a multifaceted (a) and a parabolic reflectarray (b). Sketch and cross section along the offset plane.

plane, perpendicular to the XZ or the YZ planes (see **Figure 3**). In more sophisticated multifaceted structures, the panels follow several planes, secant to the equivalent reflector [21, 22]. Each reflectarray panel p is fully characterized from its shape, dimensions, center, and a Panel Coordinate System (PCS_p).

Another solution relies on the use of parabolic reflectarrays with doubly curved surfaces that follow exactly the equivalent reflector curvature (see **Figure 3(b)**). In a single-offset configuration, the parabolic reflectarray is fully characterized according to the parameters: D, H, F, θ_0.

3. Analysis techniques of reflectarray antennas

After defining the antenna optics, the next steps of the reflectarray design consist of the calculation of the required phase shifts on the reflectarray surface and the accurate characterization and design of the reflecting elements in each cell. In this section, we start showing the fundamentals of these procedures for a planar single-faceted reflectarray, and then, we present the extension of the analysis techniques for the cases of non-planar multifaceted reflectarrays and curved reflectarrays.

3.1 Conventional single-faceted reflectarrays

The required phase distribution on the reflectarray surface must be calculated according to the specifications of the antenna radiation patterns (e.g., a collimated beam, a shaped beam, etc.). For the sake of simplicity, let us consider a reflectarray designed to produce a collimated beam. According to phased-array theory, the phase of the tangential reflected electric field in each reflectarray cell i to radiate a collimated beam in the direction (θ_b, φ_b) can be obtained as:

$$\phi(x_i, y_i) = -k_0 \sin\theta_b (\cos\varphi_b x_i + \sin\varphi_b y_i), \tag{2}$$

where k_0 is the propagation constant in vacuum, and (x_i, y_i) are the coordinates of the cell i, typically expressed in the local coordinate system of the reflector. Then, the phase delay to be introduced by each cell is computed as the difference between Eq. (2) and the phase of the incident field coming from the feed, resulting in:

$$\Phi_{Ri} = k_0\left(d_i - \left(x_i\cos(\varphi_b) + y_i\sin(\varphi_b)\right)\sin(\theta_b)\right), \tag{3}$$

where d_i is the distance from the phase center of the feed to the cell i. The phase delays obtained on the reflectarray surface are typically normalized to a range of 360°, to simplify the reflectarray design. To generate a shaped or contoured beam, the phase values given by Eq. (3) can be replaced by an appropriate phase distribution obtained by a phase-only pattern synthesis technique [15, 16].

The definition and electromagnetic (EM) characterization of the reflectarray cells to provide the required phase-shift values represents a key aspect of the reflectarray design. The amplitude and phase response of the reflectarray cells as a function of some of its geometrical parameters (element size, delay-line length, rotation angle, etc.) must be accurately analyzed, taking into account the incidence angle of the electric field, the polarization of the incident field, and the operating frequency band.

The EM characterization of the reflectarray cells is typically carried out under the local-periodicity assumption, which considers an infinite periodic array model for the analysis of each cell [1, 2]. This approach accounts for the mutual coupling between the elements, at the same time as providing an accurate and efficient characterization of the reflectarray cells [23]. The EM analysis of the reflectarray cell under periodic boundary conditions can be implemented by different numerical techniques, such as Spectral-Domain Method of Moments (SD-MoM) [23], Finite-Difference Time-Domain method (FDTD) [24], or Finite Element Method (FEM) [25]. All these techniques characterize the EM behavior of the reflectarray cell by a reflection matrix R, which relates the tangential components of the reflected electric field $\left(E_x^{ref}, E_y^{ref}\right)$ on the cell with the tangential components of the incident field $\left(E_x^{inc}, E_y^{inc}\right)$:

$$\begin{pmatrix} E_x^{ref} \\ E_y^{ref} \end{pmatrix} = \begin{pmatrix} R_{XX} & R_{XY} \\ R_{YX} & R_{YY} \end{pmatrix} \begin{pmatrix} E_x^{inc} \\ E_y^{inc} \end{pmatrix}. \tag{4}$$

As shown in Eq. (4), the matrix R consists of four complex coefficients, which are divided into two co-polar coefficients (R_{XX} and R_{YY}) and two cross-polar coefficients (R_{XY} and R_{YX}). The module of R_{XX} and R_{YY} represents the reflection losses of the cell for each linear polarization, while the phase shift introduced in each polarization is given by the phases of R_{XX} and R_{YY}. On the other hand, the cross-polar coefficients R_{XY} and R_{YX} provide the levels of cross-polarization generated at the cell.

The configuration of the reflectarray cell must be optimized to ensure a phase variation range of 360° and low reflection losses at the operating frequency. Once the required phase distributions on the reflectarray have been obtained (either by Eq. (3) or by a pattern synthesis technique), the next step of the design consists of adjusting the geometric parameters of the reflecting elements in each cell (i.e., element size, rotation angle, etc.), so that the phases of the R_{XX} and R_{YY} coefficients match the objective phases previously obtained.

The radiation patterns of the designed reflectarray antenna can be computed from the tangential components of the reflected electric field on the reflectarray cells [1]. The reflected tangential electric field on the reflectarray cell i can be obtained by multiplying the incident tangential electric field components and the reflection matrix R_i obtained from the EM analysis of the cell i, as shown in Eq. (4). For this purpose, the incident field on each reflectarray cell can be computed from the far field radiation pattern of the feed, which is normally modeled by a simple analytical $\cos^q(\theta)$ function, where q controls the shape of the feed pattern (a higher q corresponds to a narrower beamwidth with higher directivity, while a lower q indicates a wider beamwidth). Note that when the reflectarray is not in the far field region of the feed, the incident field must be computed from the near field radiated by the feed. Then, the co-polar and cross-polar components of the radiation patterns are obtained by Fast Fourier Transform (FFT) algorithms to improve the numerical efficiency [1]. Finally, the antenna gain, which includes the spillover, illumination, and dissipative losses, is computed by dividing the squared radiated field by the input power in the feed-horn.

3.2 Multifaceted reflectarrays

In the case of multifaceted reflectarrays, the composite non-planar surface that approximates a parabolic profile compensates partially the phase-shift introduced to the impinging wave to produce a collimate beam pattern. The rest of the required phase-shift is provided by the radiating element. In this sense, the phase required in a multifaceted structure to generate a high-gain beam pattern can be calculated by extending Eq. (3), defined for a single-faceted configuration, as follows:

$$\Phi_{Ri} = k_0\left(d_i - \left(x_i' \sin\theta_b \cos\varphi_b + y_i' \sin\theta_b \sin\varphi_b + z_i' \cos\theta_b\right)\right) \tag{5}$$

Here, the coordinates (x_i', y_i', z_i') are referenced in a unique coordinate system of the whole structure, considered as a baseline (typically, the *GCS* depicted in **Figures 2** and **3** is chosen for this purpose).

To characterize a multifaceted reflectarray structure, the process to be followed is similar to the one described above for conventional reflectarrays. First, the tangential reflected fields in each reflectarray panel are calculated through the behavior of the reflectarray surface (i.e., the R_i matrix) and the impinging wave in the panel, as devised in Eq. (4). The output of this process are the values E_x^{ref} and E_y^{ref} for each i element of the panel. Second, the radiated far field of each panel $\left(\vec{E}_p\right)$ is computed with respect to its local coordinate system (PCS_p), from such tangential components [1]. Finally, the total radiated far field of the multifaceted structure $\vec{E}$ is the coherent sum of the far field contributions of each panel [26]:

$$\vec{E}\left(\vec{r}_a\right) = \sum_{p=1}^{N} \vec{E}_p\left(\vec{r}_a\right), \tag{6}$$

$\vec{r}_a$ being the direction of propagation referred to an arbitrary coordinate system and $\vec{E}_p\left(\vec{r}_a\right)$ the partial radiated field of each panel, expressed in the same coordinate system. For simplicity, the *GCS* is taken as a baseline to compute this operation, so $\vec{r}_a$ is the direction of propagation in the GCS, from now, denoted as $\vec{r}_g$.

Since the partial far fields are computed in the local coordinate system of each panel $\left(\vec{E}_{pPCS_p}\right)$, the following operations are performed over the $\vec{E}_p$ field: (a) change of basis from the spherical system of PCS_p to the spherical system of GCS through the transformation matrix $S_{p\leftarrow G}\left(\vec{r}, \vec{r}_p\right)$, $\vec{r}_p$ being the direction of propagation referred to the PCS_p; (b) translation from the center of the PCS_p ($\vec{C}_p$ in **Figure 2(a)**) to the center of the GCS, introducing the phase correction $k_0\vec{C}_p \cdot \hat{r}_g$. Here, $\hat{r}_g$ is the direction of space, expressed in the spherical basis of the GCS. Both operations are summarized in the following expression:

$$\vec{E}_p\left(\vec{r}\right)|_{GCS} = \exp\left(jk_0\vec{C}_p \cdot \hat{r}_g\right)\vec{E}_p\left(\vec{r}_p\right)|_{PCS_p} S_{p\leftarrow G}\left(\vec{r}_g, \vec{r}_p\right). \quad (7)$$

The computation of the total far field components of the multifaceted reflectarray can be done by the direct evaluation of Eq. (6) in each direction of space $\hat{r}_g$, typically expressed through the spectral variables $u = \sin\theta\cos\varphi$ and $v = \sin\theta\sin\varphi$. In contrast with the analysis of conventional reflectarrays, the use of FFTs to compute more efficiently the far field components cannot be performed directly in this case. This limitation arises from the fact that far field components do not form a regular mesh of spectral points concerning the GCS, due to the different field of view considered for each panel.

3.3 Curved reflectarrays

Curved reflectarrays with a parabolic surface can be considered as the natural evolution of conformal multifaceted reflectarrays. The main idea behind this design concept is that the parabolic shape of the reflectarray will produce a collimated beam in a natural way, leaving the phase adjustment provided by the printed elements to implement some advanced features, such as an optimized beam shaping [27], conversion from linear polarization (LP) into circular polarization (CP) [27, 28], generation of two separated beams in different polarizations [27], and so on.

When designing a curved reflectarray, the parabolic shape of the surface prevents defining a regular lattice of cells (i.e., with the same unit-cell dimensions along the x and y axes, P_x and P_y). In [3, 27], the process to define the final lattice of cells on the parabolic surface has been carried out as follows. First, a grid of cells with constant dimensions ($P_x = P_y$) placed in the aperture plane of the paraboloid (the area enclosed by the rim of the dish) has been computed. Then, the periodic grid is projected over the parabolic surface, resulting in a grid of cells with slightly variable sizes.

Since Eqs. (2) and (3) relating the phase to the coordinates of the reflectarray cell are no longer valid for curved surfaces, a vectorial treatment is adopted to synthesize the phase distribution of the reflectarray elements [29]. The difference between the unit vectors of the reflected ($\hat{r}$) and the incident ($\hat{i}$) rays at the reflectarray surface is denoted as the virtual normal $\vec{N}$:

$$\vec{N} = \hat{r} - \hat{i} \quad (8)$$

The virtual normal is related to the derivatives of the phase across the curved surface through the following expression:

$$\vec{N} = \begin{pmatrix} N_x \\ N_y \\ N_z \end{pmatrix} = \frac{1}{\sqrt{z_x^2 + z_y^2 + 1}} \begin{pmatrix} z_y^2 + 1 & -z_x z_y \\ -z_x z_y & z_x^2 + 1 \\ z_x & z_y \end{pmatrix} \begin{pmatrix} \partial L/\partial x \\ \partial L/\partial y \end{pmatrix}, \quad (9)$$

where z_x and z_y are the partial derivatives of the function $z(x,y)$ (the one that describes the reflectarray surface), and L is the path length function to be implemented at reflectarray points, which is proportional to the phase:

$$\Phi_{Ri}(x,y) = -\frac{2\pi}{\lambda_0} L(x,y). \tag{10}$$

The inverse relation reduces to

$$\begin{pmatrix} \partial L/\partial x \\ \partial L/\partial x \end{pmatrix} = \begin{pmatrix} 1 & 0 & z_x \\ 0 & 1 & z_y \end{pmatrix} \begin{pmatrix} N_x \\ N_y \\ N_z \end{pmatrix}. \tag{11}$$

The EM analysis of the reflectarray cells can be performed in a similar way to the case of a flat reflectarray, described in Section 2.1. Numerical techniques, such as SD-MoM, can be applied to obtain the reflection matrix R_i for the i-th cell of the parabolic reflector, considering a locally flat and periodic array structure with the unit-cell dimensions (P_x, P_y) and angles of incidence (θ_i, φ_i) associated to the cell i on the parabolic surface [3, 27]. This approximation leads to accurate results in the simulation of the curved reflectarray, as it was shown in [27].

Once the reflectarray cells are properly characterized, the radiation patterns of the curved reflectarray antenna can be obtained by advanced Physical Optics (PO) simulations. In [3, 27], a PO algorithm based on the theory presented in [30, 31] was used to simulate the parabolic reflectarray. The surface of the reflectarray was discretized in small patches, each matching a reflectarray cell. Both electric and magnetic currents were considered. The surface reflection at each reflectarray cell i is modeled by the reflection matrix R_i, computed as abovementioned, which relates the tangential components of the incident and reflected electric fields.

4. Practical designs for space communications

The next section is devoted to present several studies that evaluate the feasibility of multifaceted and parabolic reflectarrays onboard satellites for different space applications and antenna optics. These studies are focused on the evaluation of multifaceted and curved reflectarrays to provide different electrical features (such as polarization conversion or dual-band operation, among others) while enhancing the performance provided by alternative antenna solutions.

4.1 Multifaceted reflectarrays for satellite communications

The use of multifaceted reflectarrays (denoted as MFRAs, in the following subsections) in satellite communications has been investigated in several works reported in the literature. In [32], some multifaceted configurations have been proposed to provide shaped-beam coverages onboard the spacecraft. Recent studies have explored the use of multifaceted topologies to enhance the bandwidth provided by a conventional high-gain deployable reflectarray, while they maintain high integrability with the platform [33, 34]. In this sense, the following subsections describe these structures designed for two potential space applications: deployable multifaceted reflectarrays for SmallSats [26] and an extremely large multifaceted aperture for a remote sensing mission [35]. In both cases, the electrical performance of each topology is evaluated using the formulation reported in Section 2.2 and compared with an equivalent single facet aperture (denoted as SFRA).

4.1.1 Deployable multifaceted reflectarrays for SmallSats

A deployable MFRA designed to be embarked onboard SmallSats has been presented in [26], and it is shown in **Figure 4**. The reflectarray aperture comprises three identical panels that can be folded along the spacecraft body. Once the satellite is in orbit, the panels are deployed, following a parabolic cylinder structure. In [33], different accommodations of this reflectarray concept have been evaluated, considering the physical limitations of a 3 U CubeSat (30 cm × 10 cm × 10 cm). These designs provide a high-gain coverage with relative bandwidths of about 20%, which means 40% more bandwidth than an SFRA.

Two multifaceted demonstrators have been designed, manufactured, and tested in [26] to study the features of the deployable multifaceted approach. They operate in Ka-band in a single linear polarization (one in X-and the other in Y-polarization), providing a high-gain pencil beam pattern. The phase-shift required in each reflectarray surface is properly adjusted using Eq. (5), resulting in the phase distribution depicted in **Figure 5(a)**. **Figure 5(b)** shows the phase distribution obtained if an SFRA is considered. Both multifaceted designs smooth the phase distribution along the sectoring plane (in which the panels are not aligned), which relaxes the constraints

Figure 4.
Sketch of the proposed deployable multifaceted reflectarray onboard SmallSat.

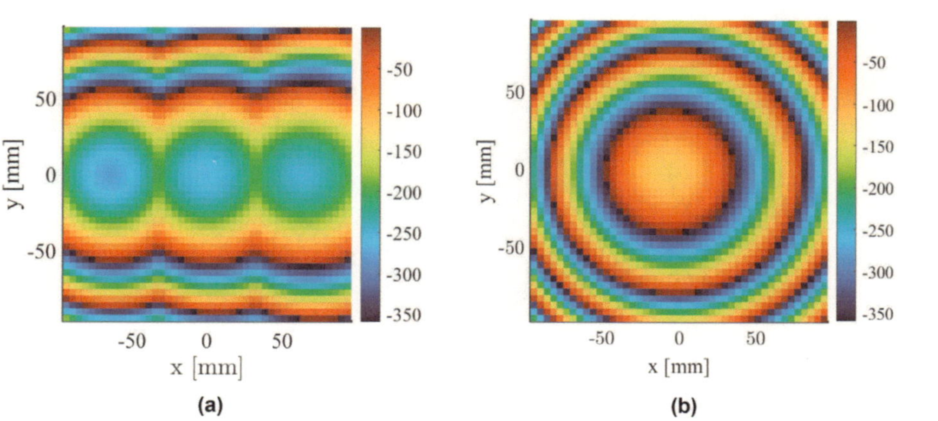

Figure 5.
Frontal view of the phase distributions in degrees [°] required in the multifaceted designs (a) and in an equivalent single facet structure (b).

imposed to the unit-cell and improves the robustness against the spatial phase delay effect explained in Section 1.2.

Figure 6 shows the measurement setup of both multifaceted designs. Each panel comprises low-profile single-layer radiating elements, depicted in **Figure 3(b)**. Such unit-cell, defined and analyzed in [36], is composed of three coplanar dipoles aligned with the x or y axis of the reflectarray panel. By the proper adjustment of their variables (w, s, l, h as depicted in **Figure 6(b)**), the unit-cell can provide the required phase (see **Figure 5(a)**) for each multifaceted design.

Good agreement has been achieved between simulation and measurements. Both predicted and measured results demonstrate the ability of the multifaceted structure to provide dual-linear polarization and enhanced performance compared with an equivalent single facet reflectarray. **Figure 7** provides the measured radiation pattern in elevation (plane in which the sectorization is applied) of one of the MFRAs compared with the SFRA. The coordinate system shown in **Figure 6(a)** is taken as a baseline. The multifaceted demonstrator exhibits higher robustness against defocusing of the main beam in-band, which results in a higher stability of the gain. According to this parameter, further analysis reported in [26] and shown in **Figure 8** reveals a 1 dB gain bandwidth of about 20% in the multifaceted designs (which is 30% enhancement regarding the SFRA). Besides, the proposed demonstrators cover the

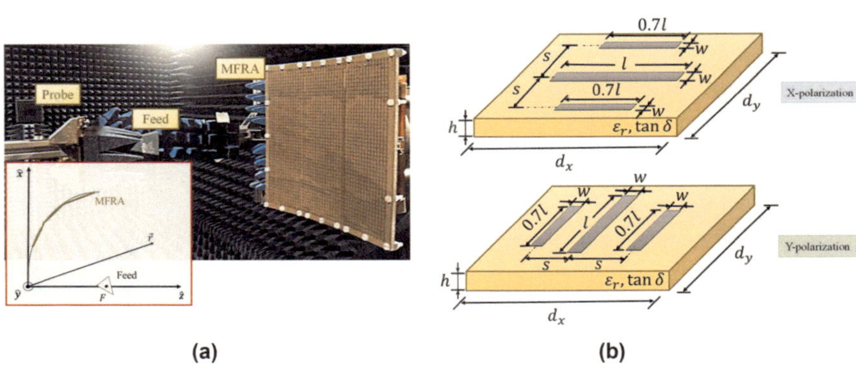

Figure 6.
Measurement setup of the multifaceted reflectarrays (MFRAs) reported in [26]. Sketch of the single-layer unit-cell used in each multifaceted demonstrator.

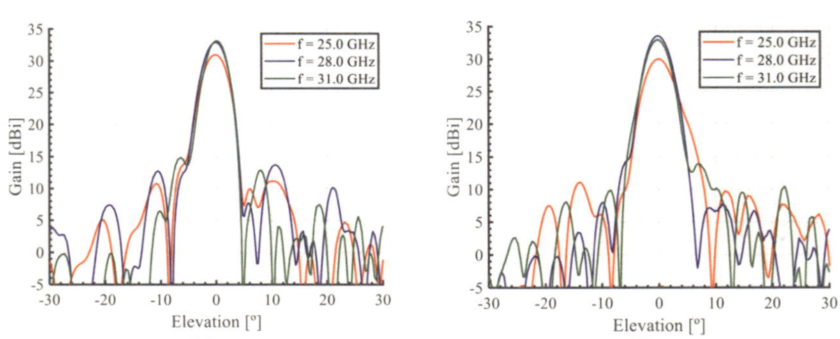

Figure 7.
Measured radiation pattern of the antenna in gain [dBi] evaluated ranging from 25 to 31 GHz. Elevation cut: (a) multifaceted reflectarray in X-polarization; (b) single-facet reflectarray.

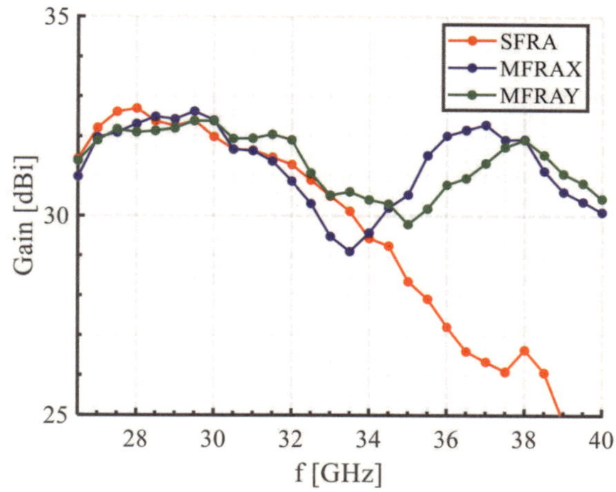

Figure 8.
Measured gain over frequency in the Ka-band (26.5–40 GHz) for the multifaceted designs (MFRAX, MFRAY) and the single facet design (SFRA).

entire Ka-band with a maximum gain ripple of 3 dB, since they exploit efficiently the broadband performance of the unit-cell.

Another critical issue in deployable reflectarrays is the effect of mechanical misalignments of the structure, due to, for instance, malfunctions in the deployment mechanism. In this sense, the effect of different misalignments in the deployable structure of **Figure 6** has been investigated in [34]. This study reveals a degradation of the pattern under the existence of gaps and errors in the tilt of the panels, which leads to a gain loss. For gaps between panels near the wavelength and angular misalignments higher than 2°, the gain loss starts to be significant.

The investigations performed in [26, 33, 34] demonstrate the advanced performance of multifaceted reflectarrays onboard small satellites with hard mechanical constraints. The multifaceted demonstrators exploit efficiently both the unit-cell performance and the antenna optics that improve the in-band performance of deployable reflectarrays based on aligned panels. Furthermore, they maintain their low-profile, low-cost, and polarization features.

4.1.2 Multifaceted reflectarray with an extremely large aperture for an InSAR mission

The feasibility of deployable and large multifaceted reflectarrays onboard satellites has been explored in [35]. Here, a large aperture reflectarray (depicted in **Figure 9**) has been designed for a remote sensing application, taking as baseline the specifications reported in [37]. The antenna operates at 35.75 GHz in dual-linear polarization: vertical (V) and horizontal (H), regarding the satellite platform. In each polarization, the antenna generates a high-gain pattern with a narrow beamwidth in azimuth and different tilt in elevation. The multifaceted reflectarray comprises nine single-layer rectangular panels that conform to a total aperture of about 4 m. Such a structure is fed spatially by two independent feeds, covering each polarization. The F/D ratio of the entire antenna is about 0.9.

The design process of the reflectarray surface was performed using the procedure introduced in Section 3.1.1. First, the phase distributions required in each polarization have been calculated using Eq. (5). The phase-shifter array used in this work has been

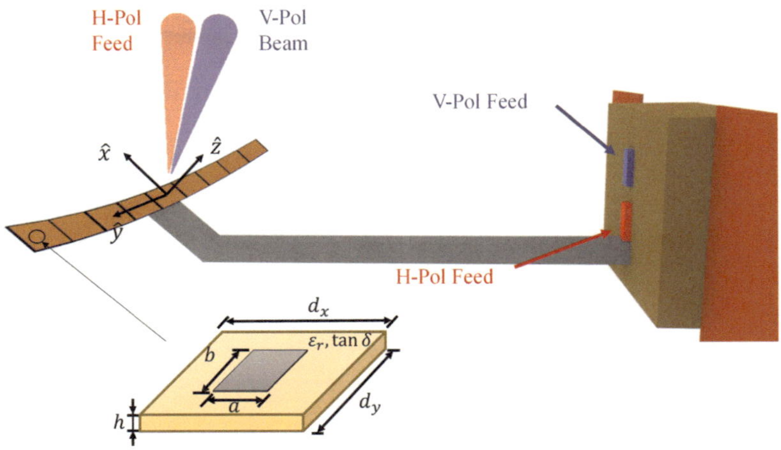

Figure 9.
Sketch of the large and deployable multifaceted aperture onboard a satellite for an InSAR mission and phase-shifter used in each reflectarray panel.

a single-layer of variable-size patches (presented in **Figure 9**). After this process, the output layouts of the MFRA and its equivalent SFRA are shown in **Figure 10**. Most of the panels in the SFRA exhibit several phase wraps (the size of the patches varies abruptly along the surface). This effect is due to the unit-cell limitation to provide the phase range required in the aperture. In contrast, the layout of the multifaceted aperture is smoother, with fewer phase wraps along the large dimension (y-axis in **Figure 10**).

The existence of several phase wraps along the surface degrades the in-band behavior of the antenna as explained in [37], and it is shown in the radiation patterns of the antennas under study, depicted in **Figure 11**. In azimuth (which corresponds to the *YZ* plane in **Figure 9**), the SFRA exhibits a significant degradation of the beam at off-center frequencies, which leads to an evident gain loss. However, the multifaceted aperture maintains the main beam shape and therefore the gain values. **Figure 12** depicts the gain of the antennas in a larger frequency range. The MFRA achieves a 1 dB gain bandwidth of about 2.5 GHz, which is about 25 times the bandwidth of its equivalent SFRA.

In conclusion, the improvements of multifaceted reflectarrays observed in electrically moderate apertures (Section 3.1.1) are more evident in large apertures. The

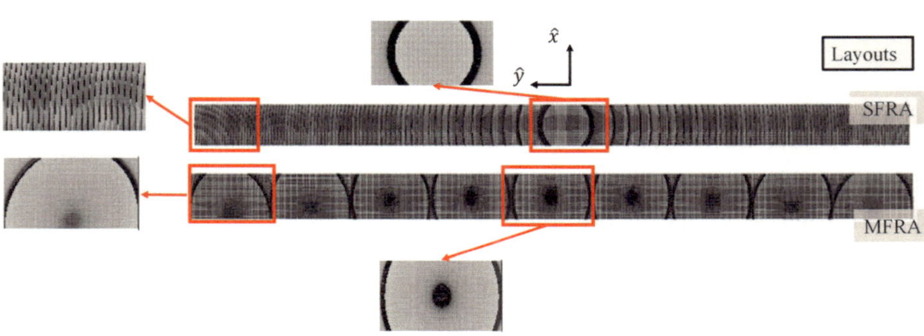

Figure 10.
Designed layouts in each panel for the multifaceted structure (bottom) and for an equivalent single-facet aperture (top) using the same unit-cell phase-shifter.

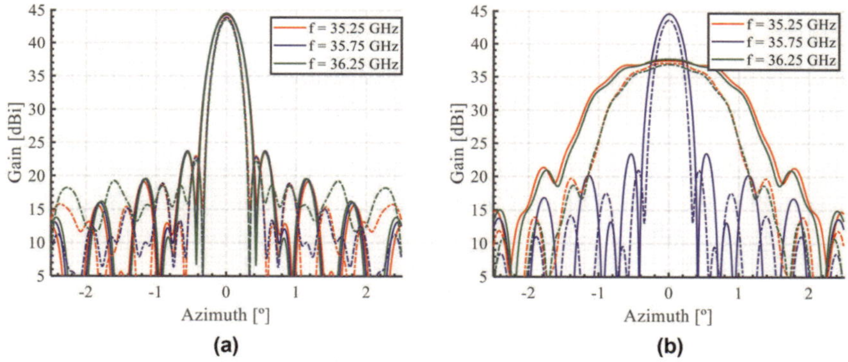

Figure 11.
Simulated radiation pattern in gain [dBi] of the large multifaceted reflectarray in 1 GHz of bandwidth. Azimuth cut: (a) multifaceted reflectarray; (b) equivalent single facet reflectarray. Solid and dotted lines correspond to the field in V- and H- polarization, respectively.

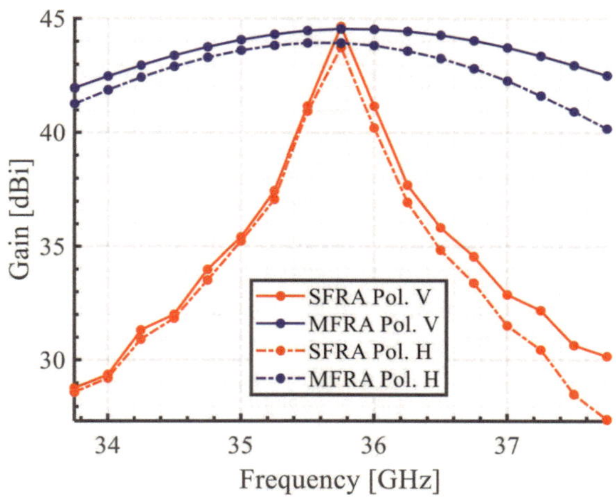

Figure 12.
Gain over frequency of the large multifaceted reflectarray (MFRA) and its equivalent single-facet reflector (SFRA).

proposed multifaceted structure maintains the performance of the antenna at the design frequency, while its equivalent single-facet configuration experiences a significant degradation of the pattern. The large multifaceted aperture exhibits high bandwidth enhancements compared with the equivalent planar aperture, underscoring the potential capabilities of these structures for space missions that require high-gain antennas.

4.2 Curved reflectarrays for satellite communications

The use of curved apertures, such as those following a parabolic profile, can result in advanced antenna configurations that provide an improved performance for satellite communications onboard GEO platforms. In this section, two antenna configurations are described: single-offset parabolic reflectarray antenna [27, 28] and dual-reflectarray antenna with flat subreflector and parabolic main reflector [3].

4.2.1 Single-offset parabolic reflectarrays

Curved reflectarrays based on a parabolic reflecting surface with printed elements have been proposed to achieve an improved performance for high-gain multibeam antennas onboard communication satellites [27, 28]. The parabolic reflectarrays are able to operate in dual-band for simultaneous transmission (Tx) and reception (Rx) in Ka-band (Tx at 20 GHz, and Rx at 30 GHz). At each band, the antennas generate multiple adjacent spot beams that form a cellular coverage over the Earth surface [38, 39]. The beams are produced at two close frequencies (f_1, f_2) and two orthogonal polarizations (right-handed CP, RHCP, and left-handed CP, LHCP); see **Figure 13(a)**.

Nowadays, four parabolic reflectors are commonly used onboard the existing GEO satellites to produce the complete four-color coverage by operating in a single-feed-per-beam scheme [38, 39], where each feed produces a high-gain spot beam at both Tx and Rx bands, as shown in **Figure 13(b)**. On the other hand, the parabolic reflectarrays proposed in [27, 28] are able to produce two adjacent spot beams per feed (in orthogonal CP) at both Tx and Rx bands, as depicted in **Figure 13(c)**. Thus, two parabolic reflectarrays are enough to produce the complete four-color coverage, reducing the number of antennas and feed-chains onboard the satellite by 50%.

In [27, 28], the generation of two spaced beams in orthogonal CP per feed can be achieved by means of the variable rotation technique (VRT) [40] applied to the reflectarray elements. The parabolic reflectarray is able to focus the beams thanks to its parabolic surface and to deviate the orthogonal CP beams in opposite directions by the application of VRT simultaneously at Tx and Rx frequencies [41, 42]. Two parabolic reflectarrays have been recently manufactured and tested [27, 28], by implementing the previous design concept. The parabolic reflectarray reported in [27] (see **Figure 14(a)**) is the largest one (0.9-m in diameter) and presents the additional advantage with respect to [28] of generating orthogonal beams at Tx and Rx bands.

The dual-band reflectarray cell used in [27] to be operated by VRT consists of a dual-layer structure with two circular arcs to operate at Tx and two orthogonal stacked sets of three parallel dipoles to operate at Rx, as shown in **Figure 14(b)**.

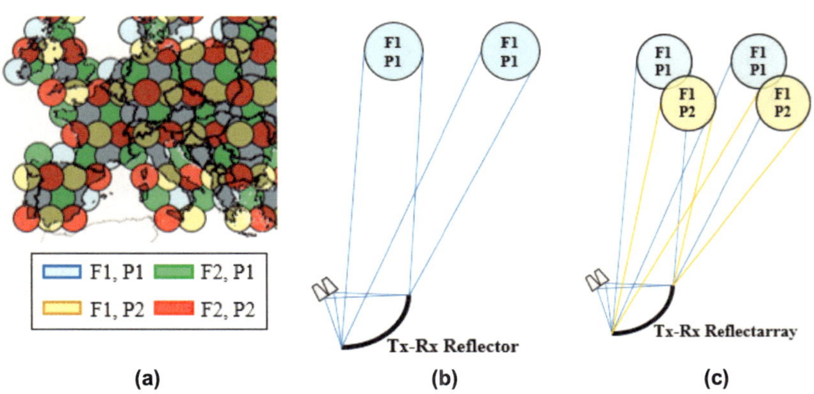

Figure 13.
(a) Sketch of a four-color cellular coverage generated from a GEO satellite, and operating principle of (b) conventional reflectors (four reflectors are necessary to produce the complete coverage at Tx and Rx) and (c) parabolic reflectarrays proposed in [1, 2] (two parabolic reflectarrays to produce the complete coverage at Tx and Rx).

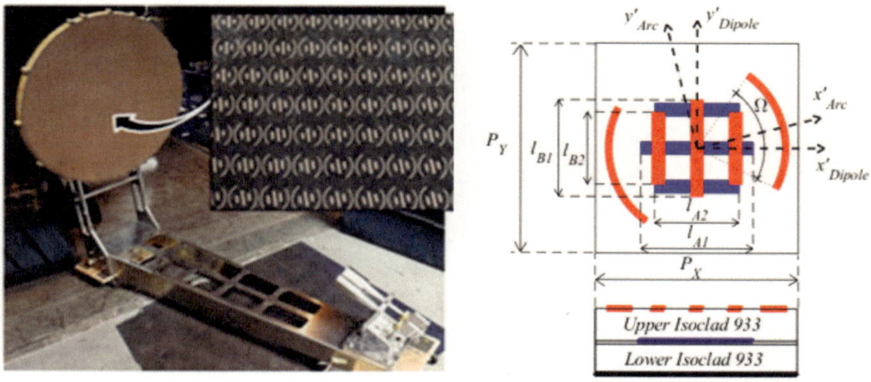

(a) (b)

Figure 14.
(a) Parabolic reflectarray demonstrator (0.9.M in diameter) reported in [27]. (b) Top and lateral views of the dual-layer reflectarray cell.

The unit-cell has been defined and characterized in [42], including the experimental validation of an in-house electromagnetic code based on SD-MoM, which has been used for the efficient EM analysis of the cell under periodic boundary conditions. The reflectarray cell is able to provide an opposite phase shift in each CP at both Tx and Rx frequencies by independently adjusting the rotation angle of the arcs and the dipoles and by tuning their lengths: variables l_{A1}, l_{A2}, l_{B1}, and l_{B2} for the dipoles, and angle Ω for the arcs (see **Figure 14(b)**).

The parabolic reflectarray antenna in [27] has been designed by applying the procedures and techniques described in Section 2.3 to analyze the reflectarray cells and obtain the appropriate dimensions and rotation angles of the reflectarray elements in each cell. The measured radiation patterns of the manufactured 0.9-m antenna demonstrator confirmed the validity of the design technique and EM simulation tools. **Figures 15(a)** and **(b)** show the contours of the measured beams generated by the parabolic reflectarray at 19.7 GHz (central Tx frequency) and 30 GHz (central Rx frequency), respectively. The contours are depicted at 40.6 dBi (end-of-coverage –EOC– gain) and 20.6 dBi (20 dB below the EOC gain). The antenna produces six spot beams that alternate in polarization (RHCP, LHCP), at both Tx and Rx bands, when only three feeds are used to illuminate the reflectarray. Thus, each feed is able to generate two adjacent beams in orthogonal CP. Moreover, the Tx and Rx spots are orthogonally polarized, that is, the same spot is generated in different polarizations at Tx and Rx. **Figure 15(c)** and **(d)** show the cut of the radiation patterns in the plane 'u = −0.036' (which contains two adjacent beams generated by one of the feeds) at Tx frequencies and at Rx frequencies, respectively. The measured beams are close to the specifications set by the masks in terms of minimum gain and side-lobe levels.

The results reported in [27, 28] are very satisfactory, since they confirm that curved reflectarrays can be a potential candidate for advanced multibeam antennas onboard Ka-band communication satellites. As shown in [27, 28], a single parabolic reflectarray can generate half the required multi-spot coverage (two colors), enabling a reduction of 50% in the number of antennas and feed-chains required on board the satellite (from four parabolic reflectors to two parabolic reflectarrays).

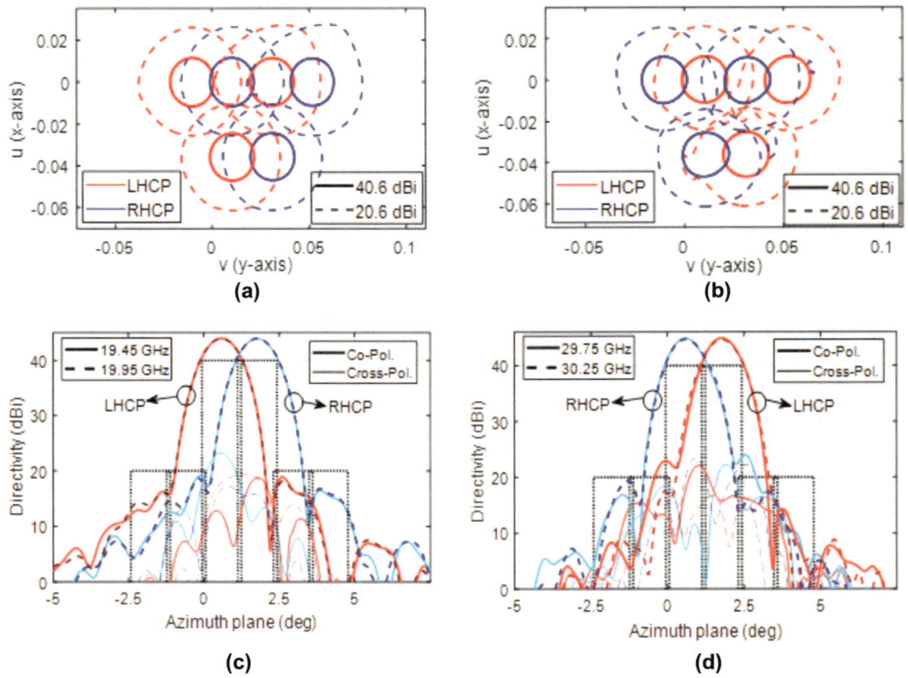

Figure 15.
Measured contours at 40.6 dBi and 20.6 dBi of the beams generated by the parabolic reflectarray illuminated by 3 feeds at (a) 19.7 GHz and (b) 30 GHz. Measured radiation patterns in the 'u = −0.036' plane at (c) Tx band and (d) Rx band, showing the two adjacent beams in orthogonal CP produced by one of the feeds.

4.2.2 Dual-reflectarray configuration with parabolic main reflector

Dual-reflectarray configurations have been proposed for the same application described in the previous section: high-gain multibeam antennas to provide four-color cellular coverage over the Earth from a GEO satellite. In [3], a dual-reflector configuration, formed by a 60-cm flat reflectarray subreflector and a 180-cm parabolic reflectarray, has been designed to generate two adjacent beams per feed in orthogonal CP at both Tx and Rx bands.

Compared to the single parabolic reflectarray of Section 3.2.1, the use of a dual-reflector configuration allows to simplify the design procedures that lead to dual-band antenna operation in dual-CP, due to the greater number of degrees of freedom that result from the use of two reflecting surfaces. In the dual-reflector antenna, the feeds operate in dual-LP (Horizontal, H, and Vertical, V), which reduces the complexity of the feed-chains with respect to the case of dual-CP feeds (eliminating the need for a LP-to-CP polarizer in the feed-chain). Then, the reflectarray subreflector is designed to deviate (without focusing) the dual-LP beams radiated by the feeds in opposite directions for each LP, simultaneously in Tx and Rx. The main reflectarray will be designed to perform as a polarizer (transforming dual-LP into dual-CP), at the same time as its parabolic surface focuses high-gain spot beams. **Figure 16** shows the operating principle of the dual-reflectarray antenna, which is able to generate two spaced beams in LHCP and RHCP from a single feed that radiates in H and V polarizations [3].

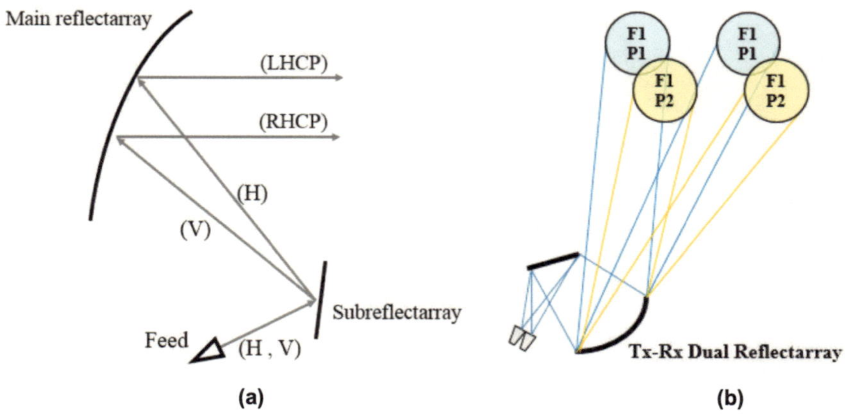

Figure 16.
(a) Operating principle of the dual-configuration, (b) functional scheme of the proposed antenna on the satellite.

The dual-band reflectarray cells used to design the subreflector are similar to those used in [43]; see **Figure 17(a)**. In this case, the unit-cell parameters have been optimized to achieve the desired phasing performance at the 20 and 30 GHz bands (for Tx and Rx). The dual-band reflectarray cells provide independent phasing capability in each LP at both operating bands. Thus, a different phase distribution can be implemented on the subreflector for each LP, which is required to deviate in different directions the two beams that are produced with the same feed (so that each feed will produce two adjacent beams in orthogonal polarizations).

The main reflectarray, implemented as a parabolic polarizing reflector, has a double mission: it is responsible for focusing the beams (thanks to its parabolic shape), at the same time as performing the LP-to-CP polarization conversion (due to the printed elements on the parabolic surface). The polarizing cells on the main reflector present the same element configuration than those used in [44] to design a flat polarizing reflector; see **Figure 17(b)**. This unit cell delivers wideband LP-to-CP conversion in the 19–30 GHz frequency range (with an axial ratio less than 2 dB), so it covers both Tx and Rx bands of the multibeam antenna.

Figure 18(a) and **(b)** show the simulated gain levels of the two-color (RHCP and LHCP) cellular coverage produced by the dual-reflectarray antenna at Tx (19.7 GHz) and Rx (29.5 GHz), respectively, when the antenna is fed by 15 horns. The average

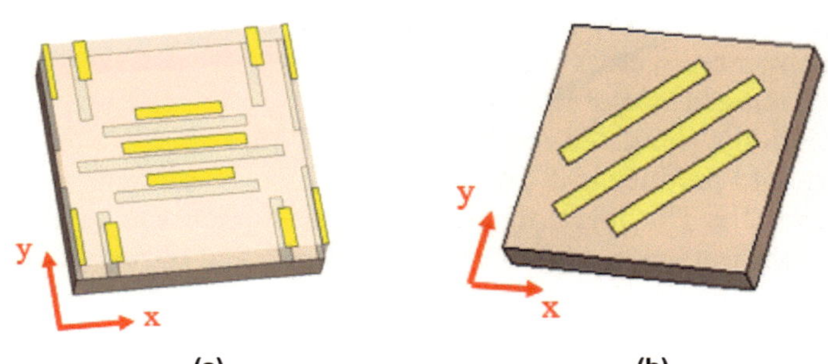

Figure 17.
Reflectarray cells used for the design of each reflector: (a) unit-cell of the flat subreflector, and (b) unit-cell of the parabolic polarizing reflector.

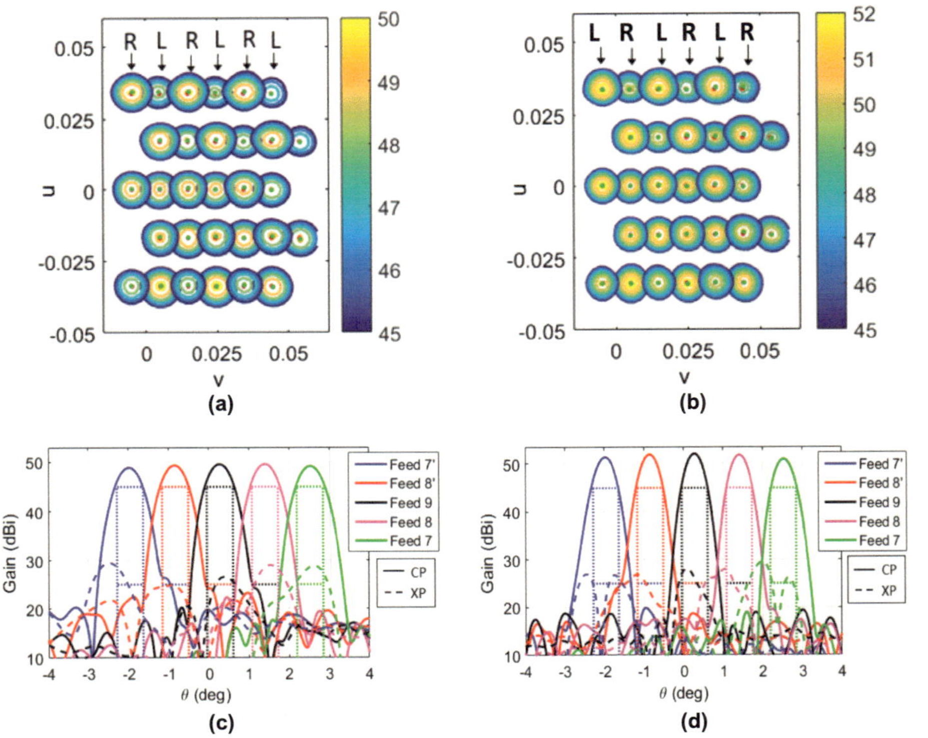

Figure 18.
Gain levels (in dBi) for the multi-spot coverage produced by the dual-reflectarray antenna in RHCP (R) and LHCP (L): (a) at 19.7 GHz and (b) at 29.5 GHz. Cuts of the radiation patterns in the plane 'u = 0' for the LHCP beams (c) at 19.7 GHz and (d) at 29.5 GHz, including the masks for the EOC gain and maximum interference levels.

maximum gain of the spots is 49.4 dBi at Tx and 51.6 dBi at Rx, and the spot separation is 0.56°. The slight differences in maximum gain between the RHCP and the LHCP beams are mainly due to the design of the subreflector, which shows a slightly better performance in V-polarization than in H-polarization. Due to the oblique incidence conditions on the subreflector cells, the phase errors resulting from the design of the printed elements affect each LP in a different way, leading to a slightly different performance for the LHCP and RHCP beams. Then, **Figure 18(c)** and **(d)** present the cuts of the simulated radiation patterns in the 'u = 0' plane at 19.7 and 29.5 GHz, respectively, for the beams generated in LHCP. The average carrier-to-interference ratio (C/I) at these frequencies is close to 20 and 15 dB, respectively. Moreover, the maximum cross-polar levels are up to −16 dB below the EOC gain for some of the extreme beams of the coverage.

In conclusion, the simulated results shown in **Figure 18** prove the capability of the dual-reflectarray to produce half the required four-color cellular coverage, simultaneously at Tx and Rx. The operating principles that support this antenna solution have been validated separately through the manufacturing and testing of reduced-size demonstrators [43, 44]. In [43], a dual-band flat reflectarray with independent operation in dual-LP at two distant frequencies was manufactured and tested, making use of reflectarray cells based on orthogonal sets of parallel dipoles for each band. In [44], a polarizing reflectarray to convert dual-LP into dual-CP with broadband operation from 20 to 30 GHz (covering both Tx and Rx bands) was demonstrated, showing an axial ratio lower than 2 dB.

5. Multifaceted and curved reflectarrays: practical insights

Antenna solutions based on flat reflectarray antennas typically exploit their low profile to obtain mechanical advantages over conventional reflector-based solutions. It is well known that flat antennas are easily foldable, have a smaller volume occupation, and can be adapted to flat surfaces on the outside of the satellite, such as the back of the solar panels. However, the practical implementation of reflectarray antennas on satellites is mainly limited to CubeSats [11, 13] due to the limitation in bandwidth associated with large aperture flat reflectarrays that would be required for high-gain antenna applications in larger satellites. The antenna solutions proposed in this chapter, based on multifaceted and curved reflectarrays, make it possible to maintain the advanced antenna performance of the reflectarrays (independent operation in frequency and/or polarization, generation of multiple beams per feed, etc.) without the bandwidth limitation associated with a large aperture.

This section provides a brief overview of the practical implications of multifaceted and curved reflectarray antennas in terms of antenna dimensions, manufacturing process, and mechanical complexity, compared with those associated with doubly curved reflectors commonly used in space. This comparison also evaluates large aperture flat reflectarrays, since they represent the classical implementation of reflectarray antennas. However, it must be noted that the main limitation of flat reflectarrays is due to their electrical performance, not due to their mechanical implementation.

First, there are no remarkable differences in the size of the projected aperture required to provide a specific gain level when using reflectors or reflectarrays. The use of dielectric materials in the reflectarray introduces ohmic losses that may be compensated by a slight increase in the projected aperture of the reflectarray, although the level of losses is typically small (less than 0.5 dB as shown in [27] at 30 GHz). In terms of weight, the carbon-fiber reflectors used in space are relatively light, and the use of similarly sized reflectarrays would not make a difference. The main difference comes from the volume occupied by the antenna system. In general, reflectors and curved reflectarrays require a larger storage volume than planar structures such as flat or multifaceted reflectarrays, as their curved surface is more difficult to fold. Nevertheless, some strategies have been proposed in the literature to solve this issue [45].

The manufacturing process of flat and multifaceted reflectarrays is based on a photo-etching process to obtain the metallization pattern in the antenna surface, the same as on printed circuit boards (PCB). The easy and low-cost manufacturing process is another benefit that has led reflectarrays to be deployed on micro-satellites. Regarding curved reflectarrays, the use of flexible dielectric substrates that adapt to the required curvature allows the same manufacturing process based on photo-etching. The manufacturing process of curved reflectarrays is indeed similar as in carbon-fiber reflectors, adding a few layers of prepreg-Quartz and Kapton with the printed pattern of the reflectarray.

Flat and multifaceted reflectarrays also enable a much simpler folding development than antennas with a curved surface. Multifaceted reflectarrays only need a hinge structure similar to that of solar panels and also to that already used in flat reflectarrays for small satellites [11, 13], with the difference that the hinges lock when they reach a certain angle, which does not pose a mechanical challenge. The planar surface of the reflectarray panels also simplifies the mounting on board the satellite. On the other hand, curved reflectarray antennas have the same difficulties in terms of folding and mounting as conventional reflectors, as they share the same parabolic structure: the use of reflectarray technology on curved surfaces does not involve a

	Flat RA	MFRA	Curved RA	Reflector
Storage volume	Low	Low	High	High
Deployment mechanism	Easy	Easy	Hard	Hard
Mounting	Easy	Easy	Standard	Standard
Electrical performance	Very limited	Advanced	Advanced	Standard

Table 1.
Main figures of merit in the practical implementation of flat, multifaceted and curved reflectarray (RA) antennas in satellite applications.

higher level of mechanical complexity. **Table 1** summarizes the main issues related to practical implementations of reflectarrays for satellite applications.

6. Conclusions

This chapter was devoted to the comprehensive discussion of different non-planar reflectarrray configurations to be onboard satellites for different space missions. Specifically, the structures have been evaluated are multifaceted and doubly curved surfaces that exploit efficiently the antenna optics and the unit-cell performance to provide advance electrical performance. Furthermore, such structures maintain low-profile and high stowage features, which make them suitable to be onboard a spacecraft.

Multifaceted reflectarrays comprise several panels that are non-aligned to each other, forming a curvature that seeks to approximate a paraboloid in one or more planes. The antenna optics of a multifaceted approach corrects partially the differential spatial phase delay, a negative effect that limits the in-band performance of conventional reflectarrays. The analysis and design of this type of structures are based on procedures used during the design of conventional reflectarrays but considering the different orientations of the panel. Several works have been proposed for the use of multifaceted reflectarrays as a high-gain solution for small satellite with challenging stowage criteria and for space applications that require electrically large apertures. In both cases, the results discussed in this chapter demonstrate the capabilities of multifaceted structures to improve the in-band performance of reflectarrays, maintaining good integrability with the spacecraft. This enhancement becomes more evident for electrically large apertures.

As for the doubly curved reflectarrays for high-gain antennas, the parabolic antenna optics make it possible to completely compensate the differential spatial phase delay. Moreover, the relatively large f/D in typical high-gain satellite antennas allows to maintain a conventional analysis of the reflectarray cells, considering a locally flat and periodic array structure, while the radiation of the complete antenna can be modeled by a combination of PO simulations and the aforementioned characterization of the reflection matrix of each cell. The main advantage provided by curved reflectarrays comes from the capability to combine the natural generation of high-gain beams by a parabolic surface with the multibeam and polarization conversion capabilities provided by reflectarray cells. Recent works have shown the potential of curved reflectarrays to halve the number of reflector antennas onboard broadband communication satellites in Ka-band. The results have been achieved by means of single offset or dual antenna configurations based on a main parabolic reflectarray antenna with simultaneous operation in transmission and reception frequencies in Ka-band.

Acknowledgements

This work was supported in part by MICIN/AEI/10.13039/501100011033 within the projects PID2020-114172RB-C21, PID2020-114172RB-C22, PDC2021-120959-C22, TED2021-130650B-C22, and RYC2021-033593-I cofounded by UE (European Union) "NextGenerationEU"/PRTR.

Author details

Borja Imaz-Lueje[1*], Daniel Martinez-de-Rioja[2], Eduardo Martinez-de-Rioja[1], Yolanda Rodriguez-Vaqueiro[3], Manuel Arrebola[4], Marcos R. Pino[4], Antonio Pino[3] and José Antonio Encinar[2]

1 Department of Signal Theory and Communications and Telematic Systems and Computing, Universidad Rey Juan Carlos, Madrid, Spain

2 Information Processing and Telecommunications Center, Universidad Politécnica de Madrid, Madrid, Spain

3 Department of Signal Theory and Communications, Universidade de Vigo, Vigo, Spain

4 Department of Electrical Engineering, Universidad de Oviedo, Gijón, Spain

*Address all correspondence to: borja.imaz@urjc.es

References

[1] Huang J, Encinar JA. Reflectarray Antennas. Hoboken, NJ, USA: John Wiley & Sons; 2008

[2] Pozar DM, Targonski SD, Syrigos HD. Design of millimeter wave microstrip reflectarrays. IEEE Transactions on Antennas and Propagation. 1997;**45**(2): 287-296. DOI: 10.1109/8.560348

[3] Martinez-de-Rioja E et al. Advanced multibeam antenna configurations based on reflectarrays: Providing multispot coverage with a smaller number of apertures for satellite communications in the K and Ka bands. IEEE Antennas and Propagation Magazine. 2019;**61**(5):77-86. DOI: 10.1109/MAP.2019.2932311

[4] Zhou M, Sørensen SB, Kim OS, Jørgensen E, Meincke P, Breinbjerg O. Direct optimization of printed reflectarrays for contoured beam satellite antenna applications. IEEE Transactions on Antennas and Propagation. 2013;**61**(4):1995-2004. DOI: 10.1109/TAP.2012.2232037

[5] Arrebola M, Encinar JA, Barba M. Multifed printed reflectarray with three simultaneous shaped beams for LMDS central station antenna. IEEE Transactions on Antennas and Propagation. 2008;**56**(6):1518-1527. DOI: 10.1109/TAP.2008.923360

[6] Chaharmir MR, Shaker J, Legay H. Dual-band Ka/X reflectarray with broadband loop elements. IET Microwaves, Antennas & Propagation. 2010;**4**(2):225-231. DOI: 10.1049/iet-map.2008.0369

[7] Deng R, Mao Y, Xu S, Yang F. A single-layer dual-band circularly polarized reflectarray with high aperture efficiency. IEEE Transactions on Antennas and Propagation. 2015;**63**(7): 3317-3320. DOI: 10.1109/TAP.2015.2429684

[8] Rengarajan SR. Reflectarrays of rectangular microstrip patches for dual-polarization dual-beam radar interferometers. Progress in Electromagnetics Research. 2013;**133**: 1-15. DOI: 10.2528/PIER12091615

[9] Martinez-de-Rioja D, Florencio R, Martinez-de-Rioja E, Arrebola M, Encinar JA, Boix RR. Dual-band reflectarray to generate two spaced beams in orthogonal circular polarization by variable rotation technique. IEEE Transactions on Antennas and Propagation. 2020;**68**(6):4617-4626. DOI: 10.1109/TAP.2020.2975294

[10] Martinez-de-Rioja E, Martinez-de-Rioja D, Lopez-Saez R, Linares I, Encinar JA. High-efficiency polarizer reflectarray antennas for data transmission links from a CubeSat. Electronics. 2021;**10**(15):1802. DOI: 10.3390/electronics10151802

[11] Hodges RE, Chahat NE, Hoppe DJ, Vacchione JD. The mars cube one deployable high gain antenna. Proceedings of the IEEE International Symposium on Antennas and Propagation (APS). 2016:1533-1534. DOI: 10.1109/APS.2016.7696473

[12] Hodges RE et al. ISARA-integrated solar array and reflectarray CubeSat deployable Ka-band antenna. Proceedings of the IEEE International Symposium on Antennas and Propagation (APS). 2015:2141-2142. DOI: 10.1109/APS.2015.7305460

[13] Chahat N, Thiel E, Sauder J, Arya M, Cwik T. Deployable one-meter reflectarray for 6U-class CubeSats. In: 2019 European Conference on Antennas

and Propagation (EuCAP), Krakow, Poland. New York City, U.S.: Institute of Electrical and Electronics Engineering (IEEE); 2019

[14] Chaharmir MR, Shaker J, Legay H. Broadband design of a single layer large reflectarray using multi cross loop elements. IEEE Transactions on Antennas and Propagation. 2009;**57**(10): 3363-3366. DOI: 10.1109/TAP.2009. 2029600

[15] Encinar JA, Arrebola M, Fuente LF, Toso G. A transmit-receive reflectarray antenna for direct broadcast satellite applications. IEEE Transactions on Antennas and Propagation. 2011;**59**(9): 3255-3264. DOI: 10.1109/TAP.2011. 2161449

[16] Huang J. Bandwidth study of microstrip reflectarray and a novel phased reflectarray concept. In: Proceedings of the IEEE Antennas and Propagation Society International Symposium. Newport Beach, CA, USA: Digest; 1995. DOI: 10.1109/ APS.1995.530086

[17] Carrasco E, Encinar JA, Barba M. Bandwidth improvement in large reflectarrays by using true-time delay. IEEE Transactions on Antennas and Propagation. 2008;**56**(8):2496-2503. DOI: 10.1109/TAP.2008.927559

[18] Carrasco E, Arrebola M, Encinar JA, Barba M. Demonstration of a shaped beam reflectarray using apertura-coupled delay lines for LMDS central station antenna. IEEE Transactions on Antennas and Propagation. 2008;**56**(10): 3103-3111. DOI: 10.1109/ TAP.2008.929452

[19] Florencio R, Encinar JA, Boix RR, Losada V, Toso G. Reflectarray antennas for dual polarization and broadband telecom satellite applications. IEEE Transactions on Antennas and Propagation. 2015;**63**(4):1234-1246. DOI: 10.1109/TAP.2015.2391279

[20] Roederer A. Reflector Antenna Comprising a Plurality of Panels. Patent US 6,411,255 B2. Alexandria, Virginia: US Patent; 2002

[21] Guarriello A, Loison R, Bresciani D, Legay H, Goussetis G. Structural and radio frequency co-design and optimization of large deployable reflectarrays for space missions. IEEE Transactions on Antennas and Propagation. 2023;**71**(5):3916-3927. DOI: 10.1109/TAP.2023.3247940

[22] Legay H, Bresciani D, Labiole E, Chiniard R, Gillard R. A multi facets composite panel reflectarray antenna for a space contoured beam antenna in Ku band. Progress in Electromagnetics Research B. 2013;**54**:1-26. DOI: 10.2528/ PIERB13061407

[23] Encinar JA, Zornoza JA. Three-layer printed reflectarrays for contoured beam space applications. IEEE Transactions on Antennas and Propagation. 2004;**52**: 1138-1148. DOI: 10.1109/TAP.2004. 827506

[24] Wan C, Encinar JA. Efficient computation of generalized scattering matrix for analyzing multilayered periodic structures. IEEE Transactions on Antennas and Propagation. 1995;**43**: 1233-1242. DOI: 10.1109/8.475095

[25] Girard E, Moulinet R, Gillard R, Legay H. An FDTD optimization of a circularly polarized reflectarray unit cell. In: Proceedings of the IEEE Antennas and Propagation Society International Symposium, San Antonio, TX. New York City, U.S.: Institute of Electrical and Electronics Engineering (IEEE); 2002. pp. 136-139. DOI: 10.1109/APS.2002. 1018174

[26] Imaz-Lueje B, Pino MR, Arrebola M. Deployable multifaceted reflectarray antenna in offset configuration with band enhancement. IEEE Transactions on Antennas and Propagation. 2022; **70**(12):11686-11696. DOI: 10.1109/TAP.2022.3209753

[27] Martinez-de-Rioja D et al. Transmit–receive parabolic reflectarray to generate two beams per feed for multispot satellite antennas in Ka-band. IEEE Transactions on Antennas and Propagation. 2021;**69**(5):2673-2685

[28] Zhou M, Sørensen SB, Brand Y, Toso G. Doubly curved reflectarray for dual-band multiple spot beam communication satellites. IEEE Transactions on Antennas and Propagation. 2020;**68**(3):2087-2096

[29] Pino A, Rodriguez-Vaqueiro Y, Martinez-de-Rioja E, Martinez-de-Rioja D, González-Valdés B, Arias M, et al. Bifocal dual reflectarray with curved main surface. Electronics. 2023; **12**(12):2619. DOI: 10.3390/electronics 12122619

[30] Martínez JA, Pino A, Vega I, Arias M, Rubiños O. ICARA: Induced current analysis of reflector antenas. IEEE Antennas and Propagation Magazine. 2005;**47**(2):92-100

[31] Arias M, Pino A, Rubiños-López O. Fast far field computation of single and dual reflector antennas. Journal of Engineering. 2013;**2013**:1-19. ISSN: 2314-4912

[32] Zhou M, Sorensen SB, Mincke P, Jorgensen E. Design and optimization of multifaceted reflectarrays for satellite applications. In: Proceedings of the 8th European Conference on Antennas and Propagation. Netherlands: The Hague; 2014. DOI: 10.1109/EuCAP.2014.6902047

[33] Imaz-Lueje B, Pino MR, Arrebola M. Evaluation of multifaceted Reflectarray configurations on SmallSats. In: Proceedings of the 16th European Conference on Antennas and Propagation. Madrid, Spain. New York City, U.S.: Institute of Electrical and Electronics Engineering (IEEE); 2022

[34] Imaz-Lueje B, Pino MR, Arrebola M. Effect of mechanical tolerances in deployable multifaceted reflectarrays. In: Proceedings of the 17th European Conference on Antennas and Propagation. Florence, Italy. New York City, U.S.: Institute of Electrical and Electronics Engineering (IEEE); 2023

[35] Imaz-Lueje B, Pino MR, Arrebola M. Evaluation of large deployable reflectarray antennas in multiple flat and not aligned panels. In: 41st ESA Antenna Workshop. The Netherlands: ESA-ESTEC; 2023

[36] Florencio R, Boix RR, Encinar JA. Efficient spectral domain MoM for the design of circularly polarized reflectarray antennas made of split rings. IEEE Transactions on Antennas and Propagation. 2019;**67**(3):1760-1771. DOI: 10.1109/TAP.2018.2888956

[37] Hodges RE, Chen JC, Radaway MR, Amaro LR, Kayatian B, Munger J. An extremely large Ka-band reflectarray antenna for interferometric synthetic aperture radar: Enabling next-generation satellite remote sensing. IEEE Antennas and Propagation Magazine. 2020;**62**(6): 23-33. DOI: 10.1109/MAP.2020.2976319

[38] Schneider M, Hartwanger C, Wolf H. Antennas for multiple spot beams satellites. CEAS Space Journal. 2011;**2**:59-66

[39] Fenech H, Amos S, Tomatis A, Soumpholphakdy V. High throughput satellite systems: An analytical approach.

IEEE Transactions on Aerospace and Electronic Systems. 2015;**51**:192-202. DOI: 10.1109/TAES.2014.130450

[40] Huang J, Pogorzelski RJ. A Ka-band microstrip reflectarray with elements having variable rotation angles. IEEE Transactions on Antennas and Propagation. 1998;**46**(5):650-656

[41] Zhou M, Sørensen SB. Multi-spot beam reflectarrays for satellite telecommunication applications in Ka-band. In: Proceedings of the 10th European Conference on Antennas and Propagation (EuCAP 2016), Davos. New York City, U.S.: Institute of Electrical and Electronics Engineering (IEEE); 2016. pp. 1-5

[42] Martinez-de-Rioja D, Florencio R, Encinar JA, Carrasco E, Boix RR. Dual frequency reflectarray cell to provide opposite phase-shift in dual circular polarization with application in multibeam satellite antennas. IEEE Antennas and Wireless Propagation Letters. 2019;**18**(8):1591-1595

[43] Martinez-de-Rioja E, Encinar JA, Barba M, Florencio R, Boix RR, Losada V. Dual polarized reflectarray transmit antenna for operation in Ku- and Ka-bands with independent feeds. IEEE Transactions on Antennas and Propagation. 2017;**65**:3241-3246. DOI: 10.1109/TAP.2017.2689059

[44] Martinez-de-Rioja E, Encinar JA, Pino A, Rodriguez-Vaqueiro Y. Broadband linear-to-circular polarizing reflector for space applications in Ka-band. IEEE Transactions on Antennas and Propagation. 2020;**68**:6826-6831. DOI: 10.1109/TAP.2020.2975617

[45] Chahat N, Hodges RE, Sauder J, Thomson M, Peral E, Rahmat-Samii Y. CubeSat deployable Ka-band mesh reflector antenna development for earth science missions. IEEE Transactions on Antennas and Propagation. 2016;**64**(6): 2083-2093

Chapter 4

Reconfigurable Smart Antennas

Faouzi Rahmani, Abdelmounaim Belbachir Kchairi, Nihade Taher, Moustapha El Bakkali, Naima Amar Touhami and Tajeddin Elhamadi

Abstract

Antennas are essential and crucial components of radar and wireless communication systems, but sometimes their inability to adapt to new operating conditions can hamper system performance. These limitations can be mitigated or eliminated, and reconfigurable antennas can bring new levels of functionality to any system by enabling it to modify its behavior in response to changing system requirements or environmental conditions. Depending on the physical property that makes the antenna reconfigurable, the antennist needs to choose very precise reconfiguration techniques and also the necessary polarization circuits. In accordance with the reconfigurability functions they offer, reconfigurable antennas can be classified into one of four categories: frequency reconfiguration, radiation pattern reconfiguration, polarization reconfiguration and hybrid reconfiguration. This classification will be presented in this chapter, preceded by the principle and main reconfiguration techniques encountered in the literature, as well as the polarization circuits of the active elements required.

Keywords: reconfigurable antenna, polarization reconfiguration, frequency reconfiguration, radiation pattern reconfiguration, hybrid reconfiguration

1. Introduction

The idea that predates the invention of reconfigurable antennas (known as agile, active or smart antennas) is the integration of electronic components on antennas, which dates back to the 1960s [1, 2]. The antenna can be considered an active antenna once it has performed an active function on the RF signal, either before transmission or after reception. The term reconfigurable antennas (RA) was first used in 1998 [3].

In the world of antennas, the reconfigurable antenna design has attracted a great deal of attention [4–6]. These antennas have garnered attention due to their ability to adapt in terms of frequency selection, polarization selectivity and pattern reconfiguration, making them highly applicable in diverse wireless communication systems. By allowing for different radiation patterns on a physical aperture at a consistent operating frequency, reconfigurable pattern antennas offer versatile control over electromagnetic waves through mechanisms like adjusting bias voltage, enabling multifunctional electrical tuning.

According to the physical property that makes the RA, the antennist needs to choose very precise reconfiguration techniques and also the necessary polarization circuits. In accordance with the reconfigurability functions they offer, reconfigurable antennas can be classified into one of four categories: frequency reconfiguration, radiation pattern reconfiguration, polarization reconfiguration, and hybrid reconfiguration (any combination of the previous three configurations). This classification will be presented in this chapter, preceded by the principle and main reconfiguration techniques encountered in the literature, as well as the polarization circuits of the active elements required.

2. Reconfigurable antenna: reconfiguration principle and techniques

2.1 Reconfiguration principle

Reconfiguring an antenna involves modifying the distribution of currents or electromagnetic fields in its effective aperture, resulting in reversible adjustments to the antenna's impedance and radiation characteristics, either discretely or continuously. These modifications are facilitated by methods such as switching, material tuning and structural alterations. System control is then used to achieve the antenna's predicted performance [7].

An RA extends the capabilities of a traditional antenna by modifying its configuration, that is, by changing one or more of its main characteristics (such as operating frequency, bandwidth, polarization and radiation pattern) by mechanical, electrical or other means, according to need and environmental context. In addition, reconfigurable antennas offer more advantages over conventional antennas, which could be decisive in terms of wireless technology lifetime and integration. What is more, a single RA may be sufficient instead of several traditional antennas. Reconfigurable antennas can significantly reduce system size, cost and complexity while improving overall performance.

2.2 Reconfiguration techniques

There are several techniques for reconfiguring the antenna. Some approaches rely on mechanical modification of the antenna structure, while others use substrates with tunable properties. In many cases, localized active components are used, making it possible to modify current lines or impedance almost instantaneously. Other techniques, not covered in this section, rely on the reconfigurability of feed networks or the appropriate excitation of antenna arrays.

The use of electrically or optically controlled switches has provided new approaches to achieving reconfigurability, thanks to significant advances in microelectronics and photoconductors. PIN diodes, varactor diodes, field-effect transistors, MEMS and optical switches are the localized active components most frequently used in antenna reconfiguration. These components can be integrated either in the radiating element or in the feed network. Reconfiguration can be achieved by modifying the effective electrical length of the antenna, creating short circuits or switchable open circuits, deactivating or activating parasitic elements, or modifying the shape or size of a slot.

2.2.1 Mechanical deformation

Antenna reconfiguration can also be achieved by physically modifying the antenna's radiating structure. Structural modification of the antenna's radiating parts enables it to be tuned. By deforming the structure, the antenna's radiation pattern, polarization or operating frequency can be altered. This technique has the advantage of avoiding switching mechanisms, bias lines, or the integration of optical fibers and laser diodes. On the other hand, this technique depends on the limitation of the device to be physically reconfigured.

Simply rotating a directional antenna, such as an antenna array or a simple patch antenna, can also achieve the same results mechanically. However, this requires rotating not only the directional antenna itself but also the feed line and perhaps even the electrical components supporting it. The design of a de-rotating platform is necessary, but can prove difficult. An example of a mechanically reconfigurable antenna pattern using a meta-surface was presented by Zhu et al. [8]. The pattern configuration of the antenna shown in **Figure 1** is achieved mechanically. In this design, it is only the planar meta-surface layer that needs to be rotated and is separated from the antenna. The rotation mechanism can be very simple.

2.2.2 Active materials

With regard to reconfiguration strategies based on smart materials, liquid crystals, ferrites and ferroelectrics can be used to modify the antenna substrate. A material can be modified by changing its permittivity or permeability under different voltage levels (for liquid crystals) or a statically applied electric/magnetic field (for ferroelectrics/ferrites). Although liquid crystals are frequently used at optical frequencies, their ability to modify the dielectric constant also makes them useful for reconfigurable antennas at microwave frequencies. Despite their low power consumption, they require a temperature of between 20 and 35°C to maintain the liquid crystal state [9].

The main advantages of ferrite-based reconfigurable antennas are their high permittivity ($\varepsilon_r > 10$) and permeability ($\mu > 1000$) values, which make it possible to reduce antenna size and achieve a wide tuning range. To achieve the widest tuning range in a bulk material, a highly sophisticated polarization network is required, resulting in higher DC power consumption. Ferroelectric thin films have lower losses and require less bias voltage than ferrites [10].

Liquids, by their very nature, can take the shape of any container. This characteristic is widely used in the design of various reconfigurable liquid dielectric antennas. **Figure 2** shows some examples of this type of antenna in various shapes.

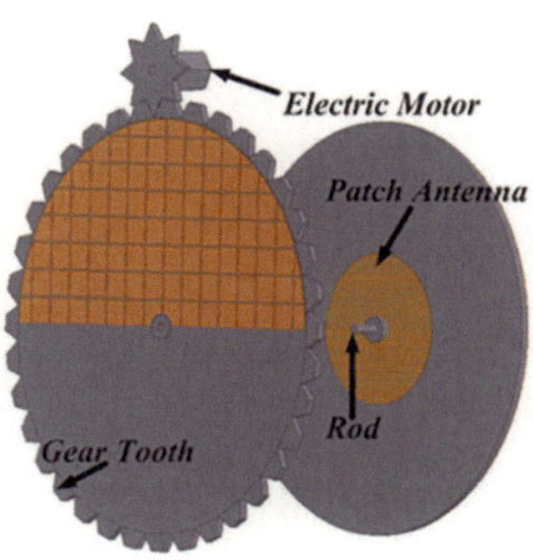

Figure 1.
Proposed schematic of a mechanically reconfigurable antenna [8].

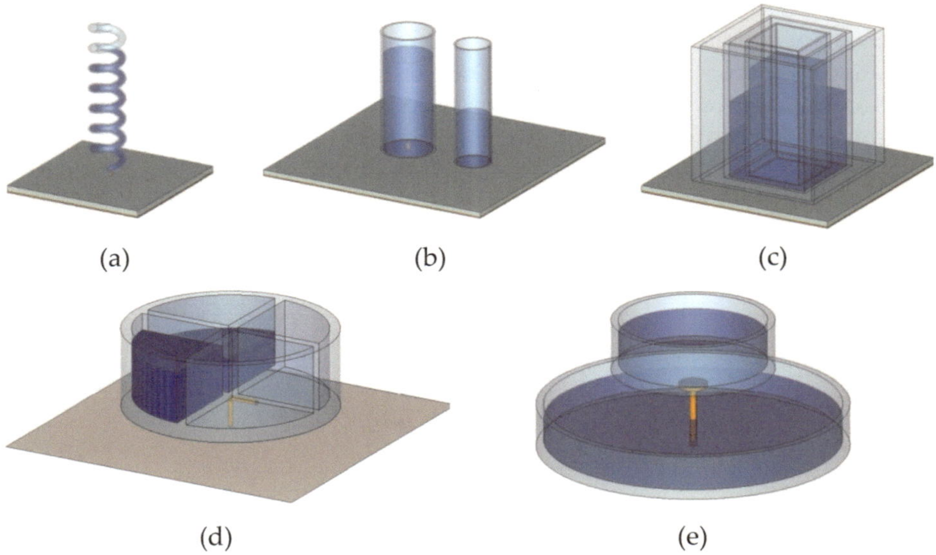

Figure 2.
Liquid dielectric antennas with different geometries: (a) [11], (b) [12], (c) [13], (d) [14], (e) [15].

2.2.3 Optical switch

Photoconductive optical switches with optical control can be used to avoid polarization lines and have the advantage of being very fast. The advantage of using fiber-optic cables to activate optical switches is that they are electromagnetically transparent and therefore do not interfere with antenna radiation patterns. They also provide thermal and electrical insulation between the antenna and the polarizing circuit. A photoconductor is a switch whose electrical resistance varies according to the intensity of the light it receives. When the laser (or a laser diode) illuminates a switch, it lights up and behaves like a short-circuited conductor. Otherwise, it acts as an insulator, that is, an open circuit [16, 17].

As shown in **Figure 3**, in the ON state, the switch can be modeled as a resistor R_{ON} connected in parallel with a parasitic capacitor $C_{p,ON}$. On the other hand, in the OFF state, the resulting equivalent circuit consists of two parallel capacitances C_{OFF} and $C_{p,OFF}$ [18].

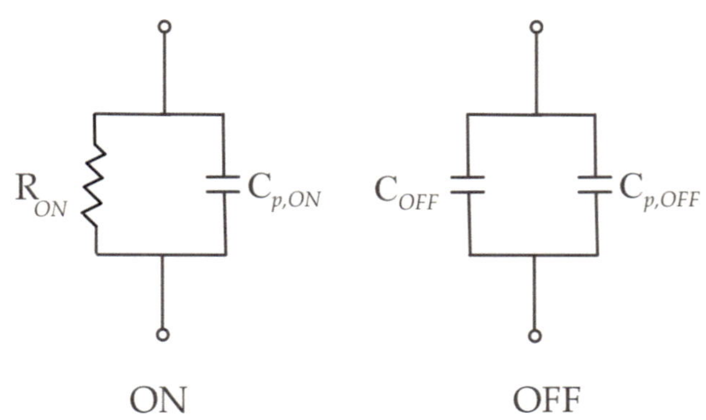

Figure 3.
Circuit equivalent to the ON and OFF states of a photoconductive switch.

2.2.4 RF-mems

Micro-Electro-Mechanical Systems (MEMS) have been the subject of research and development over the last few decades. MEMS is a microsystem component that creates a short circuit or open circuit by mechanically deforming microstructures, usually micro-membranes [19].

RF MEMS offer an alternative to semiconductor switches such as PIN diodes and FETs. They are manufactured using micromachining techniques. They provide the same basic switching function such as PIN diodes and FETs, but with the main advantage of lower insertion loss, particularly at higher microwave frequencies in the 8–100 GHz range [20]. RF MEMS components are of two main types: resistive or ohmic contact switches (metal-to-metal contact) and capacitive contact switches (metal-to-dielectric contact). To lower the RF MEMS beam, a force must always be applied to the moving part of the component. This force can take several forms (electrical, magnetic and thermal).

The schematic shown in **Figure 4** illustrates the operating principle of a MEMS-RF microswitch with ohmic contact. When the beam is in the up-position, the MEMS-RF is blocked and the radio-frequency signal is not transmitted to the output. When a voltage is applied to the actuating electrode, a metal-to-metal contact is established between input and output, and the signal is transmitted *via* a low resistance (R_{ON} due to the contact).

The same operating principle applies to capacitive MEMS-RF switches (**Figure 5**). They use a thin dielectric layer to design a variable capacitor that can take two different values between input and output. In the ON state, the signal is transmitted *via* a capacitive (metal-dielectric) contact. Note that the series capacitance (C_{ON}) is large ($C_{ON} \gg C_{OFF}$).

2.2.5 Field-effect transistors

While the PIN diode represents a simple, inexpensive and easy-to-mount control element in coplanar or microstrip lines, field-effect transistors (FETs) also offer a

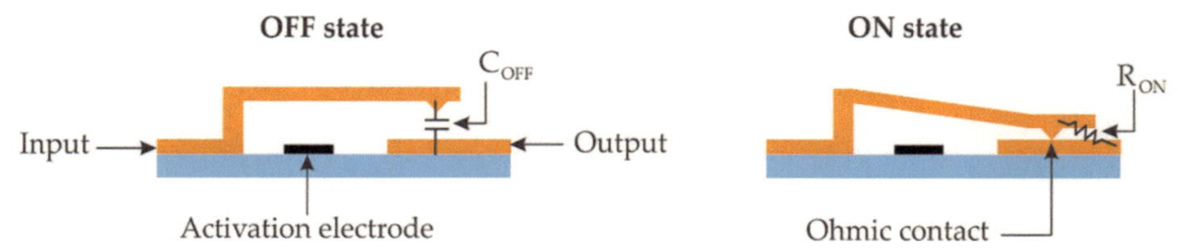

Figure 4.
Schematic of a MEMS-RF with ohmic contact.

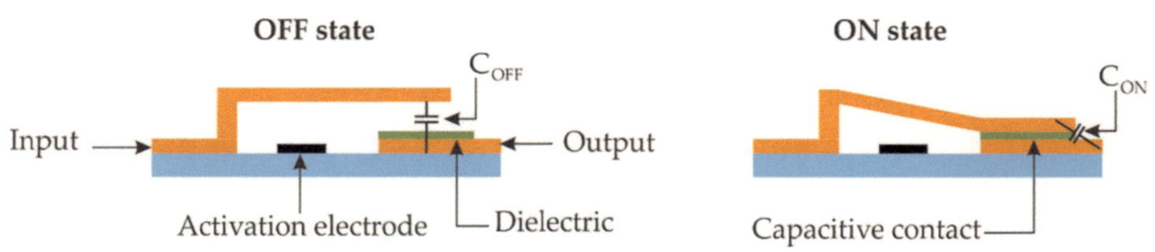

Figure 5.
Schematic of a MEMS-RF with capacitive contact.

number of advantages. The most common type of field-effect transistor used at RF and microwave frequencies is the Metal Semiconductor Field Effect Transistor (MESFET). An FET can be implemented in a variety of different structures. Gallium arsenide (GaA) is the semiconductor generally used in MESFETs designed for RF and microwave applications, as it offers high electron mobility and the possibility of fast operation.

Figure 6 shows a schematic representation of the fundamental structure of a GaAs MESFET. The MESFET consists of three poles: source, drain and gate. Three metal-semiconductor junctions serve as contact pads for the source, gate and drain. The nature of these junctions, whether ohmic or rectifying, depends on the relative work functions of the metal and semiconductor involved. When aluminum is deposited on heavily doped (N^+) GaAs, it forms an ohmic contact; whereas, if deposited on lightly doped GaAs, it forms a rectifying junction, commonly referred to as a Schottky junction. The main distinctions between a PN junction and a Schottky junction are that the Schottky junction has a lower avalanche breakdown voltage, a higher reverse-bias leakage current and a lower forward-bias turn-on voltage [20].

The active MESFET is opened or closed by applying the appropriate DC bias voltage to the gate. For greater clarity, the elements of the MESFET's equivalent circuit are shown on an outline of its structure to demonstrate their physical origin, as shown in **Figure** 7. The FETs used for the switching function are mainly: Junction FET (JFET), Metal Oxide Semiconductor FET (MOSFET), Metal Semiconductor FET (MESFET) and High Electron Mobility Transistors (HEMT).

2.2.6 Varactor diode

A varactor diode is a semiconductor device that behaves like a variable-capacitor capacitor (**Figure 8**). The junction capacitance varies nonlinearly with the inverse voltage applied across the varactor diode. The varactor diode's junction capacitance (C_T) decreases with increasing reverse bias voltage (V_R). This diode offers a continuous variation of antenna characteristics as a function of bias voltage, unlike PIN diodes which offer a discrete variation.

The nonlinear behavior described here finds utility in various applications, including voltage-controlled oscillators, continuous phase shifters, modulation, frequency multipliers (harmonic generators) and more. Varactors fall into two main categories,

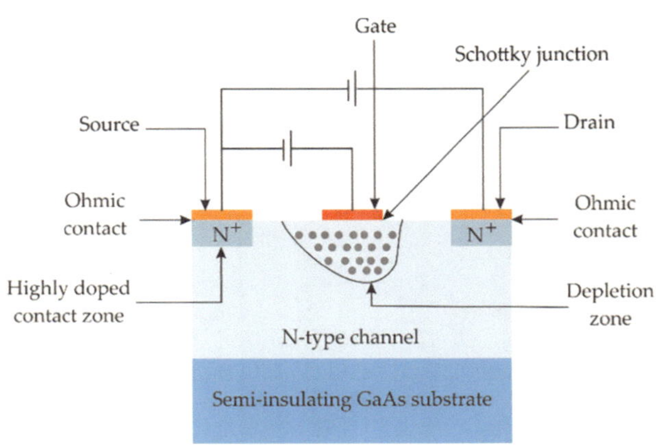

Figure 6.
GaAs MESFET structure.

DOI: http://dx.doi.org/10.5772/intechopen.1006030

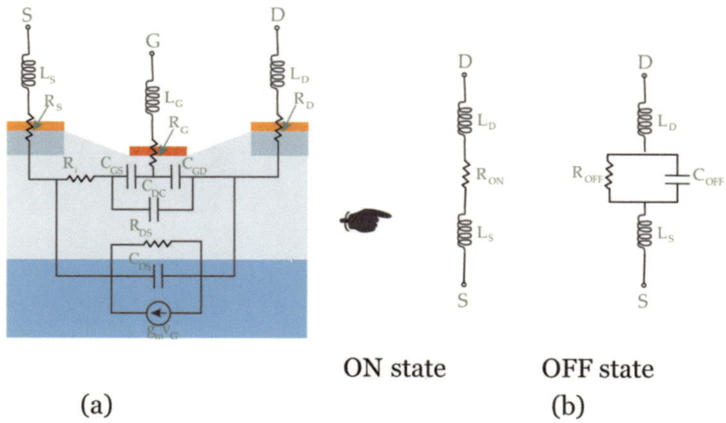

Figure 7.
Location components of (a) MESFET equivalent circuit and (b) ON and OFF equivalent circuit.

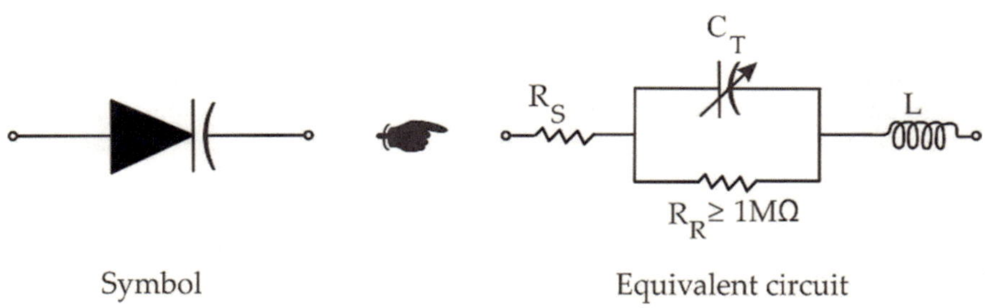

Figure 8.
Symbol and equivalent circuit of a varactor diode.

the p-n junction varactor and the Schottky varactor, typically employed at microwave and millimeter-wave frequencies, respectively [21]. It should be noted that as the number of varactor diodes used increases, so does the complexity of the polarization network. These diodes are also often used for frequency reconfiguration. In the literature, the combination of PIN diodes and varactor diodes is frequently used to achieve high-frequency overlap.

These diodes are also often used for frequency reconfiguration. In the literature, the combination of PIN diodes and varactor diodes is frequently used to achieve high-frequency overlap.

2.2.7 PIN diode

A PIN diode has internal impedance that depends on the voltage applied (bias) across its terminals. For microwaves, it is a semiconductor device that functions as a variable resistor at high frequencies [22]. Typically, the microwave resistance changes from a few kilo-ohms (R_P) in reverse bias to a few ohms (R_S) in forward bias, depending on the diode type and manufacturer. A PIN diode can therefore be used as a switch for high-frequency signals. When forward-biased, the diode has a low impedance (R_S) and therefore behaves like a closed circuit (ON state). However, when reverse-biased, it has a very high impedance (R_P) in parallel with a very low capacitance (C_T) of a few picofarads. It therefore behaves like an open circuit (OFF state),

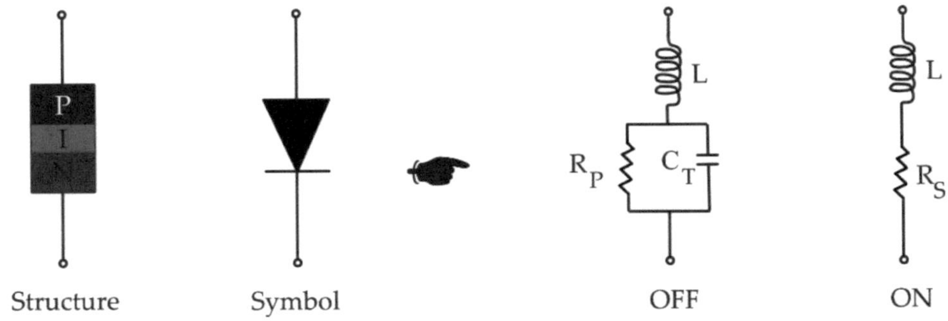

Figure 9.
Structure, symbol and equivalent circuits of a PIN diode in ON and OFF states.

as shown in **Figure 9**. Note that the parasitic capacitance (C_P) is not taken into account, as its value is almost zero. Because of their reliability, simplicity, low cost, very low control voltage, absence of moving parts, high tuning speed and linear characteristics at high frequencies, PIN diodes are most often chosen. They are often used to reconfigure frequency, polarization or radiation patterns.

PIN diodes can be used for various applications such as a switch, variable attenuator, phase shifter, limiter and duplexer. Current PIN diodes can have a cutoff frequency of around 250 GHz and are capable of operating well in the millimeter-wave region [21].

In this section, we have presented almost all the techniques used to achieve the different types of reconfigurability. The technology chosen depends on the requirements of the application in terms of size, cost, targeted performance and so on. **Table 1** shows a comparison of some of the performances of the active elements discussed above.

2.3 Diode biasing circuits

Semiconductor devices require low-frequency, DC bias circuits that must be separate from the RF circuit. This means that when a bias voltage is applied to the device, RF energy must not leak through the bias port [40]. Semiconductor devices are controlled by simple bias circuits. To achieve stable antenna performance, the RF

Active element	Response time (s)	Control voltage (V)	Power (mW)	Resistance R_{ON} (Ω)	Capacity C_{OFF} (fF)	Cost
Ferroelectric capacity [23–25]	10^{-10}	A few hundred (planar)	—	—	—	Low
Photoconductor [26–28]	10^{-9}	1,8–1,9	0–50	—	—	High
MEMS [29–32]	10^{-5}	0–100	0.05–0.1	0,5	<10	High
FET [33–35]	10^{-9}	0–3,5	0,05–0,1	1–5	30–180	Average
Varactor diode [36, 37]	10^{-6}	0–33	—	0–5	—	Average
PIN diode [38, 39]	$10^{-9}-10^{-7}$	0–3	5–100	0–5	20–100	Low

Table 1.
Performance comparison of the most useful active elements.

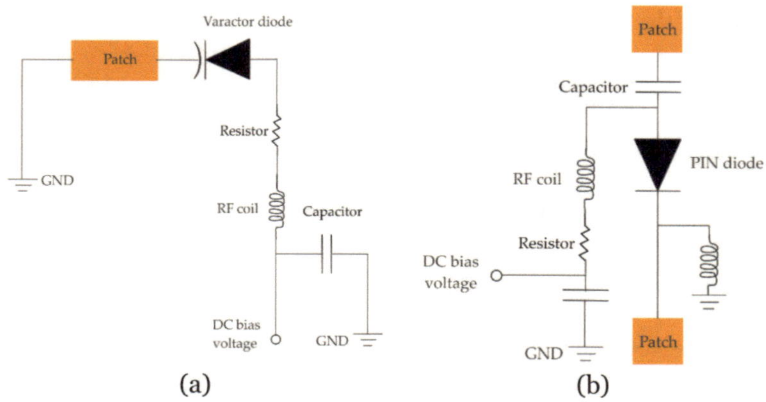

Figure 10.
DC bias circuit for (a) a varactor diode [42] and (b) a PIN diode [43].

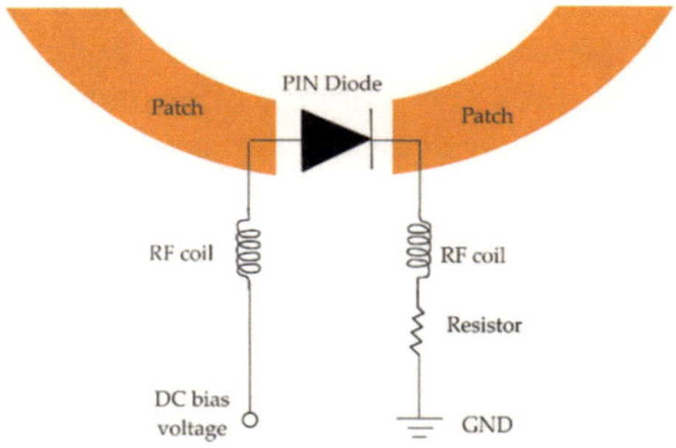

Figure 11.
DC bias circuit for a PIN diode [44].

circuit and the DC feed circuit are suitably isolated from each other using RF inductors and capacitors. To this end, the effects of the bias circuit elements on antenna performance are also examined, in order to stabilize matching and preserve antenna reconfigurability.

The main function of capacitors is to allow RF signals to pass through while preventing DC signals. Their value should ideally be as high as possible to ensure undisturbed signal transmission. These decoupling capacitors are not ideal; therefore, insertion losses and line matching are important factors to consider. The main function of the inductor in a bias circuit is exactly the opposite of that of the capacitor. The performance of RF coils, also known as RF chokes, must be high, with stable behavior over a wide frequency range [41]. Resistors are used to limit DC current to prevent excessive current from damaging the semiconductor device (**Figure 10**).

Since DC current does not affect the controlled element, the influence of capacitors can remain negligible, in the case of antennas, in DC bias circuits (**Figure 11**). Inductors can behave like capacitors as frequency increases, making it impossible to anticipate their behavior with any accuracy. There are therefore many ways to replace the inductor and ensure DC/RF isolation at very high frequencies. Consequently, an inductor can be replaced by a thin transmission line, a quarter-wave line, a quarter-wave line short-circuited by a distributed element or a quarter-wave line short-circuited by a radial stub.

3. Reconfigurable antenna types

Ideally, reconfigurable antennas should possess the capability to autonomously adjust their polarizations, operating frequencies and/or radiation patterns to accommodate evolving operational needs. Consequently, reconfigurable antennas can be classified into four main groups: polarization reconfiguration, frequency reconfiguration, radiation pattern reconfiguration and hybrid reconfiguration.

3.1 Polarization reconfiguration

Polarization reconfiguration involves changing the polarization of the antenna. The aim of these antennas is to modify the vector orientation of the emitted field without altering the shape of the radiation pattern or resonant frequencies. In general, the direction and/or phase of the currents flowing on the radiating element must be modified without altering the paths or pattern.

Polarization reconfigurable antennas (PRAs) can change the polarization state of transmitted and received microwaves by conveniently switching the DC control voltage. This can significantly improve not only the information capacity and frequency reuse of wireless communication systems but also the link quality [45]. Throughout this subsection, we will confine ourselves to listing two works that have adopted the use of PIN diodes and RF-MEMS as switches in order to achieve bias reconfiguration.

Kovitz et al. have developed a reconfigurable circularly polarized antenna [6]. A new wideband patch antenna array with reconfigurable circular polarization (CP) was designed using an E-shaped patch antenna element (**Figure 12**). RF-MEMS are used to switch between right-hand circular polarization (RHCP) and left-hand circular polarization (LHCP). A bandwidth of 17% was achieved using particle swarm optimization and confirmed by measurements for the isolated element using RF-MEMSs (RMSW100HP). The bandwidth achieved is clearly superior to the current state-of--the-art in single-layer, single-feed CP patch element design with similar substrate thickness. To overcome beam wrinkling, a new rotating element configuration is implemented to force symmetry of the pattern.

Another example has been proposed by Shi et al. [46]; this is a simple magneto-electric dipole antenna with reconfigurable polarization (**Figure 13**). The proposed antenna mainly comprises a metal cavity, three pairs of horizontal rectangular

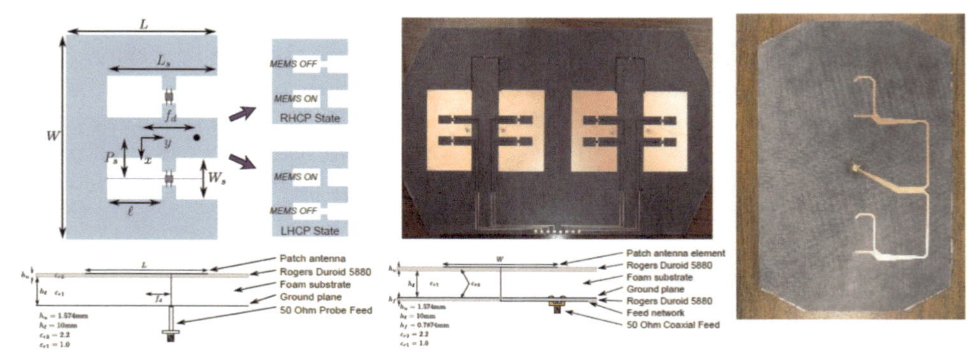

(a) (b) (c)

Figure 12.
(a) E-shaped antenna element configuration with reconfigurable CP, (b) linear antenna array with rotating elements with reconfigurable CP and (c) feed network for the linear antenna array [6].

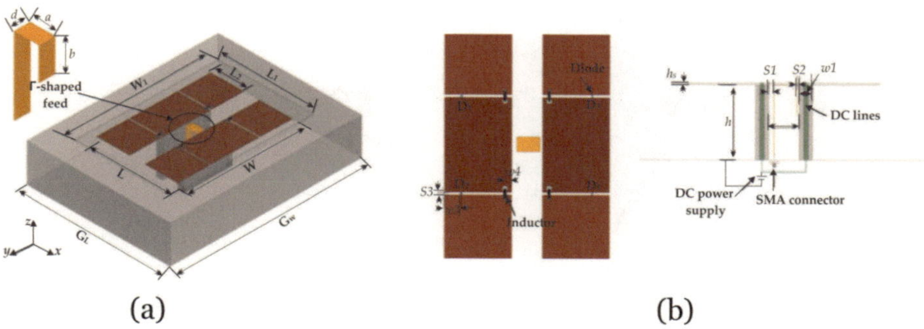

Figure 13.
Geometry of the proposed antenna: (a) perspective view, (b) top and cross-sectional views [46].

patches, a Γ-shaped feed and four PIN diodes. The horizontal rectangular patches can be connected/disconnected by independently controlling the ON/OFF state of the PIN diodes, enabling linear polarization, RHCP and LHCP to be presented. Experimental results demonstrate that the proposed dipole antenna has a stable gain of 8.2 ± 0.8 dBi and an effective wide bandwidth of 33.9% from 1.74 to 2.45 GHz.

3.2 Frequency reconfiguration

The recent development of wireless communication systems has shown a significant increase in the demand for devices that can operate in multiple frequency bands. This has accelerated the development of antennas capable of operating on multiple frequency bands. Compared with traditional antennas, frequency-reconfigurable antennas will be more advantageous and more valuable. Frequency-reconfigurable antennas can be used for a variety of communication standards, the ability to adapt to changing environmental conditions, efficient use of power and spectrum, elimination of expensive hardware and lower system costs.

The most useful and practical method of frequency reconfiguration is the electrical method, in which antennas are reconfigured using radio-frequency microelectromechanical systems (RF-MEMS) [47], varactors [48] or PIN diodes [5, 49]. These components act as switches, directing surface currents to modify the topology of the antenna's radiating elements, can be easily integrated into the antenna, and offer good isolation and minimal losses. On the other hand, the disadvantages of this method include the nonlinear nature of the switches, the need for bias lines to control switching states, associated losses, additional complexity, etc. In this subsection, some examples of frequency-reconfigurable antennas using electrical switches will be cited, and we will limit ourselves to the three types of switches mentioned previously.

Cetiner et al. have designed a frequency-reconfigurable annular slot antenna [47]. Three RF-MEMS actuators were used and monolithically integrated into the antenna architecture to ensure frequency reconfigurability (**Figure 14**). By enabling/disabling RF-MEMS actuators positioned within the antenna geometry and along the microstrip feed line, adjustments are made to the operating frequency band. The reconfigurable ring slot antenna is designed to accommodate two adjustable operating frequencies, with center frequencies at $f_1 = 2,4$ GHz and $f_2 = 5,2$ GHz, adapted to IEEE 802.11 WLAN standards.

A groundbreaking design featuring a frequency-reconfigurable, optically transparent, and unidirectional antenna was presented by Sayem et al. [48]. The innovative

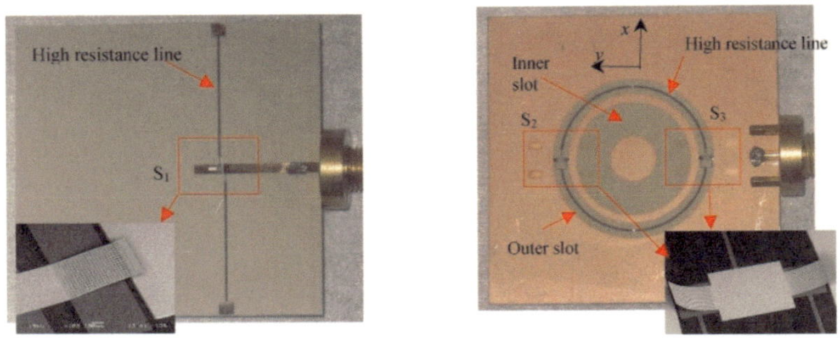

Figure 14.
Photograph of the Reconfigurable ring slot antenna: (a) integrated microstrip feedline with a single-arm MEMS actuator, (b) integrated ring-slot with two double-arm MEMS actuators [47].

design integrates a varactor diode into a folded dipole radiator to achieve continuous frequency tuning. This radiator is supported by a water-filled reflector, ensuring a unidirectional radiation pattern akin to microstrip patch antennas. The radiating element of the antenna is crafted from transparent conductive fabric, boasting approximately 72% optical transparency and occupying only a small portion of the overall antenna structure. Positioned beneath the dipole radiator, the water-filled reflector resides within a rectangular box made of flexible, transparent, hydrophobic polymer known as polydimethylsiloxane (PDMS). Despite the integration of a small varactor diode within the folded dipole radiator, the antenna's overall transparency and flexibility remain largely unaffected (**Figure 15**). The design underwent rigorous numerical analysis and experimental validation, confirming its efficacy. Experimental results showcase continuous frequency tuning ranging from 2.38 to 2.67 GHz, with an average 10 dB return loss bandwidth of approximately 170 MHz. Furthermore, the antenna maintains a favorable front-to-back ratio across its entire operating frequency range.

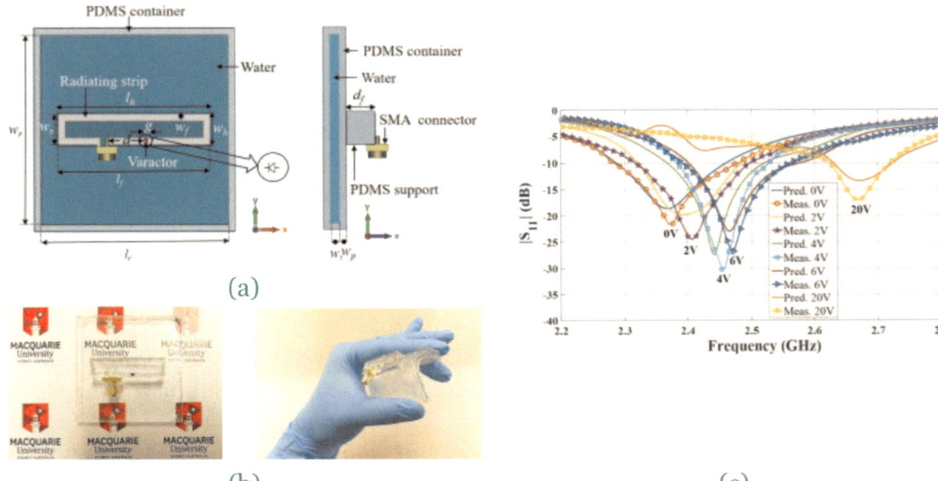

Figure 15.
(a) Geometry of the proposed antenna, (b) photograph of the manufactured prototype, (c) predicted and measured S11 at different bias states [48].

Taher et al. have realized a compact ring-shaped UWB circular monopole antenna with a reconfigurable rejected band characteristic [5]. Three slots with three PIN diodes inserted between the poles of each slot were etched on the circular ring-shaped radiating element to obtain reconfigurable characteristics (**Figure 16**). When the PIN diodes are ON, the circular patch cleverly behaves like a circular ring radiator, enabling the antenna to operate throughout the UWB frequency band (3.1–12.3 GHz). On the other hand, activating or deactivating the PIN diodes enables the rejected bands to be activated or deactivated, thus introducing reconfigurability. The antenna has two rejected bands covering the Sub 6 GHz 5G and WLAN operating with frequency bands of (3.4–3.8), (3.8–4.9) and (5.1–5.8) GHz, respectively.

A compact antenna designed for cognitive radio applications, with flexible characteristics and reconfigurable bandwidth, was presented by Tran et al. [49]. The antenna comprises a monopole serving as the primary radiator, excited by a coplanar waveguide for a streamlined single-layer design. To manipulate the operating bandwidth, they added an extra stub to the feedline (**Figure 17**), keeping the lower frequency fixed while adjusting the upper frequency. However, the inclusion of a stub disturbs the current distribution, leading to a noticeable degradation in matching performance. By controlling the length of the stub, they can modify the rejected band, making it easier to vary the upper frequency. To validate this approach, they used two stubs connected to the supply line *via* PIN diodes. By toggling the ON/OFF states of the PIN diodes type MADP-042305-130,600, they obtained three distinct bandwidth

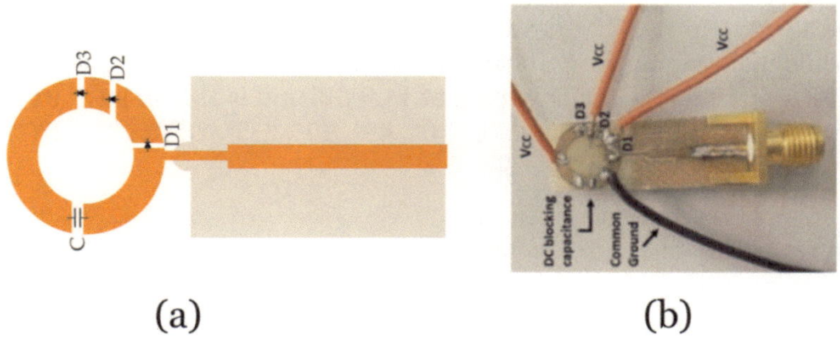

Figure 16.
Proposed band-rejected reconfigurable antenna: (a) geometry, (b) photograph of prototype [5].

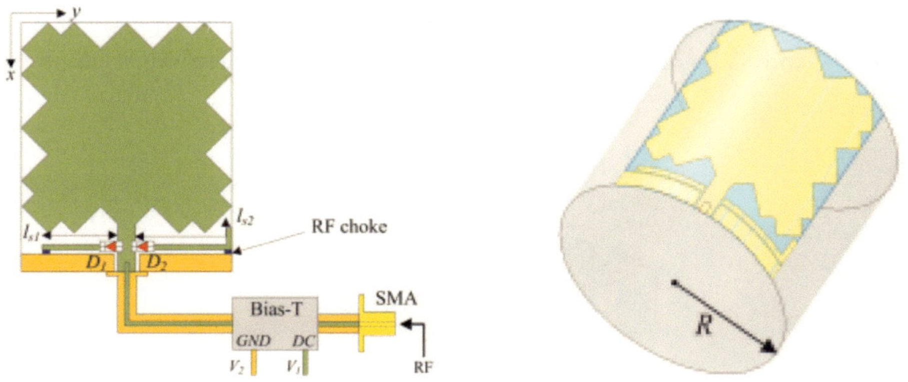

Figure 17.
Proposed BW reconfigurable antenna geometry [49].

states, with fractional bandwidth percentages of 18.9, 33.2 and 64.3%, respectively. Importantly, the antenna maintains good performance even when bent over curved surfaces.

3.3 Radiation pattern reconfiguration

Reconfigurable radiation pattern antennas, as their name suggests, modify the shape, direction and gain of radiation patterns to favor or disfavor certain directions. This reconfigurability can be achieved using a number of techniques, which will be presented throughout this subsection.

3.3.1 Mechanical reconfiguration

Unlike other reconfiguration techniques, mechanically reconfigurable antennas do not require the integration of active elements, polarization systems or scanning mechanisms. On the other hand, the performance flexibility of this type of antenna is limited, and it is difficult to provide multifunctional reconfigurable features (**Figure 18**).

A study introduces a straightforward mechanical approach to achieve beam steering in Fabry–Perot resonator antennas (FPRAs) [50]. The efficient beam steering with minimal power consumption is a crucial aspect of modern antenna designs. The method involves mechanically rotating the upper phase gradient metasurface (PGM) to alter the aperture phase distribution, resulting in continuous azimuthal plane beam steering while maintaining a significant elevation angle. Measured results demonstrate that the FPRA attains at 5.65 GHz a gain of 14.9 dBi and is able to steer its beam continuously in the azimuthal plane (φ = 0°, 45°, 90°, 135°, 180°, 225°, 270° and 315°) at an elevation angle of around θ = 50°.

3.3.2 Reconfiguration using active materials

Reese et al. first proposed a liquid crystal-filled lens antenna for beam steering (**Figure 19**) [51]. By using an array of electrodes, the orientation of the sector with the highest permittivity can be altered, thus adjusting the direction of the beam. One key benefit of this antenna design is its simplicity, as it eliminates the need for phase shifters or switching networks, presenting a straightforward approach to beam steering. The proposed antenna can steer the main beam in predetermined directions of −30°, 0° and +30° within the V-band (50 to 75 GHz). Furthermore, it offers the

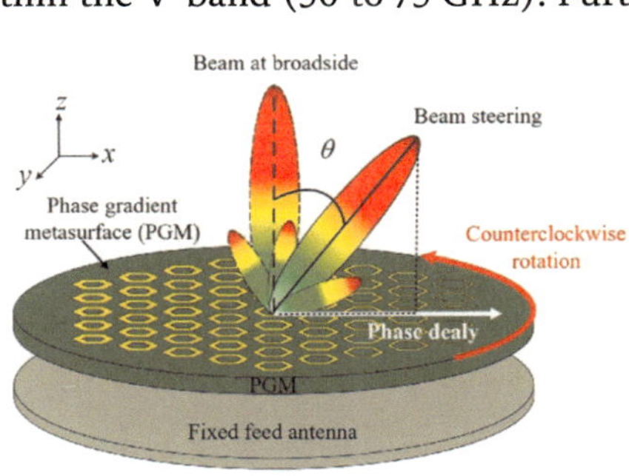

Figure 18.
Perspective view of the proposed mechanical rotation PGM. PGM, phase gradient metasurface [50].

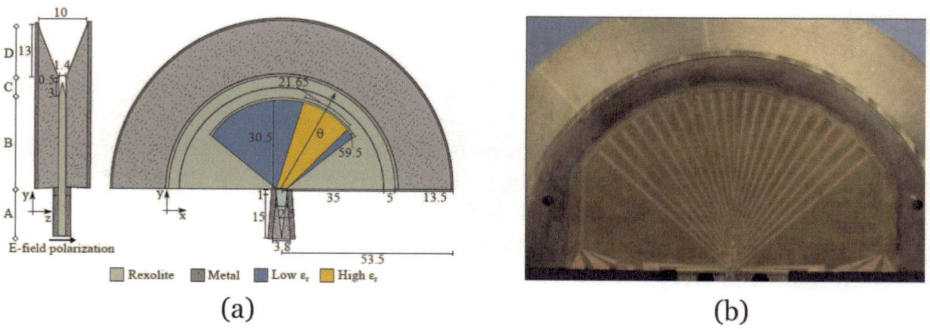

Figure 19.
(a) Geometry of the proposed lens antenna and (b) photograph of the fabricated electrode array [51].

flexibility to adjust the beam width and generate multiple beams. The measured gain falls between 13 and 15 dBi.

3.3.3 Optical reconfiguration

Optical reconfiguration relies on employing photoconductive switches crafted from semiconductor materials like silicon or gallium arsenide. This method of reconfiguration eliminates the necessity for metal wires, instead utilizing polarization lines to minimize interference and guarantee strong isolation from electrical switches.

Zhang et al. have developed an optically controlled planar Yagi-Uda antenna with a reconfigurable radiation pattern [52]. The antenna consists of a director, a controlled dipole, a reflector and the substrate (**Figure 20**). The two parasitic elements loaded with PIN photodiodes (SHARP PD60 T) in the middle have the ability to vary the radiation pattern by transforming their physical structures. The direction of maximum radiation from the two ON/OFF states of the PIN photodiodes differs by 180∘. Efficient unidirectional radiation patterns with forward/backward ratios of 12 and 8 dB are obtained. Experimental results show that the resonant frequency is 800 MHz, the relative bandwidth is ~15 and 12.5%, and the maximum gain is 6.3 and 5.4 dBi for the two operating states, respectively.

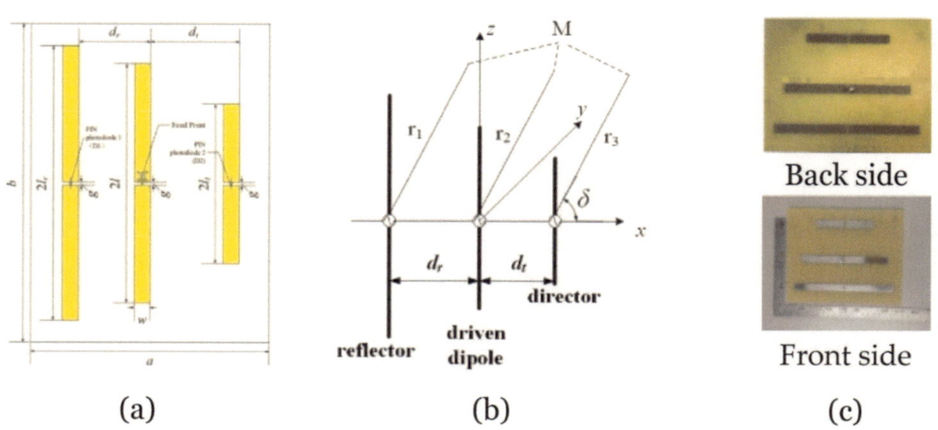

Figure 20.
(a) Proposed antenna geometry, (b) proposed antenna equivalent radiation model, and (c) photograph of fabricated prototype [52].

3.3.4 Electrical reconfiguration

3.3.4.1 Switching via *RF-MEMS*

Figure 21 shows an example of a reconfigurable radiation pattern antenna using RF-MEMSs [53]. By electrostatically modulating the unit cells *via* RF-MEMSs switches in three different configurations ("0000–0000", "1010–1010" and "1111–1111"), the resonant frequency was tuned. At a fixed frequency of 8.2 or 8.3 GHz, the phase constant β was effectively reconfigured from a negative to a positive value so as to sweep the radiation beams in three different directions (θ = 0°, +15° and −15°).

3.3.4.2 Switching via *varactor diode*

Tian et al. have proposed a reconfigurable, fixed-frequency patch antenna controlled by a single varactor diode [54]. The antenna is based on a coupled-mode patch antenna that can sweep the main beam as a function of frequency. A slot etched into the ground plane and loaded with a varactor diode is tuned to be resonant, capacitive or inductive (**Figure 22**). Two types of varactor diode, SMV1413 and SMV1405, were used to cover the entire tuning band. The frequency response of the antenna, affected by the loaded slot, shifts from the center frequency of 2.35 GHz with a bandwidth of 4.26% to the band centered on 2.3 GHz with a bandwidth of 4.35%.

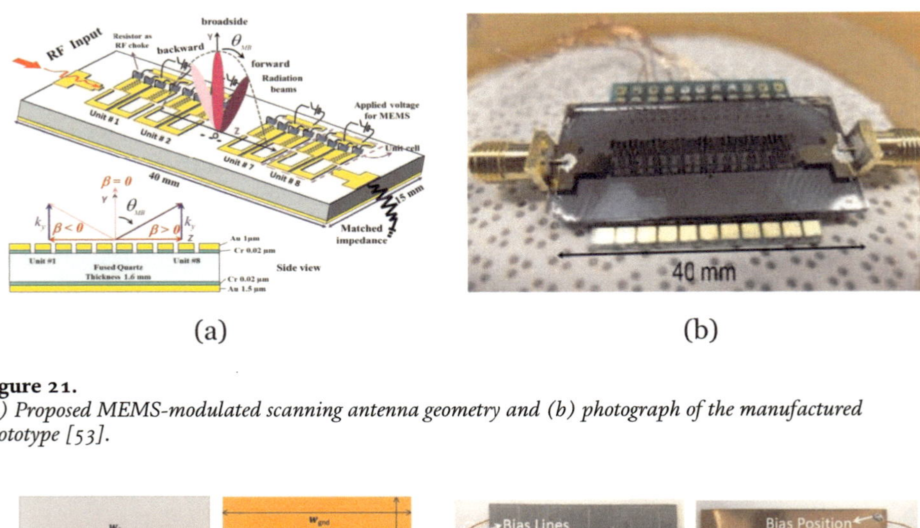

(a) (b)

Figure 21.
(a) Proposed MEMS-modulated scanning antenna geometry and (b) photograph of the manufactured prototype [53].

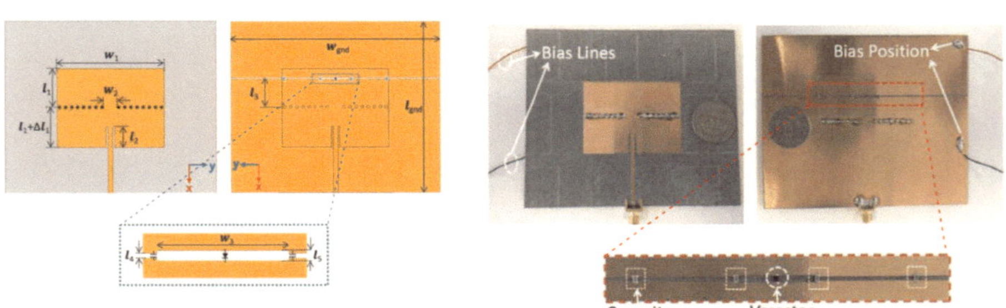

(a) (b)

Figure 22.
(a) Proposed antenna schematic and (b) fabricated prototype [54].

The phases of the radiating slits are manipulated by applying polarization, resulting in both forward and reverse beam scanning. The maximum scanning range of 66° is achieved at 2.29 GHz, where the beam scans continuously from −34° to +32°. The proposed antenna is capable of maintaining high gain and efficiency, comparable to that of an ordinary patch antenna of the same size and substrate.

3.3.4.3 Switching via *PIN diode*

Rahmani et al. have proposed an antenna consisting of two rectangular radiating elements fed by a microstrip line (**Figure 23**). By switching two MPP4203 PIN diodes on or off, the main radiation beam can be switched between three operating modes [55]. In this way, a wide-angle beam switching property ranging from (±30°) to (±38°) in the plane of elevation can be achieved using a simple, low-cost technique. Depending on operating states 1, 2 and 3 of the proposed antenna, the main beam is directed in the azimuthal plane at 0°, 180° or 0° and 180°, respectively (**Figure 24**). In addition to this electron scanning feature of the main lobe, the proposed antenna features a wide bandwidth that covers the entire spectrum between 5 and 6 GHz.

As shown in **Figure 25**, a beam-switched antenna for wireless communication systems has been proposed by Rahmani et al. [4]. The proposed reconfigurable antenna is powered by a coaxial cable and comprises a circular ground plane, and six fish-shaped radiating elements. Built on a Rogers RT5880 substrate, the antenna

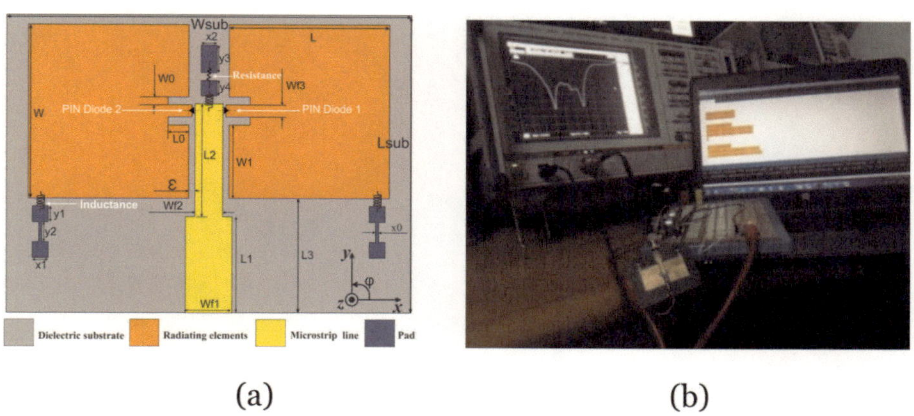

Figure 23.
(a) Proposed antenna schematic and (b) S11 parameter measurement bench [55].

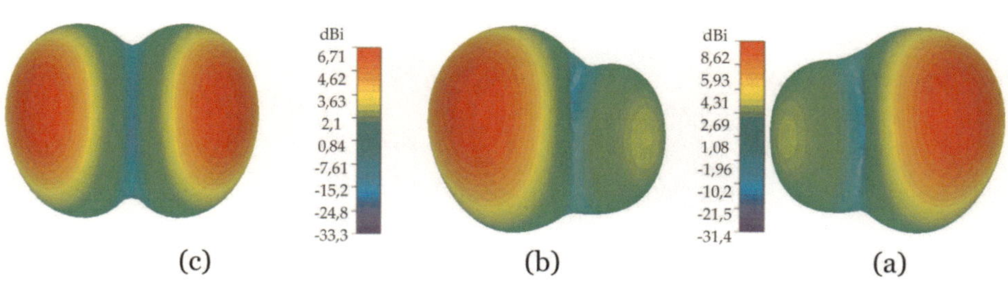

Figure 24.
Simulated 3D radiation patterns at 5.8 GHz: (a) state 1, (b) state 2, (c) state 3 [55].

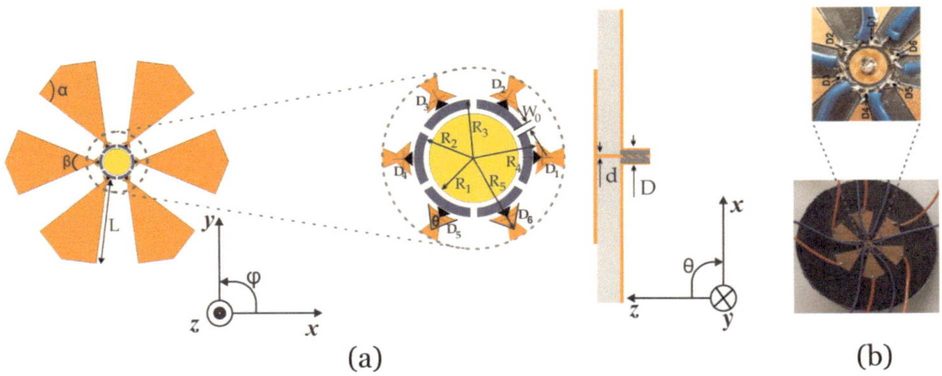

Figure 25.
(a) Proposed antenna schematic and (b) fabricated prototype [4].

features six reconfigurable operating states, at the same frequency, by switching six PIN diodes on and off to modify the beam direction. Depending on antenna operating states S1, S2, S3, S4, S5 and S6, the main beam is directed in the azimuth plane at 0°, 60°, 120°, 180°, 240° or 300°, respectively (**Figure 26**). The proposed antenna exhibits a bandwidth of 9.62% (simulated), 9.31% (theoretical) and 9.07% (measured). It achieves a maximum gain of 9.57 dBi across all operational states and boasts an efficiency ratio ranging from 85–95% over the operating frequency (5.54–6.10 GHz).

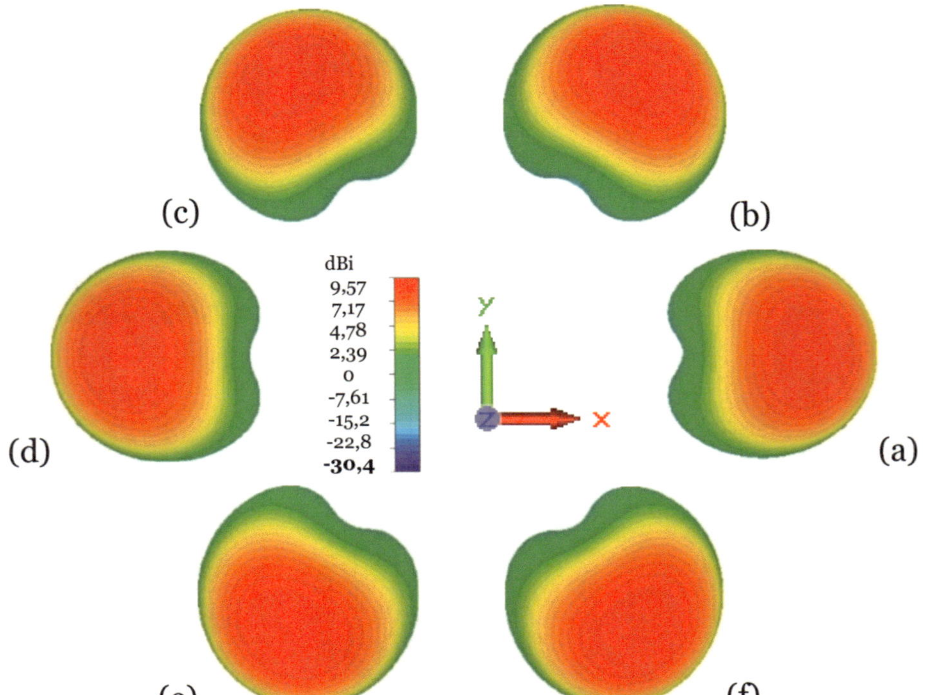

Figure 26.
Simulated 3D radiation patterns at 5.82 GHz: (a) state 1, (b) state 2, (c) state 3, (d) state 4, (e) state 5, (f) state 6 [4].

3.3.5 Comparison of different reconfiguration techniques

Table 2 below summarizes the advantages and disadvantages of the different reconfiguration techniques. Electrical reconfiguration is the most common technique for tuning antenna characteristics using active switches [4, 5, 48, 49, 53–56]. However, optical switches are less popular than active switches due to their lossy behavior and complex activation mechanism [26, 52, 57]. Despite all the advantages of mechanical reconfiguration, this type of antenna presents limited flexibility and difficult multifunctional reconfiguration [50, 58, 59]. Reconfiguration based on active materials is better suited to applications in the terahertz frequency range [51].

3.3.6 Hybrid reconfiguration

Reconfigurable antenna parameters typically include frequency, polarization and radiation pattern. Single-parameter reconfigurable antennas with very good performance have been carefully studied and reported in the literature [40–55]. With the development of wireless communication technologies, multiparameter reconfigurable antennas, which allow at least two of the three parameters to be reconfigured simultaneously, are highly desirable for compact, multifunctional communication systems.

Figure 27 illustrates, for example, an antenna that can be reconfigured in polarization and pattern [60]. The proposed antenna is mainly composed of 3 × 3 square metasurfaces, a ring-shaped pixel structure, a square microstrip patch, a metallic ground plane and a simple feed network. The proposed antenna can create steerable

Reconfiguration technique	Advantages	Disadvantages
Mechanical reconfiguration	No need for active elements No need for polarization systems	Slow response Requires a power source
Reconfiguration using smart materials	Low profile Low weight Very high frequency	Low efficiency Limited application
Optical reconfiguration	No need for polarization lines No intermodulation distortion	Loss behavior Complex activation mechanism
Electrical reconfiguration	Easy to install Low cost of ownership	Complex structure Polarization systems required

Table 2.
Advantages and disadvantages of different reconfiguration techniques.

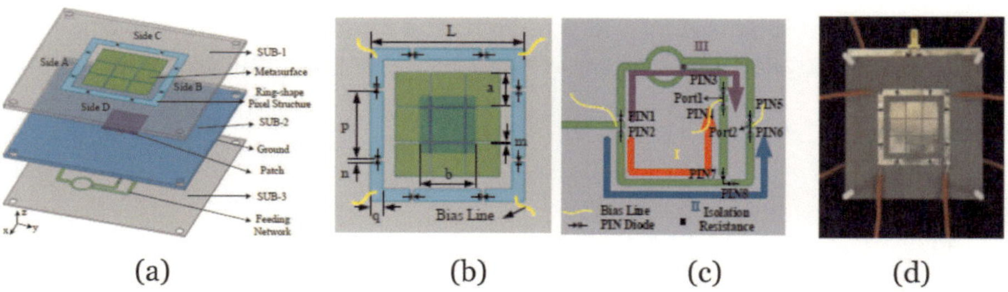

Figure 27.
Proposed antenna structure: (a) 3D view, (b) top view, (c) bottom view, (d) antenna photograph [60].

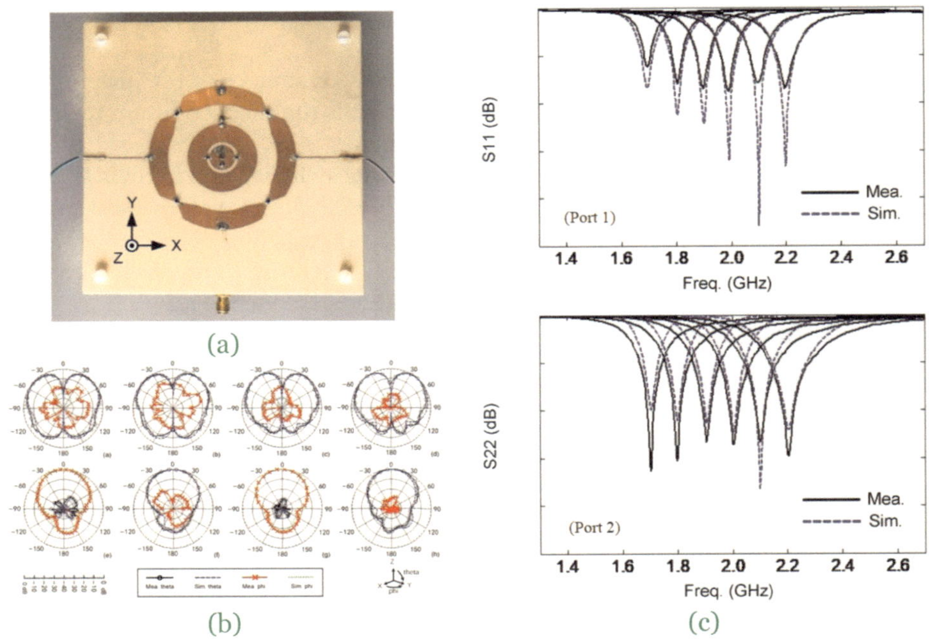

Figure 28.
(a) Antenna photograph, (b) proposed antenna normalized radiation patterns, (c) reflection coefficients: port 1, port 2 [61].

beam directions of ±40° with horizontal polarization, ±20° with vertical polarization, and ± 25° with circular polarization in the y-z plane, in the C-band at 5 GHz center frequency, by loading a ring-shaped pixel structure around the metasurface.

A six-varactor controlled antenna is developed to offer frequency reconfigurability and various radiation patterns [61]. This design integrates two separate reconfigurable structures into a single antenna, enabling the transmission of two orthogonal patterns for Multi-Input Multi-Output (MIMO) applications. The composite right/left annular transmission line (CRLH-TL) operates in mode 1, producing a wide pattern, while the internal circular radiator, loaded with split-ring resonators (SRR), operates in mode 0, emitting an omnidirectional pattern. This arrangement facilitates pattern diversity. Surface-mounted varactors enable extended frequency tuning from 1.7 to 2.2 GHz, covering the 1.71–2.17 GHz LTE band (**Figure 28**). In addition, the antenna design achieves low mutual coupling between the two radiators. A prototype of the antenna confirms the validity of the proposed concept, with measurements closely matching the simulation results. Analysis of a dual-mode MIMO system, using these antennas in a three-dimensional channel model, highlights the improvement in network compactness and spectral efficiency over systems using single-mode elements.

4. Conclusion

In this chapter, we have discussed the main approaches useful for ensuring the dynamic modification of one or more fundamental properties of an antenna. In a second step, some frequency, polarization, radiation pattern and hybrid reconfigurable antennas reported in the literature have been carefully analyzed and discussed. A comparison of the performance of different agile localized elements and also of the advantages and disadvantages of different reconfiguration techniques has been presented and discussed.

Electrical reconfiguration, employing active switches, stands out as the prevalent method for adjusting antenna properties. In contrast, optical switches, though available, are less favored due to their inherent loss and intricate activation process. Despite the benefits of mechanical reconfiguration, antennas of this kind suffer from restricted flexibility and challenging multifunctional adaptability. Reconfiguration utilizing active materials proves more suitable for terahertz frequency range applications.

The usual parameters of reconfigurable antennas involve frequency, polarization, and radiation pattern. Single-parameter reconfigurable antennas exhibit excellent performance. As wireless communication technologies advance, there is a growing demand for multi-parameter reconfigurable antennas. These antennas enable simultaneous reconfiguration of at least two out of the three parameters, offering compactness and versatility for multifunctional communication systems.

Prospects for future research work in the field of reconfigurable antennas include:

The design of miniaturized reconfigurable antennas for ISM bands;

The design of multifunctional hybrid reconfigurable antennas;

Development of flexible reconfigurable antennas;

Development of machine learning algorithms to maximize the lifetime of WSNs in which sensor nodes are equipped with reconfigurable directional antennas;

The development of machine learning algorithms to reduce or eliminate the risks associated with certain categories of security attacks on WSNs in which sensor nodes are equipped with reconfigurable directional antennas.

Author details

Faouzi Rahmani[1*], Abdelmounaim Belbachir Kchairi[2], Nihade Taher[1], Moustapha El Bakkali[1], Naima Amar Touhami[1] and Tajeddin Elhamadi[1]

1 Laboratory of Intelligent Systems Design (ISD), Faculty of Sciences, Abdelmalek Essaâdi University, Tetouan, Morocco

2 Laboratory of Engineering Sciences and Biosciences (ESB), Mohammedia Faculty of Sciences and Techniques, University of Hassan II, Casablanca, Morocco

*Address all correspondence to: rahmani.faouzi-etu@uae.ac.ma

References

[1] Gangi A. The active adaptive antenna array system. IEEE Transactions on Antennas and Propagation. 1963;**11**: 405-414. DOI: 10.1109/TAP.1963.1138061

[2] Andre S, Leonard D. An active retrodirective array for satellite communications. IEEE Transactions on Antennas and Propagation. 1964;**12**: 181-186. DOI: 10.1109/TAP.1964. 1138183

[3] Brown ER. RF-MEMS switches for reconfigurable integrated circuits. IEEE Transactions on Microwave Theory and Techniques. 1998;**46**:1868-1880. DOI: 10.1109/22.734501

[4] Rahmani F, Amar Touhami N, Belbachir Kchairi A, Aknin N, Taher N. Pattern reconfigurable antenna for VANET, wi-fi, and WiMAX wireless communication systems. International Journal of Antennas and Propagation. 2021;**2021**:1–12. DOI: 10.1155/2021/ 9973839

[5] Taher N, Zakriti A, Amar Touhami N, Rahmani F. Circular ring UWB antenna with reconfigurable notch band at WLAN/sub 6 GHz 5G mobile communication. Microsystem Technologies. 2022;**28**:965-972. DOI: 10.1007/s00542-021-05246-9

[6] Kovitz JM, Rajagopalan H, Rahmat-Samii Y. Design and implementation of broadband MEMS RHCP/LHCP reconfigurable arrays using rotated E-shaped patch elements. IEEE Transactions on Antennas and Propagation. 2015;**63**:2497-2507. DOI: 10.1109/TAP.2015.2417892

[7] Balanis CA. Modern Antenna Handbook. 1st ed. Wiley; 2008. pp. 369-398. DOI: 10.1002/97804 70294154.ch8

[8] Zhu HL, Cheung SW, Yuk TI. Mechanically pattern reconfigurable antenna using metasurface. IET Microwaves, Antennas & Propagation. 2015;**9**:1331-1336. DOI: 10.1049/iet-map.2014.0676

[9] Motovilova E, Huang SY. A review on reconfigurable liquid dielectric antennas. Materials. 2020;**13**:1863. DOI: 10.3390/ ma13081863

[10] Petosa A. An overview of tuning techniques for frequency-agile antennas. IEEE Antennas and Propagation Magazine. 2012;**54**:271-296. DOI: 10.1109/MAP.2012.6348178

[11] Ren Z, Qi SS, Hu Z, Shen Z, Wu W. Wideband water helical antenna of circular polarization. IEEE Transactions on Antennas and Propagation. 2019;**67**: 6770-6777. DOI: 10.1109/TAP.2019. 2922846

[12] Xing L, Zhu J, Xu Q, Yan D, Zhao Y. A circular beam-steering antenna with parasitic water reflectors. IEEE Antennas and Wireless Propagation Letters. 2019; **18**:2140-2144. DOI: 10.1109/LAWP. 2019.2938872

[13] Zou M, Shen Z, Pan J. Frequency-reconfigurable water antenna of circular polarization. Applied Physics Letters. 2016;**108**:014102. DOI: 10.1063/ 1.4939455

[14] Wang M, Chu QX. A wideband polarization-reconfigurable water dielectric resonator antenna, antennas wirel. Propagation Letters. 2019;**18**: 402-406. DOI: 10.1109/LAWP.2019. 2892503

[15] Sun J, Luk KM. A wideband low cost and optically transparent water patch antenna with omnidirectional conical

beam radiation patterns. IEEE Transactions on Antennas and Propagation. 2017;**65**:4478-4485. DOI: 10.1109/TAP.2017.2730250

[16] Kittel C. Introduction to Solid State Physics. 8th ed. Hoboken, NJ: Wiley; 2004

[17] da Costa IF, Cerqueira SA, Spadoti DH, da Silva LG, Ribeiro JAJ, Barbin SE. Optically controlled reconfigurable antenna array for mm-wave applications, antennas wirel. Propagation Letters. 2017;**16**:2142-2145. DOI: 10.1109/LAWP.2017.2700284

[18] Bettoumi I. Etude des Commutateurs Hyperfréquences à Base de Matériaux à Changement de Phase (PCM) [thesis]. France: Limoges University; 2022. p. 139

[19] Young DJ, Kim H. Microelectromechanical systems (MEMS). In: Burghartz JN, editor. Guide to State-of-the-Art Electron Devices. Chichester, UK: John Wiley & Sons, Ltd; 2013. pp. 239-250. DOI: 10.1002/9781118517543.ch18

[20] Free CE, Aitchison CS. RF and Microwave Circuit Design: Theory and Applications. 1st ed. NJ, USA: Wiley; 2021. DOI: 10.1002/9781119332237

[21] Ishii TK. Handbook of Microwave Technology. San Diego: Academic Press; 1995

[22] Mennai A. Conception et réalisation de commutateurs RF à base de matériaux à transition de phase (PTM) et à changement de phase (PCM) [thesis]. France: Limoges University; 2016. p. 207

[23] Modelski JW, Yashchyshyn Y. Microwave ferroelectric and reconfigurable antennas. In: 2012 IEEE MTT-S International Microwave Workshop Series on Millimeter Wave Wireless Technology and Applications. Nanjing, China: IEEE; 2012. pp. 1-2. DOI: 10.1109/IMWS2.2012.6338192

[24] Rao JBL, Patel DP, Krichevsky V. Voltage-controlled ferroelectric lens phased arrays. IEEE Transactions on Antennas and Propagation. 1999;**47**: 458-468. DOI: 10.1109/8.768780

[25] Jiang H, Patterson M, Brown D, Zhang C, Pan K, Subramanyam G, et al. Miniaturized and reconfigurable CPW square-ring slot antenna loaded with ferroelectric BST thin film varactors. IEEE Transactions on Antennas and Propagation. 2012;**60**:3111-3119. DOI: 10.1109/TAP.2012.2196918

[26] Reji V, Manimegalai CT. Light controlled frequency reconfigurable antenna for wireless applications. International Journal of Microwave and Wireless Technologies. 2023;**15**:143-149. DOI: 10.1017/S1759078722000137

[27] Huang YC, Wu JW. A reconfigurable microstrip antenna based on photoconductive switches. In: 2023 IEEE MTT-S International Microwave Workshop Series on Advanced Materials and Processes for RF and THz Applications (IMWS-AMP), Chengdu, China: IEEE; 2023. pp. 1-3. DOI: 10.1109/IMWS-AMP57814.2023.10381265

[28] Lv JM, Wang ZX, Wu JW. Optically controlled 2-bit reconfigurable antenna in C-band. In: 2023 IEEE MTT-S International Microwave Workshop Series on Advanced Materials and Processes for RF and THz Applications (IMWS-AMP), Chengdu, China: IEEE; 2023. pp. 1-3. DOI: 10.1109/IMWS-AMP57814.2023.10381113

[29] Blard F. Conception et réalisation de références de tensions alternatives à base

de MEMS [thesis]. France: Paul Sabatier University-Toulouse III; 2011

[30] Goldsmith CL, Forehand DI. Temperature variation of actuation voltage in capacitive MEMS switches. IEEE Microwave and Wireless Components Letters. 2005;**15**:718-720. DOI: 10.1109/LMWC.2005.856827

[31] Singh T. Design and finite element modeling of series-shunt configuration based RF MEMS switch for high isolation operation in K–Ka band. Journal of Computational Electronics. 2015;**14**: 167-179. DOI: 10.1007/s10825-014-0636-2

[32] Huff GH, Bernhard JT. Integration of packaged RF MEMS switches with radiation pattern reconfigurable square spiral microstrip antennas. IEEE Transactions on Antennas and Propagation. 2006;**54**:464-469. DOI: 10.1109/TAP.2005.863409

[33] Aboufoul T, Alomainy A, Parini C. Reconfiguring UWB monopole antenna for cognitive radio applications using GaAs FET switches. IEEE Antennas and Wireless Propagation Letters. 2012;**11**: 392-394. DOI: 10.1109/LAWP.2012. 2193551

[34] Yang X, Lin J, Chen G, Kong F. Frequency reconfigurable antenna for wireless communications using GaAs FET switch. IEEE Antennas and Wireless Propagation Letters. 2015;**14**:807-810. DOI: 10.1109/LAWP.2014.2380436

[35] Quddious A, Abbasi MAB, Antoniades MA, Vryonides P, Fusco V, Nikolaou S. Dynamically reconfigurable UWB antenna using an FET switch powered by wireless RF harvested energy. IEEE Transactions on Antennas and Propagation. 2020;**68**:5872-5881. DOI: 10.1109/TAP.2020.2988941

[36] Shynu S.V, Augustin G, Aanandan C.K, Mohanan P, Vasudevan K. Triple slot arm loaded reconfigurable dual frequency microstrip antenna using varactors, 2005 IEEE Antennas and Propagation Society International Symposium, IEEE, Washington, DC, USA, 2005: pp. 609–612. DOI: 10.1109/APS.2005.1552085

[37] Kehn MNM, Quevedo-Teruel O, Rajo-Iglesias E. Reconfigurable loaded planar inverted-F antenna using varactor diodes. IEEE Antennas and Wireless Propagation Letters. 2011;**10**:466-468. DOI: 10.1109/LAWP.2011.2153174

[38] Yi X, Huitema L, Wong H. Polarization and pattern reconfigurable cuboid quadrifilar helical antenna. IEEE Transactions on Antennas and Propagation. 2018;**66**:2707-2715. DOI: 10.1109/TAP.2018.2816785

[39] He X, Gao P, Zhu Z, You S, Wang P. A flexible pattern reconfigurable antenna for WLAN wireless systems. Journal of Electromagnetic Waves and Applications. 2019;**33**:782-793. DOI: 10.1080/09205071.2019.1576547

[40] Mongia R, Bhartia P, Bahl IJ. RF and Microwave Coupled-Line Circuits. 2nd ed. Boston: Artech House; 2007

[41] Alam MS, Abbosh A. Planar pattern reconfigurable antenna with eight switchable beams for WiMax and WLAN applications. IET Microwaves, Antennas & Propagation. 2016;**10**:1030-1035. DOI: 10.1049/iet-map.2015.0647

[42] Guo C, Deng L, Dong J, Yi T, Liao C, Huang S, et al. Variode enabled frequency-reconfigurable microstrip patch antenna with operation band covering S and C bands. Progress In Electromagnetics Research M. 2020;

88:159-167. DOI: 10.2528/PIERM19110204

[43] Prakash T, Chaudhary RK, Gangwar RK. Pattern-reconfigurable antenna in azimuth plane using SP3T reconfigurable switching network. IET Microwaves, Antennas & Propagation. 2021;**15**:62-68. DOI: 10.1049/mia2.12029

[44] Jin G, Li M, Xu Y, Yang J, Liao S. Differentially fed six-beam switchable reconfigurable antenna. IET Microwaves, Antennas & Propagation. 2020;**14**:573-577. DOI: 10.1049/iet-map.2019.0622

[45] Hu J, Luo GQ, Hao ZC. A wideband quad-polarization reconfigurable metasurface antenna. IEEE Access. 2018;**6**:6130-6137. DOI: 10.1109/ACCESS.2017.2766231

[46] Shi Y, Cai Y, Zhang XF, Kang K. A simple tri-polarization reconfigurable magneto-electric dipole antenna. IEEE Antennas and Wireless Propagation Letters. 2018;**17**:291-294. DOI: 10.1109/LAWP.2017.2786945

[47] Cetiner BA, Crusats GR, Jofre L, Biyikli N. RF MEMS integrated frequency reconfigurable annular slot antenna. IEEE Transactions on Antennas and Propagation. 2010;**58**:626-632. DOI: 10.1109/TAP.2009.2039300

[48] Md SAS, Simorangkir RBVB, Esselle KP, Thalakotuna DN, Lalbakhsh A. An electronically-tunable, flexible, and transparent antenna with unidirectional radiation pattern. IEEE Access. 2021;**9**:147042-147053. DOI: 10.1109/ACCESS.2021.3114004

[49] Tran H-H, Kim-Thi P, Nguyen TT-L, Thao HTP. A flexible and bandwidth reconfigurable antenna for cognitive communication. IEEE Access. 2023;**11**: 89561-89567. DOI: 10.1109/ACCESS.2023.3306451

[50] Liu Y, Zhu L, Zhang W, Wang W. Mechanical azimuthal beam-steering Fabry–Perot resonator antenna with large deflection angle. IET Microwaves, Antennas & Propagation. 2024;**18**:1-9. DOI: 10.1049/mia2.12471

[51] Reese R, Jost M, Polat E, Tesmer H, Strobl J, Schuster C, et al. A millimeter-wave beam-steering lens antenna with reconfigurable aperture using liquid crystal. IEEE Transactions on Antennas and Propagation. 2019;**67**:5313-5324. DOI: 10.1109/TAP.2019.2918474

[52] Zhang Y, Lin S, Yu S, Liu G.J, Denisov A. Design and analysis of optically controlled pattern reconfigurable planar Yagi–Uda antenna, IET Microwaves, Antennas & Propagation. 12 (2018) 2053–2059. DOI: 10.1049/iet-map.2018.5204

[53] Luo Y, Kikuta K, Han Z, Takahashi T, Hirose A, Toshiyoshi H. An active metamaterial antenna with MEMS-modulated scanning radiation beams. IEEE Electron Device Letters. 2016;**37**:920-923. DOI: 10.1109/LED.2016.2565559

[54] Tian H, Jiang LJ, Itoh T. A compact single-element pattern reconfigurable antenna with wide-angle scanning tuned by a single varactor. Progress in Electromagnetics Research C. 2019;**92**: 137-150. DOI: 10.2528/PIERC19021407

[55] Rahmani F, Amar Touhami N, Belbachir Kchairi A, Lamsalli M, Taher N, Ennasar MA. Sub-6 GHz adjustable broadband radiation pattern microstrip antenna for wireless communication system. Progress in Electromagnetics Research C. 2022;**124**: 97-109. DOI: 10.2528/PIERC22070701

[56] Ke YH, Yang LL, Chen J-X. Design of switchable dual-Balun feeding structure for pattern-reconfigurable endfire antenna. IEEE Antennas and Wireless Propagation Letters. 2021;**20**:1463-1467. DOI: 10.1109/LAWP.2021.3087479

[57] Vovchuk D, Mikhailovskaya A, Dobrykh D, Salgals T, Ginzburg P. Dual-band electro-optically steerable antenna. Journal of Optics. 2023;**25**:105601. DOI: 10.1088/2040-8986/acf1ae

[58] Siyu Wu HP, Chen J. A rotatable frequency- and pattern-reconfigurable antenna with a wide tunable bandwidth. International Journal of Electronics. 2024;**111**:1-15. DOI: 10.1080/00207217.2024.2302334

[59] Yang X, Xu S, Yang F, Li M, Fang H, Hou Y, et al. A mechanically reconfigurable reflectarray with slotted patches of tunable height. IEEE Antennas and Wireless Propagation Letters. 2018;**17**:555-558. DOI: 10.1109/LAWP.2018.2802701

[60] Li W, Wang YM, Hei Y, Li B, Shi X. A compact low-profile reconfigurable metasurface antenna with polarization and pattern diversities. IEEE Antennas and Wireless Propagation Letters. 2021; **20**:1170-1174. DOI: 10.1109/LAWP.2021.3074639

[61] Zhang J, Wang B, Yan S, Li W, Vandenbosch GAE. Metamaterial inspired varactor-tuned antenna with frequency reconfigurability and pattern diversity. Sensors. 2024;**24**:1956. DOI: 10.3390/s24061956

Chapter 5

Ultra Wideband Planar Antennas Based on Spoof Surface Plasmon Polaritons for Microwave and Milimeter-Wave Applications

Goksel Turan

Abstract

This chapter reviews ultra wideband (UWB) planar spoof surface plasmon polaritons (SSPP) antennas that operate in microwave to millimeter-wave bands. The techniques used to determine the UWB properties of SSPP-based antennas and their performance in various operating bands have been carefully analyzed and compared. Leaky-wave and end-fire antennas are the two types into which they are categorized. The pros and cons of various antenna design structures, sizes, radiation types, beam scanning capacities, bandwidths, and gains are evaluated. Most of the studies used gradient grooved transitions to SSPP mode and a variety of slots to achieve bandwidths of 100% or more. The end-fire antennas exhibit high efficiency, ranging from 75–98%, with an average gain change of about 5 dBi. On the other hand, the leaky-wave antennas present a total wide beam-scanning range of 90° to 142°, a high efficiency of 70–98%, and a low gain variation of about 3 dBi. The purpose of this paper is to serve as a reference for future research by emphasizing the most significant characteristics of the SSPP-based UWB antennas that are currently being investigated for potential use in the next generation of applications.

Keywords: ultra wideband (UWB), spoof surface plasmon polaritons (SSPP)-based antennas, planar antennas, end-fire antennas, leaky-wave antennas, microwave applications, milimeter-wave applications

1. Introduction

Wireless communication systems have progressed since their inception, keeping up with changing and evolving technology on a daily basis. Antennas, which are the primary connection between wireless communication systems and the outside world, have had to alter and renew themselves with diverse designs in order to remain functional, much like other circuit elements. In recent years, research has focused on basic antenna properties such as multi-band [1, 2], miniaturization [3, 4], wide band [5, 6], and high gain [7, 8], as well as application-oriented studies such as Wireless Body Area Network (WBAN) [9, 10], 5G [5, 11], and Radio Frequency (RF) energy harvesting [12, 13]. For example, multi-band antennas allow communication with operating frequencies in various bands [1, 2], while miniaturized antennas provide space and cost savings [3, 4]. Furthermore, wideband antennas can provide great data capacity [5, 6], whereas high-gain antennas can enable efficient transmission across long distances [7, 8]. Additionally, Multiple-Input Multiple-Output (MIMO) antennas for 5G spectrum applications [14], WBAN antennas for intra-body communication [9, 10], and RF energy harvesting antennas for wireless charging systems [12, 13] will continue to play essential roles for decades to come. While all of these studies are beneficial to current technologies, the ability to provide extremely high data speed and capacity will play the most important role in the antenna world for next-generation communication applications.

Ultra wideband technology (UWB) is on the verge of becoming the most prevalent wireless communication method for both present and future generations, thanks to the increasing demand for high-speed and large data transmission capacity. For instance, UWB wireless communication elements are necessary for applications such as synthetic aperture radars, indoor positioning, healthcare monitoring, vehicle collision surveillance, high sensitivity in robotic automation, and high-speed video transmission in order to transmit large quantities of data and achieve high resolution [15]. The most important component for data transmission in such systems is antennas with UWB characteristics. The Federal Communications Commission (FCC) defines the unlicensed UWB band range as 3.1 GHz to 10.6 GHz [16]. For this reason, the literature accepts antennas operating in these frequency ranges, with a bandwidth of 100% or more, as UWB antennas. However, obtaining an UWB in antennas might result in significant performance losses. While log periodic, monopole, Vivaldi, and slot antenna types can provide UWB operation, it is difficult to maintain basic antenna features such as gain stability, pattern shape, and efficiency over a significant portion of the working band. At this point, it is expected that antennas based on Spoof Surface Plasmon Polaritons (SSPP) with UWB [17–29], which have emerged in recent years, will be an excellent choice for the aforementioned systems due to their ease of fabrication, stable radiation properties, high electromagnetic wave restriction, high efficiency, and high gain.

Surface plasmon polaritons in optical bands are the origin of SSPP. Electromagnetic wave transmission with the SPP effect in optical bands occurs as a result of electron oscillation at the interface between the metallic structure with a negative dielectric constant and the dielectric part with positive permittivity. This area may confine electromagnetic waves during transmission [30]. Metals, on the other hand, cannot exhibit negative permittivity in the terahertz to microwave bands because they have the property of a perfect conductor. In order to resolve this issue, a transmission style comparable to optical bands can be enhanced with perforations such as sub-wavelength holes or corrugated designs created on the metal surface periodically, a method known as spoofing or mimicking [31, 32]. It is clear that, the geometric design is extremely important for to obtain spoof characteristics. The major advantage of structures constructed as spoofs for these bands is that they can confine the electromagnetic wave inside the high transmission zone, resulting in reduced losses and minimal crosstalk in microwave circuit design [33].

Planar Spoof Surface Plasmon Polaritons (SSPP) generally have a wideband design [34–37]. They recently found a very extensive field of study as UWB characteristic for filters [38, 39], transmission line [40], absorbers [41], power dividers [42], and antenna applications [17–29]. The SSPP operate primarily in the surface wave or slow-wave mode. To use this as an antenna and convert it to space modes, planar structures must establish a transition from wave mode to SSPP mode, a transition from wave mode to SSPP mode on microstrip or Coplanar Waveguide (CPW) must be established in planar structures. This is typically accomplished by progressively reducing the geometric design of SSPP to its ports [43]. After the transmission line to SSPP mode transition is complete [44], it can be converted into an antenna form by terminating it with a modified gradually decreasing edge [34], modulated SSPP parts [19], and adding additional patches [17]. This will enable it to transmit in a manner that is compatible with free space modes. According to the literature, there are two types of UWB antennas based on the SSPP: leaky-wave antennas [17–20] and end-fire antennas [21–29]. Although the primary characteristic of end-fire antennas is the ability to achieve high gain and stable gain in a specific direction, leaky-wave antennas can offer a wide beam scanning capability by sweeping frequency.

This chapter investigates planar antennas working in microwave to milimeter-wave bands based on SSPP with UWB properties, which have not been extensively studied in the literature. The methods used to obtain the UWB characteristics of SSPP-based antennas, which are primarily end-fire and leaky-wave types, as well as their performance across the operating bands, are thoroughly examined. In addition, the antennas have been evaluated in accordance with their essential characteristics for modern technologies. The objective is to provide a roadmap for future research by comparing the benefits and drawbacks of antenna size, beam scanning capabilities, bandwidths, gains, and so forth.

2. Basic principles of the spoof surface plasmon polaritons (SSPP) based structures

SSPP-based transmission lines mimic surface plasmon polaritons that operate in the optical and infrared ranges. **Figure 1** depicts an SPP structure, including its fields. A light-like wave source transmits energy to the structure's entry; the metal's negative permittivity and the dielectric half's positive permittivity cause the dielectric-metal interface to transmit energy as electrons move back and forth together. Simultaneously, while the electromagnetic wave travels in the x direction, the z component of the field in the transverse direction experiences rapid exponential attenuation. As a result, the wave on the SPP region is highly confined, while leakage out of the structure is kept to a minimum [30, 32]. This indicates that the electromagnetic wave gets delivered with minimal loss.

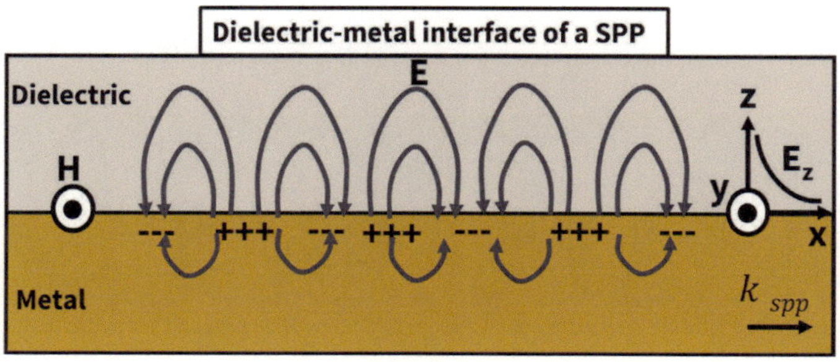

Figure 1.
An example of a SPP representation on dielectric-metal interface. This figure was adapted from [45] under Creative Commons Attribution (CC BY) license (http://creativecommons.org/licenses/by/4.0/).

Equation 1 provides the dispersion relationship of a wave transmitted using the spp. mode. The wave vector (k_{spp}) depends on the medium's dielectric constant, where ϵ_1 is the permittivity of the metal and ϵ_2 is the permittivity of the dielectric medium. This equation is obtained by calculating the formulation of the Ampere law, $\nabla \times \mathbf{H} = ik_0\mathbf{E}$, for a wave traveling in the defined direction while applying boundary conditions at $z = 0$. Here **H** is the magnetic fields, **E** is the electric fields and k_0 is the free space wave number [44].

$$k_{spp} = k_0\sqrt{\frac{\varepsilon_1\varepsilon_2}{\varepsilon_1 + \varepsilon_2}} \tag{1}$$

Until the findings of Pendry and his colleagues, it was assumed that SPP could exist in the optical and infrared operating ranges, as shown by example in **Figure 1**. The study they conducted paved the way for the mimicking of the SPPs' similar electromagnetic field distributions in optical bands for microwave to terahertz bands by modifying a metallic structure with periodic holes [31, 32].

Figure 2(a) demonstrates several example illustrations of the SSPP drawings. This process is analogous to the propagation of SPP, and the objective of all of these is to effectively restrict the electromagnetic wave at the metal-dielectric interface, as can be seen in an example in **Figure 2(b)**. The metallic structure is modified with grooves, strips with different geometries, or perforations to imitate the SPP, which is the most well-known method [46]. In addition to geometric modification of metallic structure, substrate selection is equally critical.

Figure 3 depicts an example dispersion diagram illustrating how modifying one of the SSPP geometric parameters affects the wave vector. This section considers an example of a grooved design. The dispersion diagram is an important output for SSPP

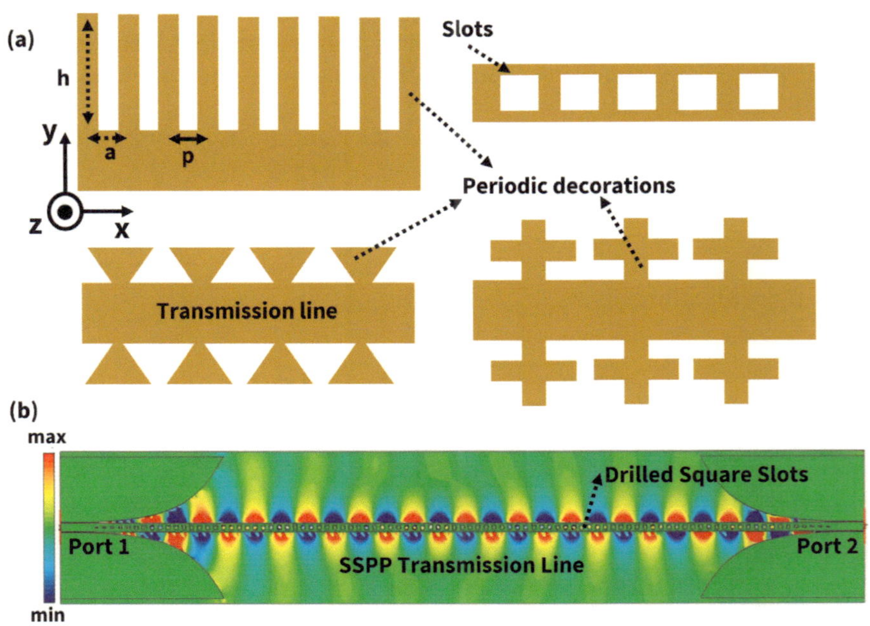

Figure 2.
(a) Several example illustrations of the SSPP drawings [46]. (b) A high-field confinement example of the SSPP [17]. Figure (a) from [46], and Figure (b) from [17] was adapted under Creative Commons Attribution (CC BY) license (http://creativecommons.org/licenses/by/4.0/).

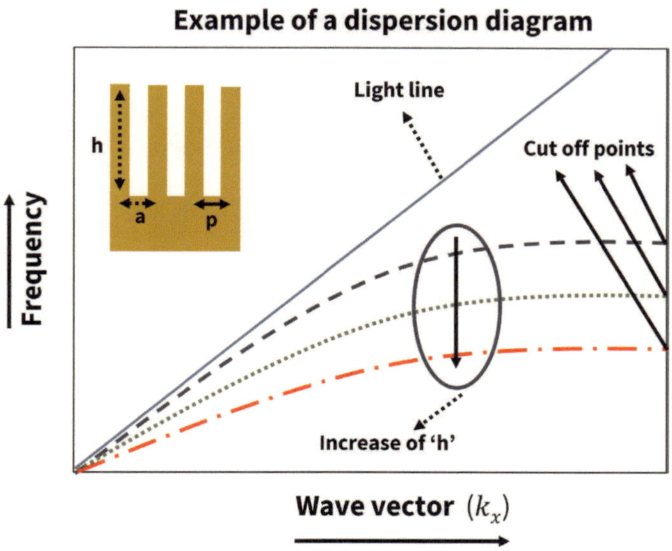

Figure 3.
A dispersion diagram example of the grooved SSPP structure. This figure was adapted from [47] under Creative Commons Attribution (CC BY) license (http://creativecommons.org/licenses/by/4.0/).

design and analysis. Because it exhibits the propagation mode's cut-off frequency, wave vector value, and whether the wave vector is lower than the airline. Most investigations employ the eigenmode solver of commercial software, such as CST Microwave Studio or ANSYS HFSS, to generate the dispersion diagram of the SSPP unit cell. According to the dispersion diagram, if the wave number is less than the airline (the change in free space wave number with frequency), the structure cannot propagate into space $(k_{sspp} < k_0)$, but can only move along an SSPP-based transmission line below its frequency cut-off. Furthermore, changing the period (p) and depth (h) helps to construct a variety of SSPP modes. As a result, the geometric design parameters should be carefully chosen based on the desired mode, application, high-field confinement, and cut-off frequency [32, 48]. Simultaneously, identifying the optimal conditions for wave confinement through field distribution control during electromagnetic wave transmission is crucial for achieving efficient transmission.

The sample dispersion diagram in **Figure 3** clearly shows that the cut-off frequency of the transmission line varies with the depth of the grooves. Although only the "depth" effect is visible, the width and period also have a significant impact. Equation 2 expresses the influence of the parameters more clearly [48]. Here, 'a' is the unit cell's width, 'h' is its depth, and 'p' is its period. Also, the depth can control the SSPPs' field confinement. When the "depth" increases, the field confinement becomes high.

$$k_x = k_0\sqrt{1 + \frac{a^2}{p^2}\tan^2(k_0 h)} \tag{2}$$

Spoof surface plasmons are unable to propagate into space because they transmit below the airline, as illustrated in the sample dispersion diagram. To achieve this, the momentum of the SSPP structure and the free space modes should match each other. This is accomplished through different methods such as the addition of periodic

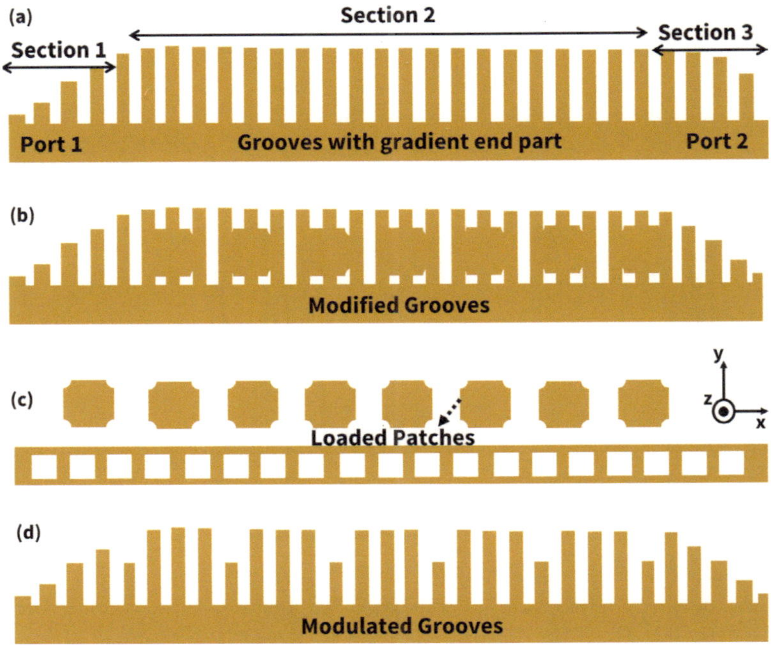

Figure 4.
Some example drawings of the SSPP-based antennas with. Figure (a) from [34], Figure (b) from [45], Figure (c) from [17], and Figure (d) from [49] was adapted under Creative Commons Attribution (CC BY) license (http:// creativecommons.org/licenses/by/4.0/).

patches [17], various tapered decreasing end shapes [25], periodically slotted structure [20], or modulations [19]. **Figure 4** illustrates the antenna designs that are based on SSPP using previously mentioned methods. If we look at **Figure 4(a)**, The SSPP structure is divided into three regions. The purpose of this initial region is to ease the transition from classical transmission lines, like microstrip and CPW, to the SSPP modes. The goal is to ensure that microstrip or CPW quasi-TEM modes are highly compatible with SSPP mode. Typically, the SSPP design uses gradient or trapezoid grooves to achieve this [32, 44]. The compatibility at the operating frequency, side-lobe levels and bandwidth are both influenced by this transition. The second section is the region where the actual SSPP mode transmission is conducted. It provides high electromagnetic wave confinement. The third section significantly influences the structure's conversion into an antenna. The SSPP modes are converted into modes that can propagate in free space with high efficiency through the modification of this part. Parasitic patch loadings of the structure, modulated SSPP designs, additional periodic slots opened on the antenna, or different connections on SSPP line can also be employed to accomplish this transformation as can be seen some example drawings in **Figure 4(a-d)**.

3. Ultra wideband planar spoof surface plasmon polartions (SSPP) based antennas

Planar antennas based on SSPP have emerged as a significant candidate for wireless communication applications, including long-distance data transmission [23], radar systems [50], and on-body communication [51], in recent years. This is due to their high gain, ease of fabrication, stable efficiency, and high electromagnetic field limitation. In contrast, the number of UWB designs with a bandwidth of 100% or greater is relatively low, as they are primarily intended to operate in the wide band. The literature has been investigated in this section by separating SSPP-based UWB antennas into two distinct groups: leaky-wave [17–20] and end-fire [21–29]. First and foremost, the methodologies and success of UWB leaky antennas based on SSPP have been investigated.

A leaky-wave antenna was realized in [17] using an SSPP structure with square openings and circular patches included to its side-fire direction. **Figure 5(a)** illustrates the antenna's finalized configuration. Square openings were periodically inserted into a microstrip transmission line to ensure that the antenna was subjected to a high electromagnetic field limitation. To facilitate the conversion from CPW to SSPP modes with a UWB matching, the square slots were arranged in a decreasing order toward the ports. The antenna's dimensions are 70 mm × 404 mm. The optimal values that exhibit UWB characteristics, low loss, and high electromagnetic wave restriction were determined by examining the impact of the square slot size change on the S parameters. In order to convert the structure into an antenna and provide that it is

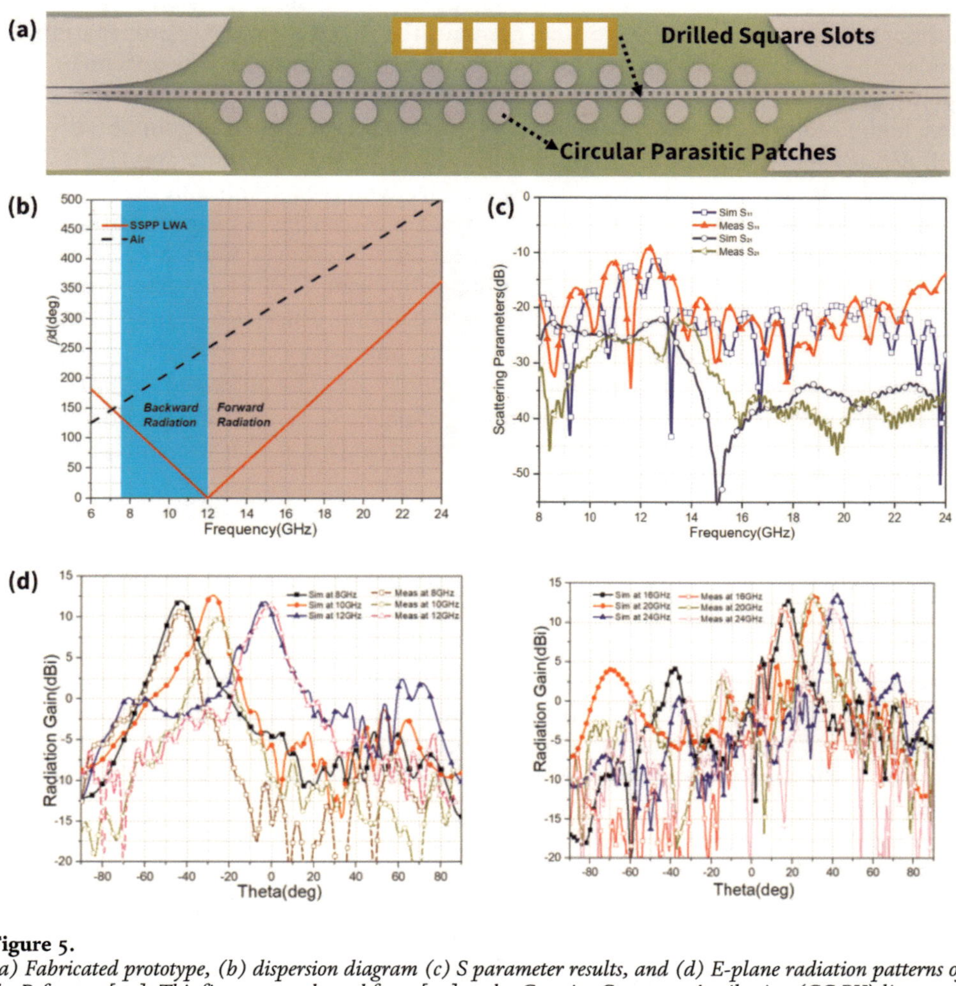

Figure 5.
(a) Fabricated prototype, (b) dispersion diagram (c) S parameter results, and (d) E-plane radiation patterns of the Reference [17]. This figure was adapted from [17] under Creative Commons Attribution (CC BY) license (http://creativecommons.org/licenses/by/4.0/).

compatible with free space modes of propagation, circular patch structures were affixed to the side of the square slot SSPP transmission line. **Figure 5(b)** illustrates the antenna's dispersion diagram, while **Figure 5(c)** illustrates the S parameter results. The antenna propagates in β_{-1} mode (β_{-1} is the first spatial harmonic phase constant) and has a beam scanning capacity of -45° to $+45^{\circ}$ as the frequency varies between 8 GHz and 24 GHz. This situation is also more clearly illustrated in **Figure 5(d)**. The antenna maintains a consistent gain of 12 dBi throughout the preponderance of the operating band, which spans from 9.5 dBi to 13.5 dBi. The most critical advantages of the antenna are its relative high gain, 100% bandwidth, and 90° back-to-forward continuous symmetrical beam scanning.

Reference [18] presents a leaky-wave antenna with a complex SSPP structure composed up of periodic peaks and gradient-shaped stubs, which provides for beam scanning between -52° and $+90^{\circ}$ degrees. The antenna's complicated SSPP structure allows for a more compact size of $0.23\lambda_L \times 3.71\lambda_L$ (λ_L represents the wavelength of the lowest working frequency). One of the key reasons for the antenna's large bandwidth is the transition section from microstrip to SSPP, which includes gradient elements within the SSPP structure. The antenna's optimum design values are determined by analyzing parameter changes in the SSPP structure using the dispersion diagram. Although the antenna's fast wave zone contains the β_{-1}, β_{-2}, and β_{-3} propagation modes, the β_{-1} mode contributes for the majority of propagation. The antenna operates from 12.5 GHz to 4.5 GHz, with a bandwidth of 109%. The antenna has a very high radiation efficiency of around 94% and a relatively stable gain of 6 dBi to 9 dBi. Although the antenna has highly wide beam scanning capability and UWB, its gain levels are significantly lower than several leaky-wave antennas in the literature.

In [19], a leaky wave antenna with multi-beam properties is proposed utilizing SSPP structures formed up of modulated linear circular slots. Conical and bidirectional multi-beam shapes are obtained by superimposing SSPP structures modified with various designs. The antenna is 20 mm × 200 mm. Analytical solutions and simulations are used in conjunction to determine antenna properties. The antenna features UWB property ranging from 7 GHz to 40 GHz. The antenna's important advantages include a total 120° beam scanning capability, multi-beam production, and analytical solution applicability; however, its disadvantage is that its gain varies greatly from 5 dBi to 17 dBi.

Reference [20] describes a leaky-wave antenna with wide beam scanning characteristics based on the propagation of higher-order SSPP modes. The antenna's SSPP structure consists of periodic holes and a thin slotted ring wrapped around them. This structure enables the antenna to realize UWB and frequency beam scanning characteristics. The size of the antenna is $0.95\lambda_L \times 5.9\lambda_L$. The antenna has achieved 120° degree beam scanning capacity, very stable radiation efficiency ranging from 70 to 90% and gain ranging from 10.5 dBi to 13.5 dBi in the operating region between 11.7 GHz and 50 GHz. The antenna has almost no negative characteristics. All the features of the antenna will be quite interesting for the next-generation UWB and stable radiation requiring applications.

Second, SSPP-based UWB end-fire antennas, which garnered greater coverage in the literature than leaky-wave antennas, are evaluated in terms of performance and other characteristics.

In [21], an SSPP-based end-fire antenna with a periodic gradient groove and flaring metal construction is presented. The transmission line modified into a gradient groove is the most important factor in generating the antenna's UWB characteristics and ensuring coherence between spatial harmonic modes and SSPP modes. The depth of the groove is determined by the dispersion diagram analysis. The antenna works at frequencies ranging from 5 GHz to 20 GHz, with a high efficiency of 85–98% and a generally steady gain of 7.5 dBi to 10.5 dBi. Unlike previous research, we conducted near-field electric field distribution measurements on the antenna, revealing the effect of SSPP modes in greater detail. However, the antenna's radiation pattern shape cannot provide full stability in the 90° end-fire direction due to pattern divisions as the operating frequency varies.

Reference [22] proposes a dual-band antenna based on SSPP, tailored as a double-layer meander with end-fire and frequency-scanning capabilities. The Vivaldi radiator provides both end-fire characteristics and efficient beam scanning, while the meanders offer high-field limitation and beam scanning operation in the fast wave zone. The antenna's gain ranges from 3.8 dBi to 8.5 dBi, and it operates at 81% efficiency in end-fire mode between 3 and 9 GHz. In addition, the antenna shows an unstable gain range of 9.5 dBi to 15.6 dBi and a broad beam scan that spans from −90° to +43° degrees at a frequency in the second operational zone between 13 and 34 GHz. In end-fire mode, the antenna exhibits UWB characteristics; nonetheless, the gain variation is rather considerable, at about 5 dBi. A wide scanning of 133° and a high gain were obtained in frequency scanning mode; however, the bandwidth did not exceed 100%.

A periodic elliptical cut and corrugated SSPP were added to an antipodal Vivaldi antenna in [23] to create an antenna that could operate between 1.8 GHz and 6 GHz. **Figure 6** illustrates the antenna's evolutionary stages. The antenna's compactness of 70 mm × 94 mm was further guaranteed by the grooves that were added to the bottom ends of the antenna arms. By observing the impact of altering the corrugated elliptical cuts' depth on the S parameters, the antenna's UWB properties were determined. **Figure 7** depicts the antenna's S parameters and gain values, respectively. The antenna's gain varies from 5.5 dBi to 9 dBi, and it operates at a bandwidth exceeding 100%. The antenna's end-fire radiation properties were retained across the operating bands. To help visualize this condition, the *E* and *H*-plane radiation patterns of the antenna are shown graphically in **Figure 8**. For applications in the bands intended for base station use, which range from 2G to 5G, one may consider the antenna's gain to be somewhat modest.

Reference [24] presents an end-fire antenna that boosts the gain of a standard low-gain Vivaldi antenna using parasitic patches based on SSPP. The reference antenna's UWB characteristics, which had a 130% bandwidth, remained unchanged after SSPP

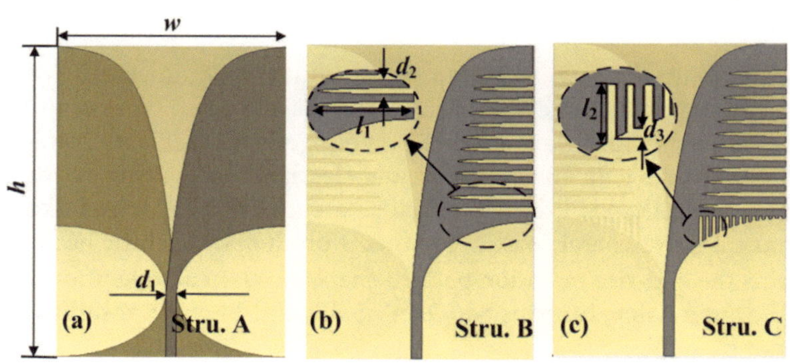

Figure 6.
Evolution stages of the Reference [23]'s antennas. This figure was adapted from [23] under Creative Commons Attribution (CC BY) license (http://creativecommons.org/licenses/by/4.0/).

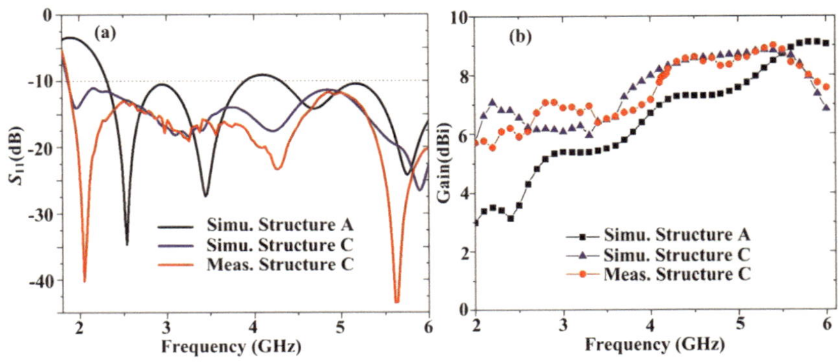

Figure 7.
(a) S parameter and, (b) gain results of the Reference [23]'s antennas. This figure was adapted from [23] under Creative Commons Attribution (CC BY) license (http://creativecommons.org/licenses/by/4.0/).

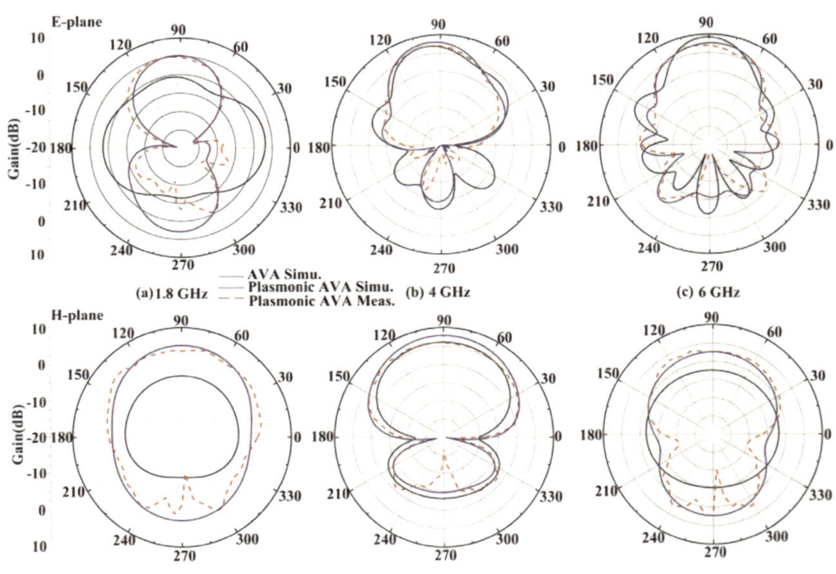

Figure 8.
Radiation pattern of the Reference [23]'s antennas. This figure was adapted from [23] under Creative Commons Attribution (CC BY) license (http://creativecommons.org/licenses/by/4.0/).

was applied. The antenna's size is 80 mm × 235 mm. The Sommerfeld-Zenneck wave mode—a situation in which the SSPP wave vector is almost equal to k_0 in the slow-wave region—and the slow-wave mode worked together to achieve the high-level accomplishment of the gain enhancement and electric field limiting on the parasitic patches. In this case, the SSPP-based parasitic patches exhibited a lens-like effect. The antenna's most notable characteristics are that it produces very little back radiation and maintains the end-fire radiation pattern shape at 90° in all operational bands. However, the antenna's gain spans between 7.5 dBi and 14.5 dBi, resulting in a significant variance of 7 dBi.

A bidirectional end-fire antenna with the SSPP wave vector mismatching method described in [25]. The antenna contains a mismatched power divider with the SSPP structure for bidirectional propagation. The SSPP part, which acts as radiator and transmission line, consists of a slot line with drilled holes and the slot line with double-side 45° tilted metallic lines. Thanks to this structure, the antenna achieves 142% UWB feature between 9.6 GHz and 40 GHz. Similar to other studies, the analysis was performed by changing the length of the tilted strip over the dispersion diagram, and the most suitable values were determined. The antenna showed an average efficiency of 93% and a gain of 9.2 dBi. The antenna maintains its end-fire radiation pattern in a very wide bandwidth. The most important feature that distinguishes the antenna from others is that it is bidirectional. This feature can make it a very big potential to be used in vehicle radar applications for collision control purposes.

In Ref. [26], an end-fire antenna with UWB characteristics is introduced, which is based on the high-order modes of SSPP. The antenna design was initiated in a progressive manner with a quad-band SSPP gradient groove structure that offers higher-order-mode propagation. The antenna arms are printed on the front and rear surfaces of this structure, in opposition. Subsequently, this structure is modified to enhance the antenna's gain with a tilted shape. Lastly, the lower portion of the arms is equipped with I-shaped strip resonators to eradicate the first two bands and enable the antenna to operate in the UWB bands. The antenna functions within the frequency range of 6.9 GHz to 39.7 GHz. Nevertheless, the antenna's gain exhibits a significant variation of approximately 9 dBi, ranging from 6.4 dBi to 15.2 dBi, in contrast to other studies. The antenna's 140% bandwidth and its ability to preserve the end-fire propagation pattern in nearly all frequencies are its most significant advantages.

An exponential SSPP-modified Vivaldi antenna with end-fire characteristics is presented in [27] for radio astronomy applications. Exponential SSPP structures achieve high electric field confinement by compensating for the near-field phase. This also results in improved gains. The antenna accomplishes end-fire propagation, a very stable gain between 9.2 dBi and 11.5 dBi, and 140% bandwidth between 3 GHz and 9 GHz. These properties make it an extremely significant candidate for applications that necessitate next-generation UWB characteristics.

An end-fire antenna based on SSPP is proposed in [28], which consists of modulated circular groove transitions. The prototype of the antenna is presented in **Figure 9(a)**. The quasi-TEM mode of the CPW feed of the antenna is converted to SSPP modes with semicircular peaks and spaced parts in order to provide high bandwidth. The antenna exhibits 146% bandwidth between 7 GHz and 45 GHz, indicating UWB characteristics that can be seen in **Figure 9(c)**. The SSPP structure's tapered design toward the output section has significantly contributed to the high suitability of the free space modes with the TM modes of SSPP. The Hanson-Woodyard relationship is the primary factor influencing the antenna's operation. This relationship suggests that the antenna's entire body structure has a beneficial impact on propagation. **Figure 9(b)** depicts the dispersion diagram of the antenna. It optimizes the values that can remain beneath the airline and provides the maximum field limitation by adjusting the depth of the grooves. The antenna's dimensions are 80 mm × 30 mm. The antenna's efficiency is 80% or higher throughout the operating spectrum, despite the fact that it lacks a consistent gain between 4.7 dBi and 10 dBi. The antenna's primary beam direction degree is not precisely 90° at the majority of frequencies; however, it can be regarded as stable due to its proximity to that level of variation. The antenna's radiation efficiency and high bandwidth are its most significant advantages.

In [29], a modified antipodal end-fire Vivaldi antenna is introduced, which is based on SSPP and features periodic arc type slotted structures. By incorporating the Vivaldi antenna into the edges, the antenna's SSPP structure facilitates both high-field

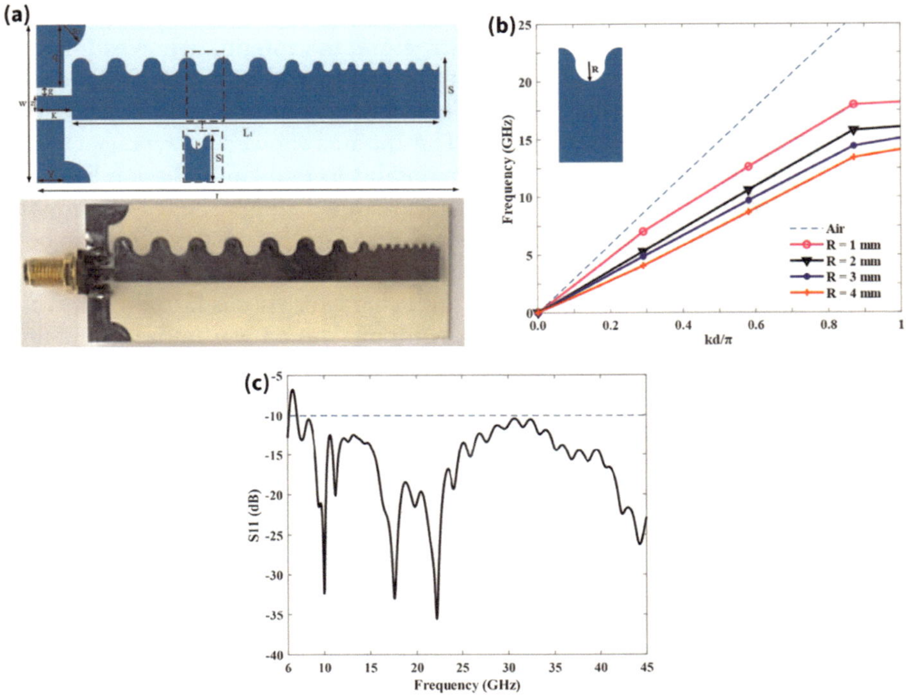

Figure 9.
(a) Fabricated prototype, (b) dispersion diagram, and (c) S parameter results of the Reference [28]. This figure was adapted from [28] under Creative Commons Attribution (CC BY) license (http://creativecommons.org/licenses/by/4.0/).

confinement and space mode matching. The trapezoidal arc type slots from the microstrip input line to the SSPP section are crucial to achieving high impedance bandwidth. The antenna reaches a 120% bandwidth within the range of 3 GHz to 13 GHz. The SSPP line dispersion diagram is employed to analyze the optimum value of the arc type beam with respect to the angular change. The antenna is capable of achieving a radiation efficiency of 90% or higher throughout the entire operating spectrum; however, its gain is relatively low and its variation is high, ranging from 4 dBi to 8 dBi. One of the antenna's essential features is that it provides full end-fire propagation in all bands, despite the fact that the Half Power Beam Width (HPBW) ngle is wider at lower frequencies than at higher frequencies. The antenna's UWB characteristics, high efficiency, and stable end-fire main beam radiation direction are its positive attributes. However, its limited gain is its negative feature.

Nowadays, the preference level of antennas in UWB wireless communication systems is determined by factors such as compactness, high gain, stable efficiency, bandwidth percentage, and low loss. The performance comparison of the antennas in the literature is summarized in **Tables 1** and **2** taking these properties into account. The performance of leaky-wave antennas is compared in **Table 1**, which is based on SSPP. Reference [18] is distinguished by its high efficiency, wide beam scanning capacity, and compact size; however, its gain levels are relatively low in comparison to antenna studies in the same category. The bandwidth of Reference [19] is the maximum, but it exhibits a large gain variation of 12 dBi. Compared to its counterparts, Reference [17] is relatively large and has a low beam scanning capacity. The majority of Reference [20]'s features, with the exception of its size, can be regarded as optimal.

Refs	[17]	[18]	[19]	[20]
Operating freqs (GHz)	8–24	12.5–42.5	7–40	11.7–50
BW(%)	100	109	140	124
Gain (dBi)	9.5–13.5	6–9	5–17	10.5–13.5
Gain Variation (dBi)	4	3	12	3
Efficiency	—	90–98%	—	70–90%
Freq. Scan. Capacity	90°	142°		129°
Sizes (W × *L*)	$1.9\lambda_L \times 10.8\lambda_L$	$0.2\lambda_L \times 3.7\lambda_L$	$0.5\lambda_L \times 4.7\lambda_L$	$\lambda_L \times 5.9\lambda_L$

Table 1.
Characteristics comparisons of the previously reported SSPP-based leaky-wave UWB antenna studies.

Refs	[21]	[22]	[23]	[24]	[25]	[26]	[27]	[28]
Operating freqs (GHz)	5–20	3–9	1.8–6	3–14	9.6–40	6.9–39.7	3–9	7–45
BW(%)	120	100	108	130	142	141	100	146
Gain (dBi)	7.5–10.5	3.8–8.5	5.5–9	7.5–14.5	4.5–11	6.4–15.2	9.2–11.5	4.7–10
Gain Variation (dBi)	3	4.7	3.5	7	6.5	8.8	2.3	5.3
Efficiency(%)	85–98	75–85	—	—	80–99	80–97	—	80–98
Radiation Type	Direc.	Direc.	Direc.	Direc.	Bi-direc.	Direc.	Direc.	Direc.
End-fire Direction Stability	No	Yes	Yes	Yes	Yes	Yes	No	Yes
Sizes (W × *L*) ($\lambda_L \times \lambda_L$)	1.2 × 2.6	0.6 × 2.7	0.4 × 0.6	0.8 × 2.9	0.4 × 5.5	0.4 × 2.5	0.9 × 1	0.7 × 1.8

Table 2.
Characteric comparisons of the previously reported SSPP-based end-fire UWB antenna studies.

Table 2 illustrates the performance comparison of end-fire antennas based on their critical features. Reference [21] and Reference [27] are particularly noteworthy in this table due to their low gain variations and consistent radiation characteristics. However, they are unable to consistently maintain their end-fire radiation patterns at 90° direction across all operating bands. The antenna with the most compact structure among the studies is Reference [23]. Despite the fact that studies [25, 26, 28] exhibit high bandwidths and radiation stability, their gain variations are substantial. Furthermore, the bidirectional end-fire pattern of the study [25] distinguishes it from the other studies in terms of its propagation. It is evident that the antennas listed in this table exhibit superior bandwidth and gain compared to other non-SSPP-based end-fire pattern antennas that have been published in recent years [8, 52–55].

The primary issues with UWB antennas based on SSPP are high gain variation and compactness, as is evident from these comparisons. It is obvious that these issues will be a unique candidate for next-generation microwave and millimeter-wave bands communication systems if they are more effectively resolved in future research.

4. Conclusions

This study presents a comprehensive examination of UWB planar antennas based on SSPP that operate in the microwave and millimeter wave bands, as reported in the literature. It has been noted that antennas are typically classified into two groups: end-fire or leaky-wave. The methods employed for obtaining UWB characteristics and their critical features, including gain, stability, beam scanning capacity, bandwidth, and efficiency, have been considered in the investigation and comparison of antennas. It has been observed that the design parameters are typically analyzed using a dispersion diagram to attain ultra-wide operating bands in the antennas that have been examined. It has also been recognized that the transition structure from transmission line to SSPP mode is the most critical aspect of UWB characteristics.

The leaky-wave antennas that were investigated can be identified by their wide beam scanning capacity, high gain, and low gain variations; however, they suffer from compactness issues. The end-fire antennas stand out for their high radiation efficiency and consistent directionality, but their primary drawback is their high gain variations. It was found that the bandwidths varied between 100% and 140% in the literature studies on the two types of antennas. The leaky-wave antennas have a maximum gain of 17 dBi along their operational bands, whereas the end-fire antennas have a maximum gain of 15.2 dBi. Both types of antennas have a minimum efficiency of 70% and 75%, respectively, and a maximum efficiency of up to 98%. Furthermore, the leaky-wave antennas shows a total wide beam-scanning range of 90° to 142°. It is apparent that SSPP-based UWB planar antennas, which possess distinctive features in comparison to other UWB antennas in the literature, will be a highly exciting candidate for next-generation microwave-millimeter wave imaging applications, long-distance data communication, and radar systems.

Abbreviations

SSPP	spoof surface plasmon polaritons
WBAN	wireless body area network
RF	radio frequency
MIMO	multiple-input multiple-output
UWB	ultra wideband antenna
FCC	federal communications commission
CPW	coplanar waveguide
HPBW	half power beam width

Author details

Goksel Turan
Department of Electrical and Electronics Engineering, Eskisehir Osmangazi University, Eskisehir, Turkey

*Address all correspondence to: goksel_turan@outlook.com

References

[1] Hussain M, Ali EM, Awan WA, Hussain N, Alibakhshikenari M, Virdee BS, et al. Electronically reconfigurable and conformal triband antenna for wireless communications systems and portable devices. PLoS One. 2022;**17**(12):e0276922. DOI: 10.1371/journal.pone.0276922 [Accessed: 28 September 2024]

[2] Turan G, Odabasi H. A simple dual-band quasi-Yagi antenna with defected ground structures. Turkish Journal of Electrical Engineering and Computer Sciences. 2022;**30**(1):158-166. DOI: 10.3906/elk-2104-165 [Accessed: 28 September 2024]

[3] Tariq S, Hussain Q, Alzaidi MS, Ghoniem RM, Alibakhshikenari M, Althuwayb AA, et al. Frequency selective surfaces-based miniaturized wideband high-gain monopole antenna for UWB systems. AEU-International Journal of Electronics and Communications. 2023; **170**:154841. DOI: 10.1016/j.aeue.2023.154841 [Accessed: 28 September 2024]

[4] Turan G, Odabasi H. A miniaturized quasi-Yagi antenna using defected ground structure and double dog-bone driver. Microwave and Optical Technology Letters. 2022;**64**(2):358-362. DOI: 10.1002/mop.33096 [Accessed: 28 September 2024]

[5] Alibakhshikenari M, Virdee BS, Vadalà V, Dalarsson M, de Cos Gómez ME, Alharbi AG, et al. Broadband 3-D shared aperture high isolation nine-element antenna array for on-demand millimeter-wave 5G applications. Optik. 2022;**267**:169708. DOI: 10.1016/j.ijleo.2022.169708 [Accessed: 28 September 2024]

[6] Sehrai DA, Asif M, Shoaib N, Ibrar M, Jan S, Alibakhshikenari M, et al. Compact quad-element high-isolation wideband MIMO antenna for mm-wave applications. Electronics. 2021;**10**(11): 1300. DOI: 10.3390/electronics10111300 [Accessed: 28 September 2024]

[7] Alibakhshikenari M, Virdee BS, Mariyanayagam D, Vadalà V, Naser-Moghadasi M, See CH, et al. High gain/bandwidth off-chip antenna loaded with metamaterial unit-cell impedance matching circuit for sub-terahertz near-field electronic systems. Scientific Reports. 2022;**12**(1):17893. DOI: 10.1038/s41598-022-22828-3 [Accessed: 28 September 2024]

[8] Turan G, Odabasi H. A high-gain quasi-Yagi antenna using nested metaresonator structures. Microwave and Optical Technology Letters. 2023;**65**(6):1755-1761. DOI: 10.1002/mop.33615 [Accessed: 28 September 2024]

[9] Dumanli S. Pattern diversity antenna for on-body and off-body wban links. Turkish Journal of Electrical Engineering and Computer Sciences. 2018;**26**(5): 2395-2405. DOI: 10.3906/elk-1707-196 [Accessed: 28 September 2024]

[10] Godeneli K, Bengi U, Kati OA, Dumanli S. A wearable dual-mode repeater antenna for implant communications. IEEE Transactions on Antennas and Propagation. 2021;**70**(2): 868-875. DOI: 10.1109/TAP.2021.3111603 [Accessed: 28 September 2024]

[11] Alibakhshikenari M, Virdee BS, See CH, Shukla P, Moghaddam SM, Zaman AU, et al. Dual-polarized highly folded bowtie antenna with slotted self-grounded structure for sub-6 GHz 5G applications. IEEE Transactions on Antennas and Propagation. 2021;**70**(4): 3028-3033. DOI: 10.1109/TAP.2021.3118784 [Accessed: 28 September 2024]

[12] Vinnakota SS, Kumari R, Majumder B. Metasurface-assisted broadband compact dual-polarized dipole antenna for RF energy harvesting. IEEE Antennas and Wireless Propagation Letters. 2023; **22**(8):1912-1916. DOI: 10.1109/LAWP.2023.3269307 [Accessed: 28 September 2024]

[13] Sang J, Qian L, Li M, Wang J, Zhu Z. A wideband and high-gain circularly polarized antenna array for radio-frequency energy harvesting applications. IEEE Transactions on Antennas and Propagation. 2023;**71**(6): 4874-4887. DOI: 10.1109/TAP.2023.3271526

[14] Alibakhshikenari M, Virdee BS, Benetatos H, Ali EM, Soruri M, Dalarsson M, et al. An innovative antenna array with high inter element isolation for sub-6 GHz 5G MIMO communication systems. Scientific Reports. 2022;**12**(1):7907. DOI: 10.1038/s41598-022-12119-2 [Accessed: 28 September 2024]

[15] Chinmoy S, Siddiqui JY, Antar YMM. Multifunctional ultrawideband antennas: Trends, techniques and applications. CRC Press; 2019. DOI: 10.1201/9781351026543 [Accessed: 30 August 2024]

[16] Federal Communications Commission. First report and order and third notice of proposed rulemaking in the matter of revision of part 15 of the Commission's rules regarding ultra-wideband transmission systems. Federal Communications Commission. FCC Record. 2002;**17**:7435

[17] Leilei L, Wang J, Yin X, Chen ZN. Wide-angle beam scanning leaky-wave antenna using spoof surface plasmon polaritons structure. Electronics. 2018;**7** (12):348. DOI: 10.3390/electronics7120348 [Accessed: 30 August 2024]

[18] Dai XW, Ding C, Yu W, Liu L, Luo GQ. Ultrawide-angle and compact leaky-wave antenna based on complex periodic structure. IEEE Antennas and Wireless Propagation Letters. 2023;**22**(6):1486-1490. DOI: 10.1109/LAWP.2023.3247435 [Accessed: 30 August 2024]

[19] Pan Y, Cheng Y, Dong Y. Surface plasmon polariton leaky-wave antennas with wideband arbitrary multibeam radiation. IEEE Transactions on Antennas and Propagation. 2021;**70**(2): 931-942. DOI: 10.1109/TAP.2021.3111210 [Accessed: 30 August 2024]

[20] Wang J, Zhao L, Hao ZC, Cui TJ. Wide-angle frequency beam scanning antenna based on the higher-order modes of spoof surface plasmon polariton. IEEE Transactions on Antennas and Propagation. 2020;**68**(11):7652-7657. DOI: 10.1109/10.1109/TAP.2020.2993325 [Accessed: 30 August 2024]

[21] Yin JY, Zhang HC, Fan Y, Cui TJ. Direct radiations of surface plasmon polariton waves by gradient groove depth and flaring metal structure. IEEE Antennas and Wireless Propagation Letters. 2015;**15**:865-868. DOI: 10.1109/LAWP.2015.2477877 [Accessed: 30 August 2024]

[22] Zu HR, Wu B, Zhao YT, Su T, Fan YF. Dual-band antenna with large beam steering angle incorporating endfire and frequency scanning modes using double-layer SSPPs structure. IEEE Transactions on Antennas and Propagation. 2021;**70** (1):46-55. DOI: 10.1109/TAP.2021.3096556 [Accessed: 30 August 2024]

[23] Dai LH, Tan C, Zhou YJ. Ultrawideband low-profile and miniaturized spoof plasmonic Vivaldi antenna for base station. Applied Sciences. 2020;**10**(7):2429. DOI:

10.3390/app10072429 [Accessed: 30 August 2024]

[24] Liao Z, Che YZ, Luo GQ, Zhang ZH, Qian YH, Cai BG. Enhanced radiation characteristics for Vivaldi antenna using spoof surface plasmon polaritons. IEEE Transactions on Plasma Science. 2021;**49**(9):2730-2736. DOI: 10.1109/TPS.2021.3101814 [Accessed: 30 August 2024]

[25] Dong G, Liu Y, Li X, Zhang A. Ultrawideband spoof surface plasmon polariton bidirectional endfire antenna based on wave vectors mismatching. IEEE Transactions on Antennas and Propagation. 2021;**69**(11):7895-7899. DOI: 10.1109/TAP.2021.3076560 [Accessed: 30 August 2024]

[26] Liang Z, Qi SS, Wu W. Ultra-wideband planar end-fire antenna based on high-order modes of spoof surface plasmon polaritons. IEEE Transactions on Antennas and Propagation. 2022;**71**(2):1469-1477. DOI: 10.1109/TAP.2022.3227750 [Accessed: 30 August 2024]

[27] Qi H, Liu H. Wideband high gain differential Vivaldi antenna design based on exponential spoof surface plasmon polaritons metamaterial. AEU-International Journal of Electronics and Communications. 2023;**163**:154603. DOI: 10.1016/j.aeue.2023.154603 [Accessed: 30 August 2024]

[28] Arghandeh F, Abbasi-Arand B, Hesari-Shermeh M. A compact wideband antenna with high gain based on spoof surface plasmon polaritons. Scientific Reports. 2024;**14**(1):10107. DOI: 10.1038/s41598-024-54682-w [Accessed: 30 August 2024]

[29] An G, Zhu C, Zhang Y, Li W, Jiang Y, Ye L. A wideband endfire antenna incorporating modified antipodal Vivaldi resonator and spoof surface plasmon polaritons. IEEE Antennas and Wireless Propagation Letters. 2024;**23**(6):1874-1878. DOI: 10.1109/LAWP.2024.3372191 [Accessed: 30 August 2024]

[30] Junxi Z, Zhang L, Xu W. Surface plasmon polaritons: Physics and applications. Journal of Physics D: Applied Physics. 2012;**45**(11):113001. DOI: 10.1088/0022-3727/45/11/113001 [Accessed: 30 August 2024]

[31] Pendry JB, Martin-Moreno L, Garcia-Vidal FJ. Mimicking surface plasmons with structured surfaces. Science. 2004;**305**(5685):847-848. DOI: 10.1126/science.1098999 [Accessed: 30 August 2024]

[32] Tang WX, Zhang HC, Ma HF, Jiang WX, Cui TJ. Concept, theory, design, and applications of spoof surface plasmon polaritons at microwave frequencies. Advanced Optical Materials. 2019;7(1):1800421. DOI: 10.1002/adom.201800421 [Accessed: 30 August 2024]

[33] You ZH, Gao X, Cheng ZW, Wang HL, Ma HF, Cui TJ. An ultra-compact spoof surface plasmon polariton transmission line and its signal integrity analysis. Advanced Engineering Materials. 2022;**24**(12):2200621. DOI: 10.1002/adem.202200621 [Accessed: 30 August 2024]

[34] Kandwal A, Li J, Igbe T, Liu Y, Li S, et al. Broadband frequency scanning spoof surface plasmon polariton design with highly confined endfire radiations. Scientific Reports. 2020;**10**(1):113. DOI: 10.1038/s41598-019-56720-4 [Accessed: 30 August 2024]

[35] Liang Z, Qi SS, Wu W. Wideband vertically polarized planar endfire antenna based on spoof surface plasmon polaritons. IEEE Antennas and Wireless

Propagation Letters. 2023;**22**(9):2240-2244. DOI: 10.1109/LAWP.2023.3281840 [Accessed: 30 August 2024]

[36] Liu L, Jiang Y, Hu Y, Jiang D, Zhu L. Wideband millimeter-wave endfire antenna based on symmetrical spoof surface plasmon polaritons. IEEE Transactions on Antennas and Propagation. 2021;**69**(11):7386-7393. DOI: 10.1109/TAP.2021.3076199 [Accessed: 30 August 2024]

[37] Turan G. Dualband and tunable beam tilting wideband high gain compact endfire antennas based on spoof surface plasmon polaritons. In: 2023 Workshop on Microwave Theory and Technology in Wireless Communications (MTTW), Riga, Latvia. Newyork, USA: IEEE; 2023. pp. 104-108. DOI: 10.1109/MTTW59774.2023.10320050 [Accessed: 30 August 2024]

[38] Hu MZ, Zhang HC, Yin JY, Ding Z, Liu JF, Tang WX, et al. Ultra-wideband filtering of spoof surface plasmon polaritons using deep subwavelength planar structures. Scientific Reports. 2016;**6**(1):37605. DOI: 10.1038/srep37605 [Accessed: 30 August 2024]

[39] Wang M, Sun S, Ma HF, Cui TJ. Supercompact and ultrawideband surface plasmonic bandpass filter. IEEE Transactions on Microwave Theory and Techniques. 2019;**68**(2):732-740. DOI: 10.1109/TMTT.2019.2952123 [Accessed: 30 August 2024]

[40] Pan BC, Luo GQ, Liao Z, Cai JL, Cai BG. Wideband miniaturized design of complementary spoof surface plasmon polaritons waveguide based on interdigital structures. Scientific Reports. 2020;**10**(1):3258. DOI: 10.1038/s41598-020-60244-7 [Accessed: 30 August 2024]

[41] Jidi L, Cao X, Gao J, Li T, Yang H, Li S. Ultra-wideband absorber for electromagnetic waves under large incident angle based on spoof surface plasmon polaritons. Optical Materials Express. 2021;**11**(11):3917-3929. DOI: 10.1364/OME.442579 [Accessed: 30 August 2024]

[42] Xu HT, Guan DF, Yang ZB, Zhang Q, Liu L, Xu S, et al. An ultra-wideband out-of-phase power divider based on odd-mode spoof surface plasmon polariton. International Journal of RF and Microwave Computer-Aided Engineering. 2021;**31**(4):e22583. DOI: 10.1002/mmce.22583 [Accessed: 30 August 2024]

[43] Liao Z, Zhao J, Pan BC, Shen XP, Cui TJ. Broadband transition between microstrip line and conformal surface plasmon waveguide. Journal of Physics D: Applied Physics. 2014;**47**(31):315103. DOI: 10.1088/0022-3727/47/31/315103 [Accessed: 30 August 2024]

[44] Geng J, Ren C, Wang K, Zhou H, Zhang J. Spoof Surface Plasmon Polaritons Antenna. Singapore: Springer; 2022. pp. 68-73. DOI: 10.1007/978-981-16-4721-5 [Accessed: 30 August 2024]

[45] Kong GS, Ma HF, Cai BG, Cui TJ. Continuous leaky-wave scanning using periodically modulated spoof plasmonic waveguide. Scientific Reports. 2016;**6**(1):29600. DOI: 10.1038/srep29600 [Accessed: 28 September 2024]

[46] Ping R, Ma H, Cai Y. Compact and highly-confined spoof surface plasmon polaritons with fence-shaped grooves. Scientific Reports. 2019;**9**(1):12045. DOI: 10.1038/s41598-019-48616-0 [Accessed: 30 August 2024]

[47] Li S, Zhang Q, Li J, Zhao H, Yin X, Yang M. Monopulse antenna based on singular spoof surface plasmon polariton

structure for angle measurement. Electronics. 2020;**9**(12):2156. DOI: 10.3390/electronics9122156 [Accessed: 28 September 2024]

[48] Ma HF, Shen X, Cheng Q, Jiang WX, Cui TJ. Broadband and high-efficiency conversion from guided waves to spoof surface plasmon polaritons. Laser Photonics Reviews. 2014;**8**(1):146-151. DOI: 10.1002/lpor.201300118 [Accessed: 30 August 2024]

[49] Chu D, Mao Y, Li H, Bie H, Zhou Y. Dual-polarized multi-beam fixed-frequency beam scanning leaky-wave antenna. Sensors. 2023;**23**(11):5070. DOI: 10.3390/s23115070 [Accessed: 28 September 2024]

[50] He PH, Ren Y, Shao C, Zhang HC, Cui TJ. Suppressing high-power microwave pulses using spoof surface plasmon polariton mono-pulse antenna. IEEE Transactions on Antennas and Propagation. 2021;**69**(12):8069-8079. DOI: 10.1109/TAP.2021.3083836 [Accessed: 30 August 2024]

[51] Tian X, Zeng Q, Nikolayev D, Ho JS. Conformal propagation and near-omnidirectional radiation with surface plasmonic clothing. IEEE Transactions on Antennas and Propagation. 2020;**68**(11):7309-7319. DOI: 10.1109/TAP.2020.2998216 [Accessed: 30 August 2024]

[52] Li J, Xiang Z, Chen X, Xu M, Li J. Broadband bowtie-based log-periodic array antenna via GIPD process for 5G mm-wave applications. Progress in Electromagnetics Research Letters. 2024;**118**:55-61. DOI: 10.2528/PIERL23121402 [Accessed: 30 August 2024]

[53] Turan G, Odabasi H. A multiband quasi-Yagi antenna using stepped driver and slotted ground for 2.4/3.5/5.2 GHz bands applications. Microwave and Optical Technology Letters. 2023;**65**(7):2031-2038. DOI: 10.1002/mop.33643 [Accessed: 30 August 2024]

[54] Wang S, Li HC, Li Y, Guo ZC, Zhang G, Yang L, et al. A wideband end-fire stepped slot antenna with gain enhancement and filtering capability. IET Microwaves, Antennas Propagation. 2024;**18**(6):422-429. DOI: 10.1049/mia2.12472 [Accessed: 30 August 2024]

[55] Turan G, Odabasi H. A wideband quasi-Yagi antenna using symmetrical slots on the ground plane. Electromagnetics. 2023;**43**(5):349-360. DOI: 10.1080/02726343.2023.2244841 [Accessed: 30 August 2024]

Chapter 6

Ultra-Wideband Antennas: Designs, Investigations, Enhancement Techniques, Propagation Characteristics, and Practical Applications

Seyed Ramin Emadian and Masoumeh Forgonchi Nojavan

Abstract

Ultra-wideband (UWB) systems because of unique characteristics including specific frequency spectrum allocated by the Federal Communication Committee (FCC) obtained many applications in modern communication technologies like industrial IOT, outdoor and indoor positioning, healthcare and biomedical, intelligent home electronics and so forth. UWB Antennas present unique challenges, particularly in achieving size reduction and bandwidth enhancement. This chapter addresses these challenges by exploring a variety of UWB antenna types, including planar monopole and slot antennas, and examines techniques for bandwidth enhancement and size reduction. Given the potential interference with narrowband systems such as WLAN, WiMAX, and X-band satellite downlink, the chapter also delves into band notching and filtering techniques. Furthermore, the chapter discusses UWB pulse characteristics essential for impulse radio applications and provides an in-depth analysis of indoor UWB propagation characteristics. Through case studies and practical examples, the chapter offers insights into real-world UWB antenna designs and provides a comprehensive understanding of the challenges and solutions associated with UWB systems.

Keywords: UWB antennas, band-notched characteristics, bandwidth enhancement, impulse radio, time domain characteristics, indoor propagation channel

1. Introduction

Ultra-wideband (UWB) technology has emerged as a key solution for high-data-rate communications, offering advantages such as low power consumption, precise ranging, and minimal interference with existing communication systems. However, UWB antenna design presents challenges, including the need to achieve wide bandwidth, compact size, and minimal interference with narrowband systems [1–4]. The Federal Communications Commission (FCC) has allocated the 3.1–10.6 GHz frequency range for UWB, imposing stringent requirements that necessitate innovative antenna design approaches [5–9]. Integrating UWB into environments with narrow-band systems, such as WLAN, WiMAX, and X-band satellite downlinks, requires band-notching and filtering to mitigate interference.

This chapter addresses these challenges, focusing on bandwidth enhancement, size reduction, and interference mitigation. The chapter delves into various UWB antenna types, such as planar monopole and slot antennas, exploring design techniques to enhance bandwidth and reduce size. Methods such as adding parasitic elements, using defected ground structures (DGS), and optimizing geometrical parameters are discussed in detail. The integration of band-notching techniques, including stub resonators, slots, and slits, ensures UWB antennas can operate effectively alongside narrowband systems.

Additionally, the chapter covers pulse transmission, crucial for impulse radio applications, focusing on the impact of different pulse shapes—Modulated Gaussian, and SRRC pulses—on antenna performance. The chapter also analyzes UWB indoor propagation characteristics, discussing power delay profiles, RMS delay spread, and channel impulse responses.

Practical case studies illustrate real-world UWB antenna designs, showcasing the application of discussed design techniques with both simulated and measured results. This chapter provides a comprehensive exploration of UWB antenna design, addressing the key challenges of bandwidth, size, interference, and indoor propagation, while equipping readers with practical knowledge for UWB system development.

Ultra-wideband (UWB) technology, operating over a large frequency range (3.1–10 GHz), meets the growing demand for wide bandwidth in modern communication systems. Likewise, millimeter-wave (mmWave) and terahertz (THz) technologies are gaining prominence for high-data-rate applications due to their wide bandwidth capabilities. Although UWB operates at lower frequencies, both mmWave and THz technologies share a common goal: to maximize bandwidth and meet the requirements of next-generation wireless systems [10]. This illustrates a broader trend towards spectrum efficiency across various frequency domains.

2. UWB antenna's characteristics in frequency domain

In this section, we explore the critical aspects of UWB antennas in the frequency domain, focusing on parameters that define their performance and functionality. Over the past 2 decades, numerous studies have focused on the design of UWB antennas, with some utilizing modified radiation patch shapes to achieve UWB characteristics [11–15]. In [16–19], slotted radiation patches are applied to enhance the antenna bandwidth. In [20, 21], two concentric rings are employed around the main circular radiation patch to design a conformal UWB antenna. In [22], a Vivaldi structure is proposed to improve the bandwidth. In [23, 24], bowtie structures are presented to achieve super-wideband characteristics. In [25–31], metamaterials and metasurfaces are used to design UWB antennas. To enhance data rate and channel capacity and improve covering and range and positioning accuracy as well as to reduce multipath interferences, multi-input multi-output (MIMO) UWB antennas are reported [32–34].

2.1 Bandwidth enhancement techniques in UWB antenna design

One of the key issues in designing UWB antennas is achieving a broad impedance bandwidth. This subsection presents various methods to enhance the bandwidth of the antenna. These methods include the use of parasitic or resonator stubs, different types of slots and slits, application of ***via*** holes, employing defected ground structures

(DGS) and other modifications to the antenna geometry. In [35], a ribbon-shaped slot is applied to the radiation patch to enhance bandwidth. In [36], an inverted T-shaped slit is cut in the ground plane to generate a strong resonance around 15.5 GHz. In [37] two rectangular stubs are connected at the bottom of the simple rectangular patch to increase the impedance bandwidth. **Figure 1** displays the VSWR of the antenna with and without these rectangular stubs. It is observed that the impedance bandwidth of the antenna clearly improved to more than 20 GHz. In fact, by adding these rectangular stubs, the electromagnetic coupling between the lower part of the radiation patch connected to the CPW feed line and ground plane can be improved resulting in better impedance matching at higher frequencies.

In [38], a trident feed line is employed to ensure impedance matching over a wide frequency bandwidth. As illustrated in **Figure 2**, with the improved feed line structure, the VSWR of the antenna extends from 2.5 to over 25 GHz. The simulated current distribution on the radiation patch, both with and without the trident-shaped feed line, is also presented in **Figure 2**. The impedance bandwidth enhancement is achieved by increasing the connections to the radiation patch. The bandwidth improvement is clearly due to the vertical alignment of current in the tuning stub through the branches, resulting in a more uniform distribution of the magnetic current within the slot.

In [39], a truncated rectangular radiation patch, combined with two semicircular slots in the ground plane, is utilized to enhance the antenna's bandwidth characteristics. By introducing two appropriately sized triangular notches at the corners of the rectangular radiating patch, excellent impedance characteristics are achieved. In a

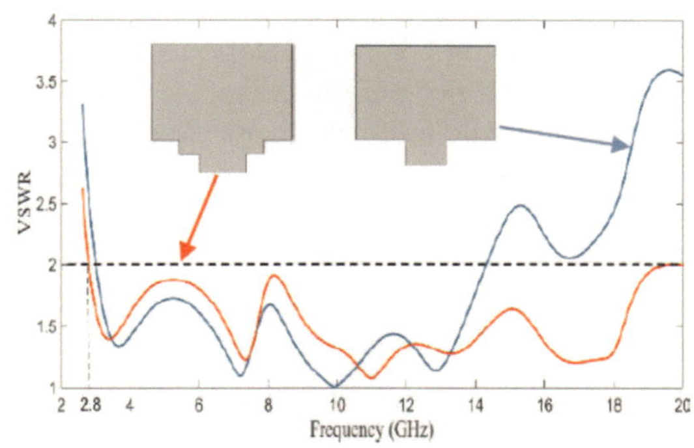

Figure 1.
VSWR of antenna in [37] with simple patch and modified patch with rectangular stubs.

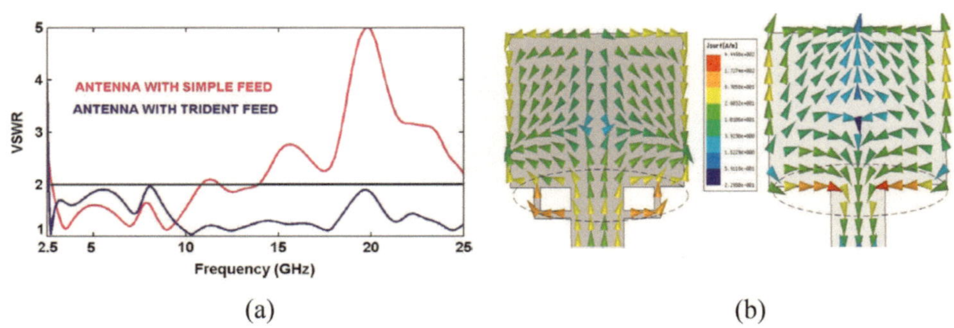

Figure 2.
(a) VSWR of the antenna reported in [38] and (b) simulated current distributions on the antenna at 15.5 GHz.

simple rectangular patch without these triangular notches, there is a sharp discontinuity at the junction between the CPW feed line and the rectangular patch, which can significantly reduce bandwidth at higher frequencies. **Figure 3** demonstrates the return loss and current distributions of antenna. The addition of these triangular notches helps balance the vertical and horizontal electrical currents on the patch (as seen from **Figure 3(b)**), generating multiple resonances and thereby improving the overall bandwidth of the modified antenna. To further enhance the antenna's bandwidth, two semicircular slots are incorporated on both sides of the CPW feed line in the ground plane. This modification improves coupling between the lower edges of the patch and the ground while also eliminating unwanted coupling that could negatively impact the antenna's impedance and radiation characteristics.

2.2 Band-notched characteristic methods in UWB antennas

Given the extremely wide frequency range of UWB systems, they can interfere with existing narrowband systems such as WiMAX (3.3–3.6 GHz), WLAN (5.15–5.35 GHz, 5.475–5.725 GHz, 5.725–5.825 GHz), and X-band downlink satellite communication systems (7.25–8 GHz), which operate within the UWB spectrum (3.1–10.6 GHz). To address this issue, compact antennas with broad operating bandwidths and band-notched characteristics are highly desirable for UWB applications. Achieving these characteristics involves introducing various slot and stub resonators, typically with lengths of a quarter wavelength or half wavelength, to create effective notch filter properties.

A single notch-band UWB antenna is presented in [40], with its geometry and reflection coefficient illustrated in **Figure 4**. In this work, a compact slit-loaded monopole antenna for Bluetooth and UWB applications is introduced. Inverted, open-ended L-shaped slits are employed not only to create an additional 2.4 GHz resonance for Bluetooth but also to obtain band-notched properties in the WLAN spectrum.

In [41–43], dual-band-notched UWB antennas are presented. In [42] two L-shaped slots in the ground plane, connected to the circle-like slot, are employed to create band-notched properties in the WiMAX spectrum, while two parasitic strips inside the circle-like slot provide filtering characteristics in WLAN frequencies. **Figure 5** shows the geometry and VSWR of the antenna. As illustrated in **Figure 5(b)**, by adjusting the lengths of the L-shaped slots and parasitic strips, the center frequency of the notched band can be controlled.

The antennas discussed above demonstrate either single or dual-band-notched characteristics. In [44], however, a UWB antenna with triple band-notched properties

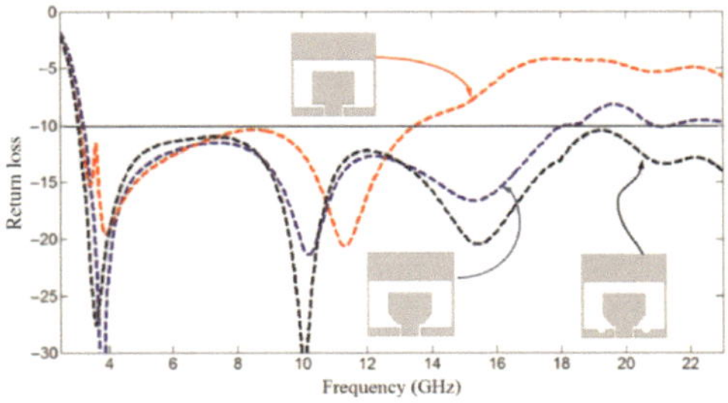

(a)

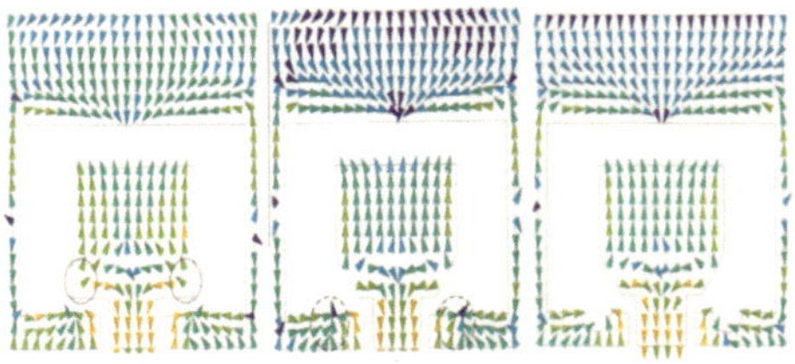
(b)

Figure 3.
(a) Return loss of the antenna presented in [39], (b) Current distribution simulation on the Antenna with simple rectangular patch, truncated patch, and semicircular slots in the ground plane.

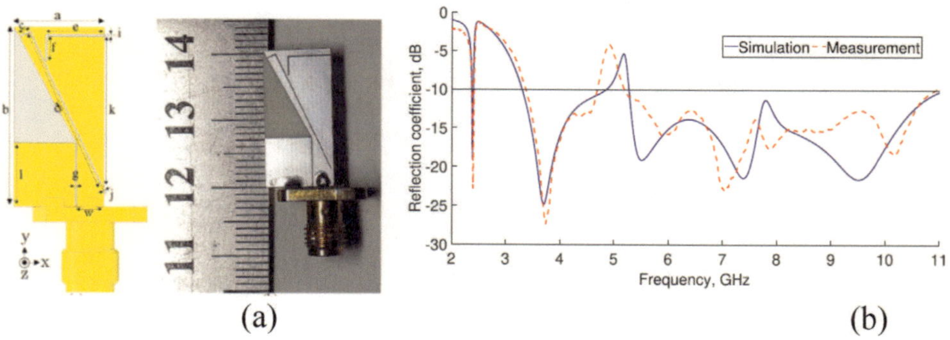

Figure 4.
(a) Compact slit loaded UWB monopole antenna with single band-notch characteristics and (b) reflection coefficient of the antenna [40].

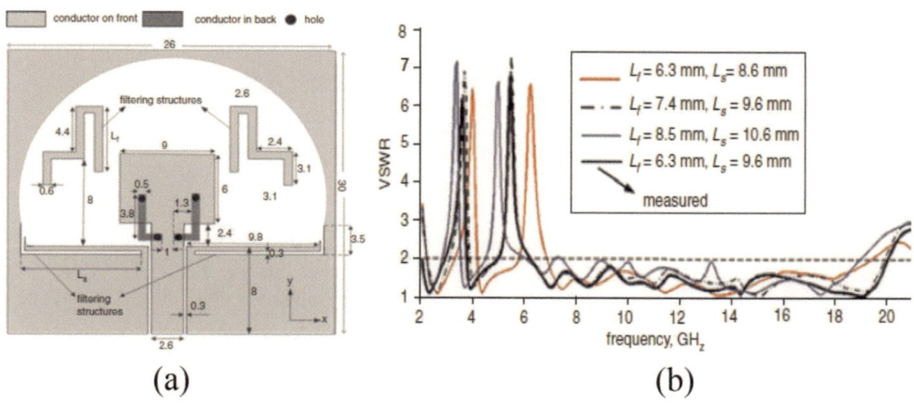

Figure 5.
(a) Geometry of antenna reported in [42] and (b) VSWR of the antenna.

is introduced. **Figure 6** illustrates the antenna's structure and VSWR. This design uses a CPW-fed rectangular patch and a circle-like slot to meet UWB requirements. A single multi-resonance structure is employed to achieve both bandwidth enhancement and band-rejection characteristics. **Figure 7** illustrates the design of the multi-resonance structure. The configuration incorporates several elements: a Y-shaped stub to enhance bandwidth, a T-shaped stub to reject the WiMAX band, two C-shaped stubs for band-notched characteristics in the WLAN range, and two Z-shaped stubs to reject frequencies in the X-band.

To better explain the band-notched behavior of the multi-resonance structure, the surface current distributions at the notch frequencies were simulated and are displayed in **Figure 8**. At the first notch frequency (3.4 GHz), the electrical current primarily concentrates on the T-shaped section of the MFC. At the second notch frequency (5.6 GHz), the current is largely focused on the C-shaped sections, and at the third notch frequency (7.6 GHz), it is predominantly distributed along the Z-shaped parts. At each notched frequency, the surface currents flow in opposing directions between the corresponding section of the MFC and the rectangular radiation patch, as indicated by the dashed lines in **Figure 8**. This opposing current flow effectively cancels out their effects, resulting in the antenna's triple band-notched characteristics.

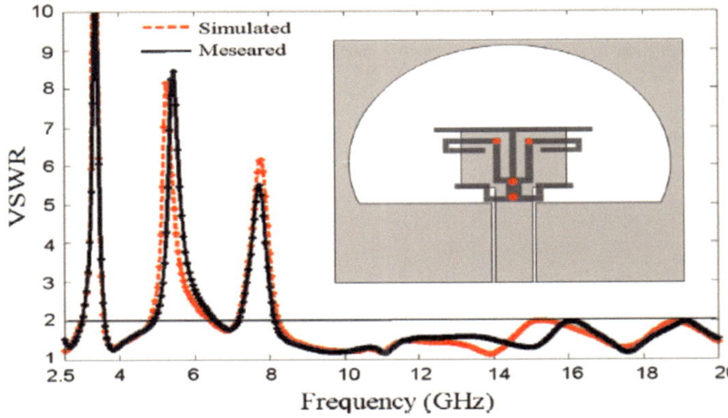

Figure 6.
Structure and VSWR of the antenna in [44].

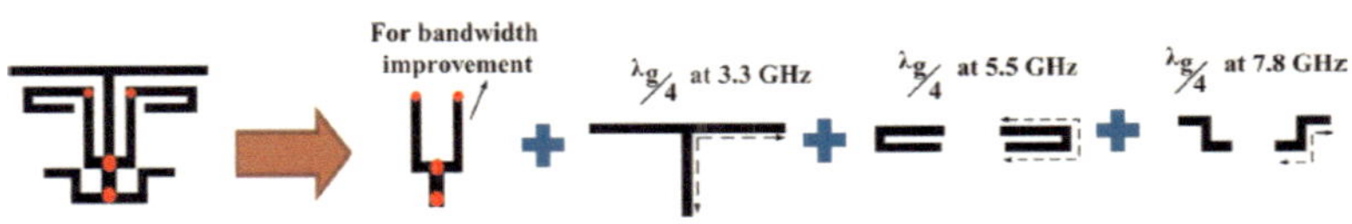

Figure 7.
Structure of the multi-resonance resonator [44].

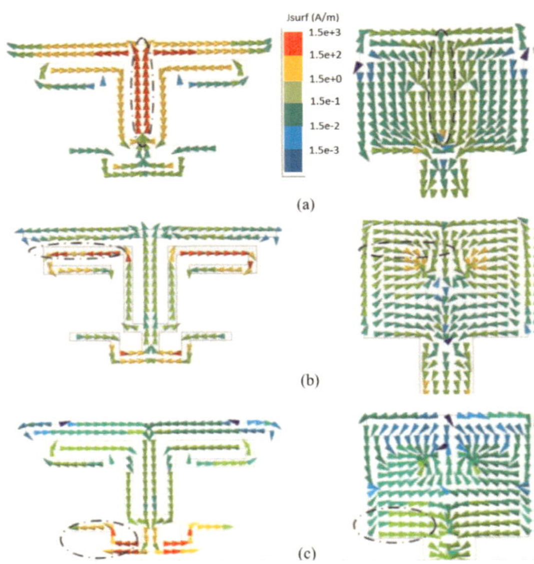

Figure 8.
Current distribution of the antenna presented in [44] at (a) 3.4 GHz, (b) 5.6 GHz and (c) 7.6 GHz.

3. UWB antenna's characteristics in time domain

In pulse-based UWB communication systems, the antennas must be able to transmit and receive extremely short pulses with minimal distortion. As a result, it is crucial to analyze the time-domain characteristics of UWB antennas during the design process. In this section, various pulses used in UWB systems are first defined and their characteristics are analyzed. Following this, the system fidelity factor (SFF), which is a metric for assessing received signals, is introduced, along with the methods for calculating it. Lastly, measurement techniques for evaluating the time-domain characteristics of the antenna system are presented.

Each pulse type is characterized by unique features that influence system performance, including bandwidth, duration, and spectral efficiency. These pulse characteristics are critical for optimizing UWB communication and minimizing interference with other systems.

The Gaussian pulse is a well-known waveform frequently utilized by researchers as an excitation signal in antenna design. This pulse exhibits smooth amplitude variations in the time domain. The Gaussian pulse is mathematically expressed by the following equation:

$$f(t) = e^{-\left(\frac{t}{\tau}\right)^2} \tag{1}$$

where τ represents the pulse width parameter, which controls the duration and shape of the pulse. This parameter is defined as follows:

$$\tau = \frac{1}{2\pi f_c} \tag{2}$$

Here, f_c denotes the center frequency of the pulse. The Gaussian pulse spectrum lacks side lobes, with the energy primarily concentrated in the main lobe. This characteristic makes it ideal for minimizing interference and ensuring efficient energy transfer in UWB systems.

To comply with FCC ultra-wideband regulations, one approach is to use a modulated Gaussian signal, which can be achieved using the following equation:

$$MG(t) = e^{-\left(\frac{t}{\tau}\right)^2} \cos(2\pi f t) \tag{3}$$

Here, f represents the center frequency of the carrier signal.

Another pulse that can be used for UWB systems is the square root raised cosine (SRRC) pulse, which is particularly useful in telecommunications for filtering, and is derived from the following equation.

$$SRRC(t) = \frac{4\beta}{\pi\sqrt{T_s}} \frac{\cos\left[(1+\beta)\pi\frac{t}{T_s}\right] + \frac{sin\left[(1-\beta)\pi\frac{t}{T_s}\right]}{4\beta\frac{t}{T_s}}}{\left[1-\left(4\beta\frac{t}{T_s}\right)\right]} \tag{4}$$

where $SRRC(t)$ is the squared root raised cosine pulse in the time domain; T_s is the symbol period; β is the roll-off factor $(0 \le \beta \le 1)$; t is time.

The efficiency of UWB impulse systems is directly proportional to the quality of the received signal relative to the reference signal. Therefore, it is crucial for antenna designers to assess the level of signal degradation in the time domain.

The most important parameter used for evaluating and comparing signals in the time domain is the system fidelity factor (SFF). The factor is defined as the maximum absolute value of the cross-correlation coefficient between two normalized pulses. This metric effectively indicates the extent of time-domain dispersion introduced by a two-antenna system to a transmitted pulse. The SFF is calculated using the following equation:

$$SSF = \frac{R_s(t)T_s(t)}{\sqrt{\int_{-\infty}^{+\infty} |R_s(t)|^2 |T_s(t)|^2 dt}} \tag{5}$$

Where is $T_s(t)$ transmitted signal and $R_s(t)$ is received signal.

The SFF evaluates the dispersion effects caused by a pair of antennas within a radiation plane. This system can include any combination of antennas—whether identical or different—and can be spaced at any distance from each other. The transmitting coefficient (S_{21}) or transfer function ($H(f)$) are then simulated using a full-wave simulator such as HFSS or CST. This approach requires a large number of mesh elements to solve Maxwell's equations, resulting in increased simulation time and the need for high-performance computing resources. In [45], a simplified method is described for obtaining the SFF from the simulation results of a single antenna, which reduces simulation time and computational complexity.

In this method by simulating a single antenna, the transmission coefficient between two identical antennas, regardless of their distance and rotation angle (in the azimuth plane), can be determined. The transfer function $H(f)$ of an antenna system can be determined using the following equation:

$$H(f) = H_{Tx}H_{Ch}H_{Rx} \tag{6}$$

Where H_{Tx} represents the transfer function of the transmitter antenna, H_{Ch} is the transfer function of the channel, and H_{Rx} is the transfer function of the receiver antenna.

Using the transfer function of two identical antennas, the transfer functions of each antenna can be derived. For an antenna system with two different antennas, the system transfer function can be obtained by combining the transfer functions of the respective elements. The transfer function is defined as the ratio of the voltage received at the terminals of the Rx-antenna to the voltage at the input of the Tx-antenna. This ratio can be calculated using Friis' Transmission Equation.

$$\frac{P_{Rx}}{P_{Tx}} = e_{pol}\left(1 - |\Gamma_{Tx}|^2\right)\left(1 - |\Gamma_{Rx}|^2\right) G_{Tx}G_{Rx}\left(\frac{\lambda}{4\pi r}\right)^2 \tag{7}$$

Where e_{pol} represents the system's polarization efficiency, Γ shows the mismatch at each antenna input, G refers to the total antenna gain, and the final terms describe the path loss (wavelength and distance between antennas).

Using the values of G and Γ from single antenna simulations, the transfer function $H(\omega)$ for a system with two identical antennas can be determined. Eq. (7) simplifies by assuming no polarization losses ($e_{pol} = 1$), as both antennas are positioned in the same plane, and by setting $\Gamma_{Tx} = \Gamma_{Rx} = S_{11}$, since the antennas are identical. As noted earlier, the system's transfer function is the ratio of the received voltage to the transmitted voltage, so $|H(\omega)|$ is the square root of the equation. Knowing that $P_{Rx} = V_{Rx}^2/Z_{Rx}$ and with both antenna impedances being equal ($Z_{Rx} = Z_{Tx}$), the amplitude of the transfer function can be expressed as follows:

$$\frac{|V_{Rx}|}{|V_{Tx}|} = \left(1 - |S_{11}|^2\right)\frac{\lambda}{4\pi r} G_A \tag{8}$$

To determine the phase of $H(\omega)$, it is necessary to separately calculate the phase of the channel and the phase distortion within the antennas. The phase shift in the channel can be derived from the universal expression e^{-jkr}, which accounts for the radial dependence of the radiated field. The phase distortion caused by each antenna ($\varnothing_{Tx}$and $\varnothing_{Rx}$) significantly impacts the pulse shape, and the resulting time delay is considerable. This phase distortion is embedded in the radiated E-field of the antennas, which can be computed using simulation tools. The overall phase of the transfer function is thus influenced by both the channel and the radiated E-field of the antennas, and is given as follows:

$$\angle \mathrm{H}(\omega) = -\varnothing_{Tx} - \mathrm{kr} + \varnothing_{Tx} \tag{9}$$

In HFSS, for example, the radiated electric field *rE* is multiplied by the radial distance to obtain the necessary value. Depending on the antenna's orientation, either the *theta* or *phi* component is used to calculate the phase in radians, denoted as *ang_rad*. To ensure accurate time-domain signal representation, simulation parameters such as Gain, Return Loss, and the phase of the radiated E-field should be calculated across a frequency range that exceeds the desired bandwidth. Additionally, both the antenna gain and the E-field phase must be computed for every angle within the plane of interest. Based on Eqs. (8) and (9), the channel transfer function can be defined as follows:

$$H_{Ch} = \frac{\lambda}{4\pi r} e^{-jkr} \tag{10}$$

The reciprocity theorem establishes the relationship between the transfer functions of two identical antennas.

$$H_{Tx} e^{-j\varnothing_{Tx}} = \frac{\omega}{2\pi c} H_{Rx} e^{-j\varnothing_{Rx}} e^{-j\frac{\pi}{2}} \tag{11}$$

By substituting Eqs. (10) and (11) into Eqs. (6) and (9)

$$|\mathrm{H}(\omega)| = \frac{\omega}{2\pi c} |H_{Rx}||H_{Ch}||H_{Rx}| \tag{12}$$

$$\angle \mathrm{H}(\omega) = \varnothing_{Rx} - \mathrm{kr} + \varnothing_{Rx} - \frac{\pi}{2} \tag{13}$$

and determining H_{Rx} and H_{Tx} through the use of Eq. (8)

$$H_{Rx} = \sqrt{\frac{2\pi c\left(1 - |S_{11}|^2\right) G_A}{\omega}} e^{j\varnothing_{Rx}} \tag{14}$$

$$H_{Tx} = \sqrt{\frac{2\pi c\left(1 - |S_{11}|^2\right) G_A}{2\pi c}} e^{j\varnothing_{Rx}} e^{-j\frac{\pi}{2}} \tag{15}$$

The transfer function for both antennas (Tx and Rx) can be determined using the provided equations. By substituting the transfer functions H_{Rx} and H_{Tx} into Eq. (6), the system transfer function is obtained. This section focuses on the free-space channel to analyze antenna properties alone, but the channel can be replaced as needed since the three transfer functions are independent. With the transmitting antenna oriented at ∅ = 0°, rotating the receiving antenna at different ∅ points allows us to characterize the system within the Rx antenna's azimuthal plane. Two measurement approaches, time-domain and frequency-domain setups, are commonly used to simulate the time-domain characteristics of UWB antennas.

In [46], a time-domain setup is used to evaluate a UWB planar dipole antenna, as shown in **Figure 9**. An SRRC pulse is generated by a Tektronix AWG7122C Arbitrary Waveform Generator (AWG). A Picosecond Pulse Labs wideband amplifier (Model 5865) boosts the output signal. The antennas, connected *via* a rotary joint, are mounted on a turntable. An ultra-wideband anti-phase power divider was used for the balance-fed dipole antennas to create a 180-degree phase difference. A reference antenna with minimal dispersion was placed 20 cm from the transmitting antenna and linked to an Agilent DSO81204A oscilloscope (12 GHz sampling bandwidth). This distance ensured far-field measurement while maintaining a good signal-to-noise ratio, as greater distances (e.g., 50 cm) proved unreliable. The amplifier output was set to 7 dBm, with an oscilloscope noise floor of −54 dBm, and the antenna input power was 0 dBm, resulting in approximately −29.2 dBm at the receiving antenna port.

Figure 10 shows the SRRC pulse generated by the AWG (in(t)) and the corresponding SRRC pulse received by the system (out(t)). The Power Spectral Density (PSD) of these pulses is also shown in **Figure 10**. The UWB SRRC pulse generated by the AWG is affected by the transient response of the transmission network (the response shown in red), which must be accounted for to properly evaluate the antenna under test. An equalized pulse in'(t) is obtained numerically using equation below:

$$in'(t) = IFFT\left[\frac{FFT(in(t))}{FFT(out(t))}\right] \quad (16)$$

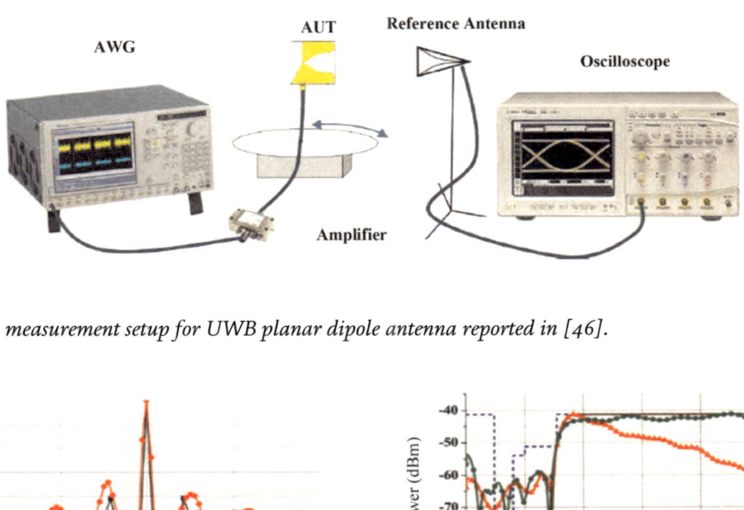

Figure 9.
Time domain measurement setup for UWB planar dipole antenna reported in [46].

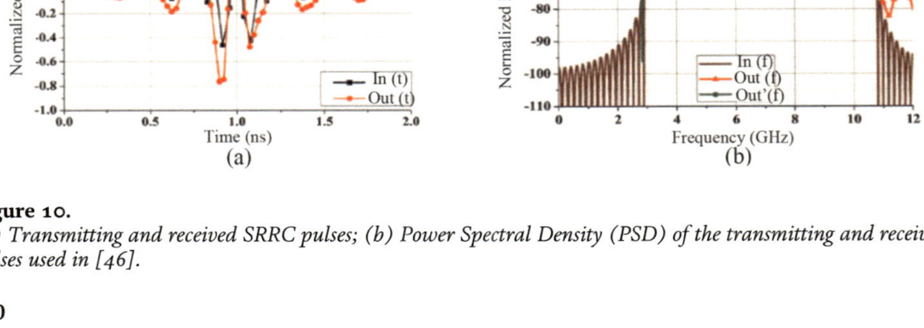

Figure 10.
(a) Transmitting and received SRRC pulses; (b) Power Spectral Density (PSD) of the transmitting and received pulses used in [46].

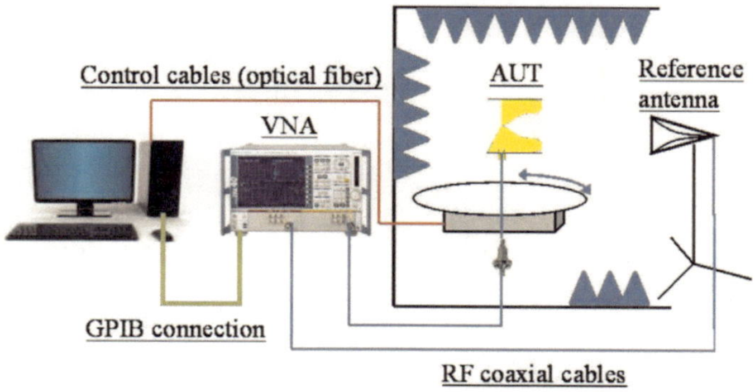

Figure 11.
Frequency domain measurement setup for UWB planar antenna reported in [46].

where the pulse is processed in the frequency domain using the Inverse Fast Fourier Transform (IFFT) and the Fast Fourier Transform (FFT). The PSD of the received equalized pulse out′(f) is displayed in green in **Figure 10(b)**.

Frequency domain measurement technique measures the transfer function of the antenna system in the frequency domain. The time domain characteristics are then obtained by applying a FFT to the measured frequency domain data.

In [45], a frequency domain measurement setup is utilized to derive the time domain characteristics of UWB antennas. S_{21} measurements are performed using a Vector Network Analyzer (VNA) within an anechoic chamber. The frequency measurement setup for time domain characteristics of UWB antenna is depicted in **Figure 11**. In [45], antennas are mounted at opposite ends of a 1.09-meter foam frame, with one antenna fixed (Tx) and the other (Rx) rotated in the azimuthal plane. Measurements are taken at each step of rotation. Each antenna is connected to a port on the HP 8720D Network Analyzer *via* a 3-meter high-quality cable. The VNA, situated outside the chamber, was calibrated at the antenna ports with an electronic calibration kit to cover the full frequency range (50 MHz–20.05 GHz). This process is simple and can utilize any input signal. The pulse intended for transmission is defined in MATLAB, and its frequency response is determined using a Fast Fourier Transform (FFT).

This frequency response is then multiplied by the antenna system's transfer function to derive the received signal in the frequency domain. The Inverse Fast Fourier Transform (IFFT) is then applied to convert the received signal back into the time domain. The following equations outline these steps.

$$R_s(\omega) = FFT(T_s(t))H(\omega) \tag{17}$$

$$R_s(\omega) = IFFT(R_s(\omega)) \tag{18}$$

The received signal is determined by replacing the simulated data with the measured data. This received signal reflects the distortion and dispersion caused by both the antennas and the transmission channel.

Due to the advantage of representing arbitrary input signals in the frequency-domain approach, in Ref. [47], the effect of a triple band-notched slot antenna on two

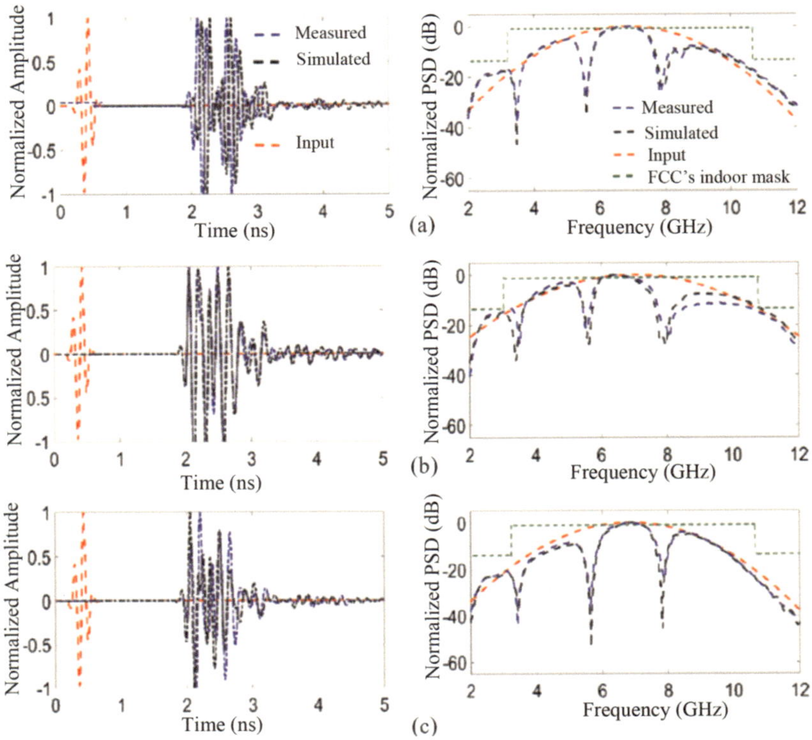

Figure 12.
Transmitting and received MG(t) pulses and their power spectral densities in (a) ∅=0°, (b) ∅ = 90°, (c) ∅ =180° [47].

modulated Gaussian (MG) and square root raised cosine (SRRC) pulses are studied. **Figures 12** and **13** show the input and received pulses, along with their PSD, in different orientations (face-to-face ∅ =0°, face-to-side ∅ =90°, and face-to-back ∅ =180°), when using MG and SRRC pulses as excitation signals, respectively. It is evident that there is a strong agreement between the results across different directions, attributed to the omnidirectional radiation characteristics of the proposed triple-band-notched antenna. However, at higher frequencies, where the radiation patterns become somewhat directive, variations in the PSD diagrams at different angles can be observed. The three nulls in the PSD diagrams result from the band-notched features of the multi-resonance band-notched structure in the antenna design. Additionally, **Figures 12** and **13** show that the PSDs of the received signals fully comply with the FCC's indoor power emission mask, demonstrating the presented antenna's suitability for indoor IR UWB applications.

4. UWB antenna's characteristics in indoor propagation channels

UWB signals have unique characteristics, such as short pulses and high bandwidth, that make them suitable for indoor environments. Understanding the behavior of UWB signals indoors is critical for applications like high-resolution radars, localization systems, and high-speed wireless communication. In indoor environments, the signal undergoes multiple interactions such as reflection, diffraction, scattering, and attenuation. These interactions can impact the accuracy of signal detection and

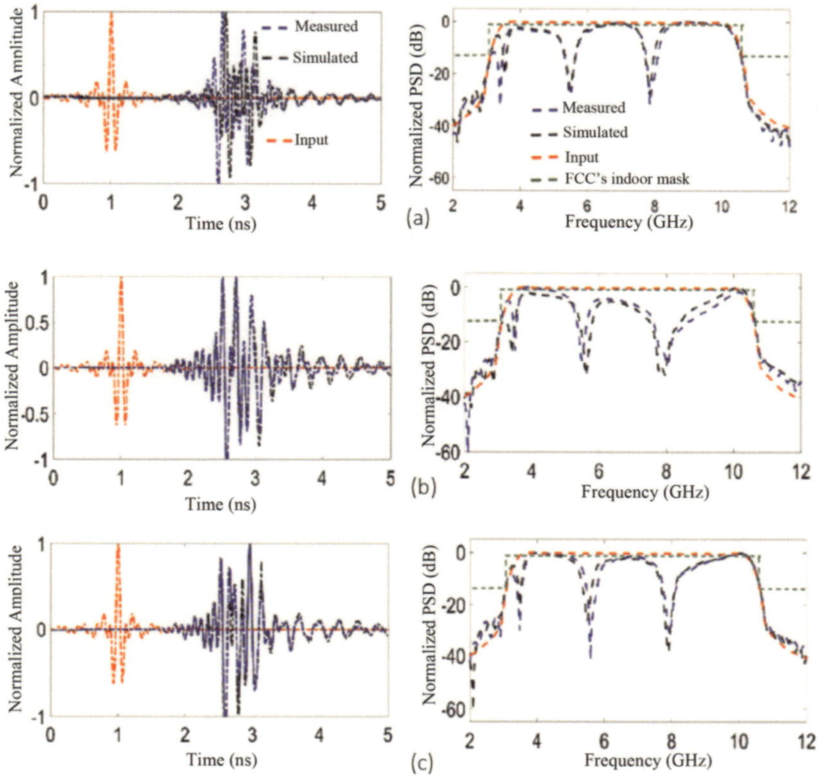

Figure 13.
Transmitting and received SRRC pulses and their power spectral densities in (a) ∅ = 0°, (b) ∅ = 90°, (c) ∅ = 180° [47].

communication performance. Multipath propagation is a significant challenge for UWB signals in indoor environments. Due to the wide bandwidth of UWB, different frequency components of the signal experience different propagation paths, leading to time-domain dispersion and multipath fading. The propagation mechanisms are shown in **Figure 14** and can be described as follows:

1. *Reflection*: This phenomenon occurs when an electromagnetic wave encounters large obstacles relative to its wavelength. If the separating surface between two materials is smooth, the reflection is specular, following Snell-Descartes and Fresnel laws. If the surface is rough, the reflection becomes diffuse, scattering part of the energy in various directions while most remains along the reflected path.

2. *Scattering*: When an electromagnetic wave encounters small obstacles relative to its wavelength, it experiences numerous random diffractions, resulting in a phenomenon known as diffusion. This causes the wave to scatter in various directions with varying attenuation. While diffusion is commonly observed outdoors, such as with tree foliage, it can also occur indoors with clusters of small objects.

3. *Diffraction*: This occurs at the edges of large obstacles relative to the wavelength, affecting the electromagnetic field beyond the line of sight. Huygens' principle is

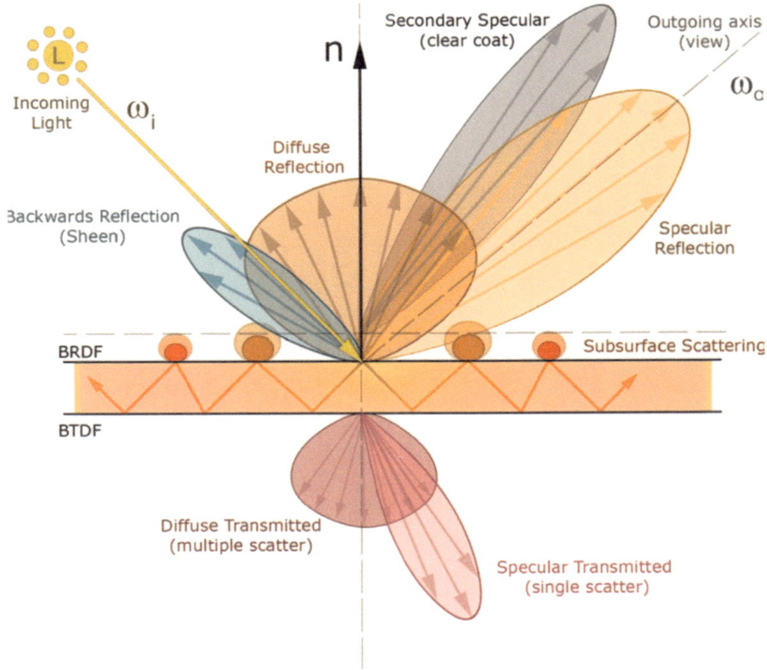

Figure 14.
Different channel multipath characteristics [48].

used to calculate the diffracted field, treating each point of the wavefront as a secondary source. The diffracted waves spread in a cone with an angle matching the incident angle.

4. *Transmission*: This occurs when part of the incident wave passes through a material that is not entirely radio-opaque. Indoor building materials typically attenuate this transmitted wave significantly. The extent of attenuation and the direction of the transmitted signal depend on the wavelength, as the material's refractive index varies with frequency. Additionally, layered materials like plasterboard can cause multiple reflections within the material.

5. *Waveguide effect*: This occurs indoors, such as between corridor walls, where successive reflections on parallel surfaces guide the wave along a specific direction. It also happens in urban settings, like between buildings in narrow streets.

Several important metrics help characterize the UWB indoor propagation channels:

- *Channel impulse response* (*CIR*): This gives insight into how multipath components affect the received signal. It shows the time delay and amplitude of each multipath component.

- *RMS delay spread*: A critical parameter that quantifies the time dispersion caused by multipath propagation. It impacts the coherence bandwidth of the channel.

- *Path loss models*: Path loss models characterize the attenuation experienced by UWB signals as they propagate through space. Models like the Friis free space model can be adapted for indoor environments, taking into account reflections and diffractions.

- *Power delay profile (PDP)*: PDP represents the power received from different multipath components over time and is important for understanding signal strength over multiple paths.

To study UWB propagation channels, both experimental and simulation-based methods are employed.

Experimental methods involve physical measurements of UWB signals in real indoor environments, using both time-domain and frequency-domain setups to characterize the channel:

- *Time-domain measurements*: These involve transmitting UWB pulses and capturing their reflections, arrivals, and delays at the receiver using high-speed oscilloscopes. Time-domain analysis provides valuable insights into multipath behavior and pulse distortions.

- *Frequency-domain measurements*: In this method, the transfer function of the propagation channel is measured using a vector network analyzer (VNA). By applying the Inverse Fast Fourier Transform (IFFT), the channel impulse response can be derived, enabling analysis in both the time and frequency domains.

While experimental methods capture all physical effects, including multipath, interference, and attenuation, they are often costly, time-consuming, and constrained by equipment limitations such as power and bandwidth.

In contrast, simulation methods are crucial for scenarios where physical measurements are impractical. They offer flexibility to modify environmental parameters, antenna characteristics, and test various scenarios without the constraints of real-world setups. Different channel models can be used to characterize UWB propagation in indoor environments. These models can be classified into stochastic and deterministic models:

Saleh-Valenzuela (S-V) Model [49] is a widely used stochastic model for UWB indoor channels, which assumes multipath components arrive in clusters. It characterizes the delays and amplitude of each path and is suitable for capturing the statistical nature of multipath. Modified S-V Model for UWB is an extension of the Saleh-Valenzuela model, adapted for UWB characteristics by considering wider bandwidths and shorter pulse durations. This approach is displayed in **Figure 15**.

Deterministic models offer several advantages over stochastic models in UWB indoor propagation studies. They provide high accuracy by simulating the exact physical characteristics of specific environments, including detailed reflection, diffraction, and scattering effects. This allows for precise predictions of the channel impulse response (CIR) and the exact timing of multipath components. Deterministic models also offer detailed time and frequency domain information and can be tailored to specific scenarios by adjusting environmental parameters. While stochastic models are more generalized and computationally less demanding, deterministic models excel in delivering detailed, accurate insights necessary for applications requiring high

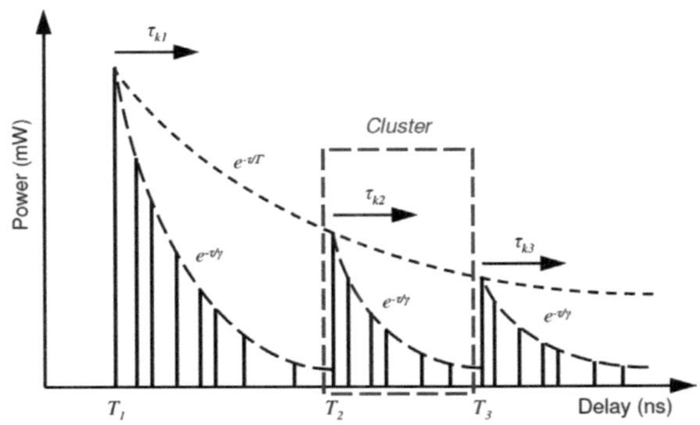

Figure 15.
Power delay profile based on Saleh and Valenzuela model [49].

precision, such as indoor positioning and network planning. Ray Tracing (RT) and Finite Difference Time Domain (FDTD) are both deterministic methods used to predict UWB signal propagation. FDTD solves Maxwell's equations directly in the time domain, making it suitable for wideband simulations. However, it also requires substantial computational power, particularly for large-scale indoor environments and higher frequencies. In [50], FDTD approach is used to study a small office propagation channel with two different antennas: a planar inverted cone antenna (PICA) and a horn-shaped self-complementary antenna (HSCA).

Antenna patterns from CST at 3.5 and 10.5 GHz are applied for channel modeling. Both PICA and HSCA exhibit omnidirectional radiation at lower frequencies. However, HSCA's pattern is more distorted at higher frequencies than PICA's.

Simulations are conducted in 2D, focusing on the transverse magnetic polarized field. For simulation, the FDTD cell size is set to $\Delta x = \Delta y = 3$ m, with a time step of $\Delta t = 7 \times 10^{-12}$ s, following the stability criterion.

The simulated power delay profile (PDP) at a distance of 3 meters from the transmitter with a threshold of 23 dB for multipath component (MC) detection exhibits PICA's PDP has more MCs than HSCA's, suggesting that higher-frequency pattern distortions reduce the number of received MCs.

However, the FDTD method demands substantial computational resources and is generally more suited for small to medium-sized environments due to its time-intensive nature. In contrast, the RT (Ray Tracing) approach offers several advantages over FDTD, particularly in large-scale environments. The RT model can accurately simulate complex scenarios such as reflections, diffractions, and transmissions by precisely modeling walls, doors, and various obstacles. It provides highly detailed results with a significantly lower computational cost and faster processing time. Additionally, RT is more flexible for real-world applications and is widely adopted in network planning and antenna positioning, making it ideal for large environments like buildings or urban areas where multiple interactions between waves and obstacles occur.

In [51], ray tracing method (RT), whose algorithms are based on the uniform theory of diffraction and geometrical optics, is applied to achieve UWB indoor channel characteristics.

To thoroughly assess channel characteristics, both line-of-sight (LoS) and non-line-of-sight (NLoS) scenarios were examined. In the NLoS scenario, an absorbing plane was placed between the transmitting antenna and the receiver grid. To optimize simulation accuracy, four and five ray bounces were deemed sufficient for LOS and NLOS cases, respectively. Additionally, the frequency-dependent radiation pattern of the discone antenna was applied at both the transmitter and receiver, taking advantage of the RT simulator's ability to track the angle of departure (AoD) and angle of arrival (AoA). A vector network analyzer (VNA) was employed to characterize the channel's behavior. Channel measurements were conducted across the entire ultra-wideband (UWB) band of 3.1–10.6 GHz, using a total of 1,601 discrete frequency points. The frequency response at each measurement point was determined using the RT simulator. In the simulation, a set of dipole antennas transmitted continuous wave (CW) carriers at the same 1,601 frequencies used in the VNA measurements.

In [52], a UWB indoor propagation channel as shown in **Figure 16** is fully investigated. The UWB channel measurement was performed in a typical laboratory setting characterized by a high density of radiofrequency equipment, electronic devices, and metallic objects. This environment exhibited a complex multipath propagation due to reflections, diffraction, and scattering. The laboratory was situated in a modern building with large glass windows, concrete floors and ceilings, and steel-framed walls. A frequency-domain measurement campaign was conducted to determine the impulse responses at 15 distinct locations on a reviver grid. Each grid point consisted of nine receiver points, as illustrated in **Figure 16**. A frequency-domain measurement campaign is applied to measure impulse response at 15 different positions of reviver grid. Each grid includes nine receiver points. The complex channel transfer function (CTF) was measured using a VNA at all receiver positions. The VNA, antennas, and LNAs were configured to cover the UWB FCC band (3.1–10.6 GHz). Hundred snapshots were captured at each receiver position. **Figure 17** presents a representative channel impulse response (CIR) for the maximum transmitter-receiver separation distance and a single position within the square grid.

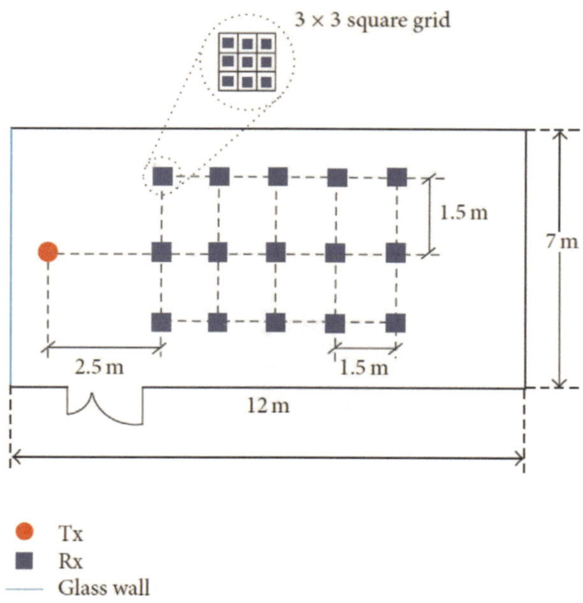

Figure 16.
Indoor propagation channel reported in [52].

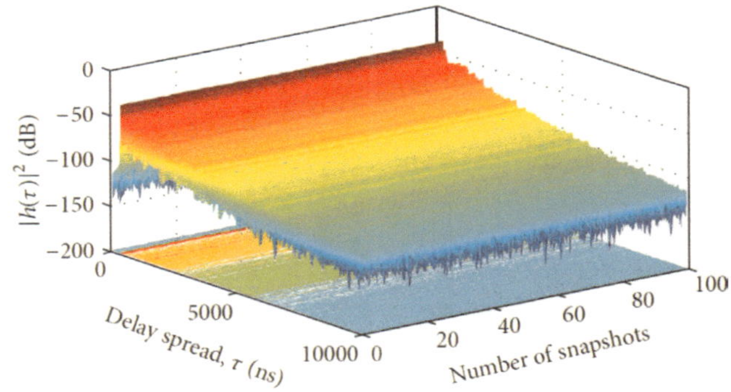

Figure 17.
Measured channel impulse response for maximum Tx-Rx separation, one grid position. Hundred snapshots shown in [52].

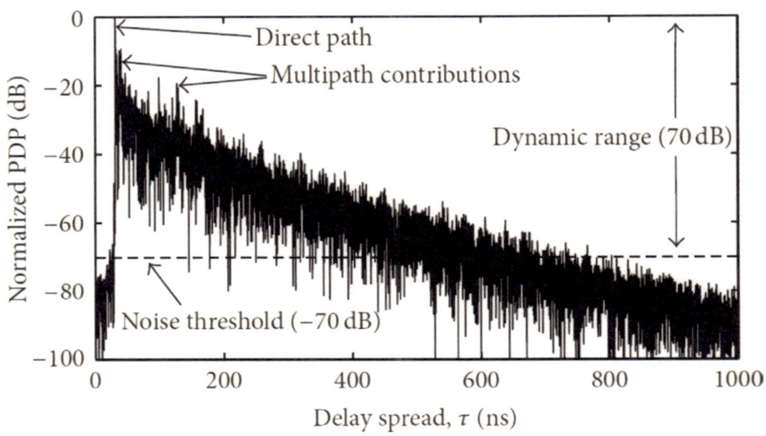

Figure 18.
*Normalized power delay profile corresponding to the channel impulse response shown in **Figure 38** [52].*

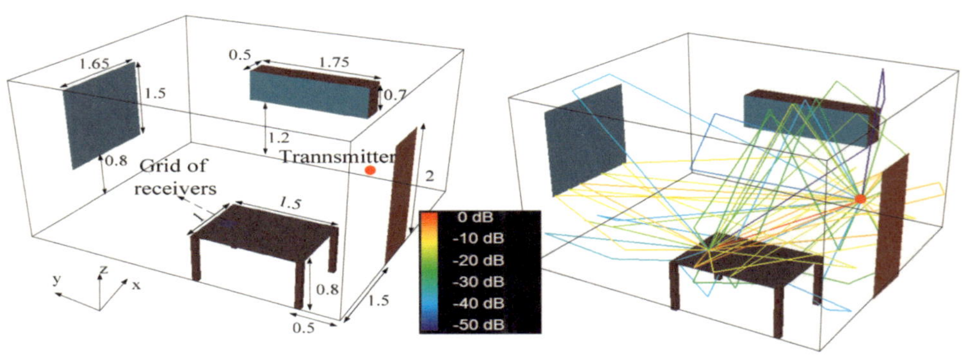

Figure 19.
Indoor propagation channel presented in [54].

A posterior averaging of all CIR snapshots was performed to reduce the impact of thermal noise, potential interference, and minimal temporal variations of the propagation channel, thereby enhancing the reliability of statistical analysis.

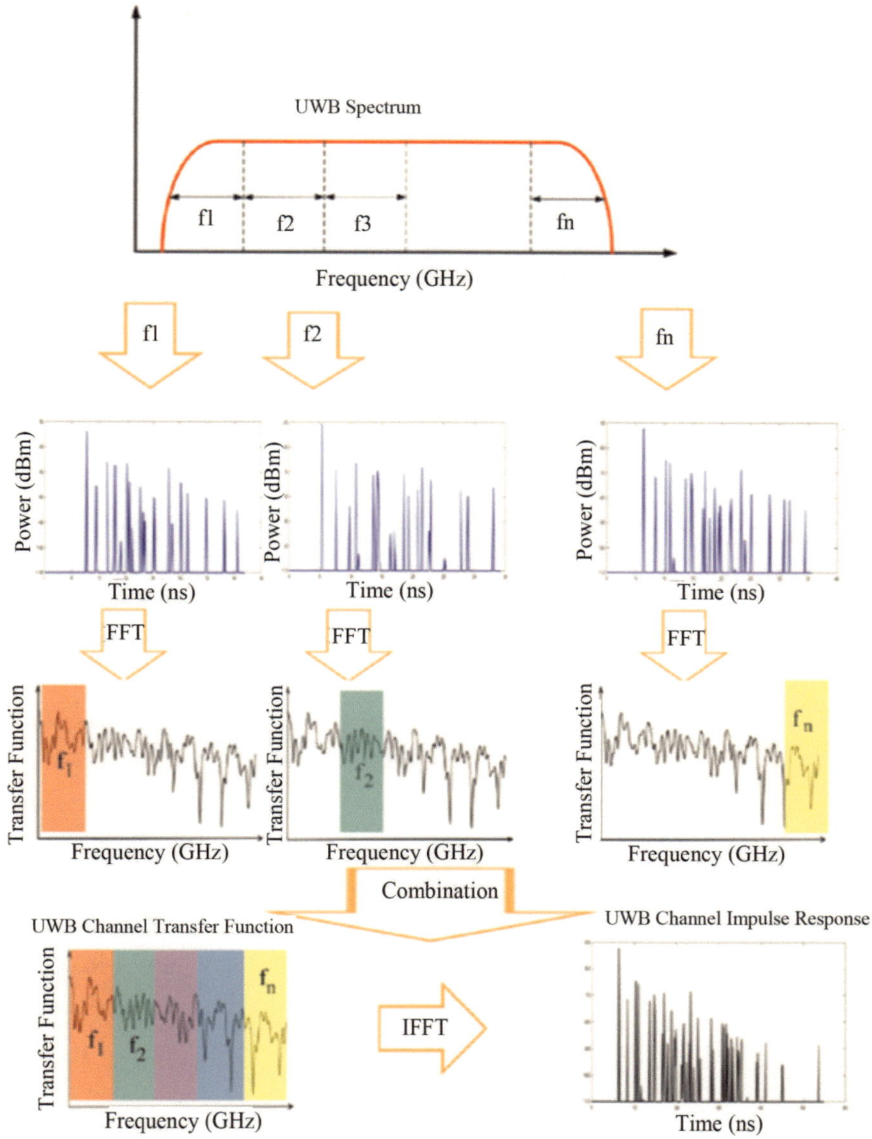

Figure 20.
Sub-band divided RT method [47].

Figure 18 illustrates the normalized power delay profile derived from the channel impulse response presented in **Figure 17**. The PDP exhibits a minimum dynamic range of 70 dB.

However, performing RT simulations at all 1,601 points, as done in [51, 52], can be computationally intensive.

In [53, 54], a sub-band divided RT model is applied to consider the indoor propagation channel when dual band-notched antennas are applied for transmitter and receiver as shown in **Figure 19**. The steps of this method are explained in the following and depicted in **Figure 20**.

1. The entire frequency bandwidth is divided into multiple sub-bands, each narrow enough to be considered frequency-flat.
2. The conventional RT method is used to calculate the time-domain delay profile at the center frequency of each sub-band.
3. The Fourier transform is applied to obtain the channel transfer function for each sub-band. Only the accurate portions are retained and combined to create a new frequency response covering the entire bandwidth.
4. The inverse Fourier transform is then applied to the combined frequency response to obtain a delay profile valid for the entire bandwidth. The whole UWB bandwidth was divided into 15 sub-bands with 500 MHz bandwidth.

References	Dimensions	Bandwidth (GHz)	Technique
[5]	10 × 15 × 0.25	3–14.55	Fork-shaped radiation patch
[6]	29 × 39 × 1.5	2.19–25	Slotted circular patch
[7]	27 × 38 × 1.6	3.17–11.61	Octogen radiation patch
[8]	23 × 128 × 1.6	3.4–12.4	Slotted circular patch
[9]	22 × 32 × 1.6	2.8–13.3	Foot-shaped patch
[11]	60 × 60 × 50	1–6	Bowtie radiation patch
[20]	38 × 38 × 5.5	2.85–8.6	Annular ring patch
[22]	20 × 30 × 0.8	2.9–13.55	Vivaldi structure
[25]	15 × 25 × 1.6	0.65–9.2	Metamaterial elements
[28]	25 × 25 × 1.6	0.7–8	Metamaterial elements
[36]	24 × 25 × 1.6	2.5–23	Forked-shaped patch
[38]	26 × 30 × 1.6	2.5–25	Trident feed line
[42]	26 × 30 × 1.6	2.1–20	Parasitic elements
[44]	27 × 27 × 1.6	2.5–20	Parasitic elements

Table 1.
Comparison of different UWB antennas.

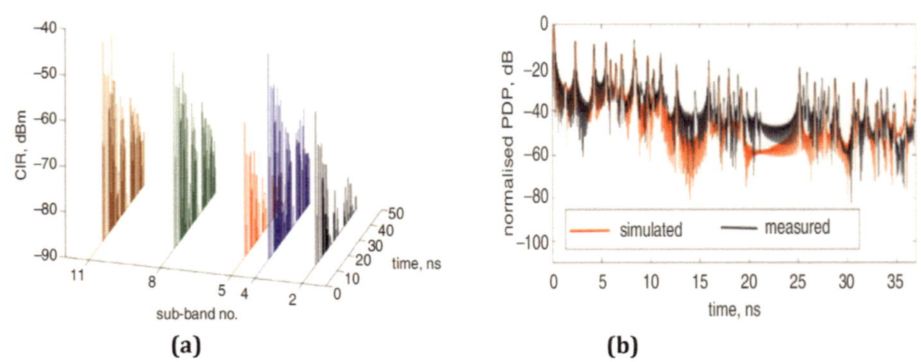

Figure 21.
(a) Channel impulse responses in different sub-bands studied in [53] and (b) simulated and measured power delay profile.

Figure 21(a) displays the channel impulse response at different sub-bands. For clarity, only the channel impulse responses (CIRs) of sub-bands 2, 4, 5, 8, and 11 are shown. **Figure 21(a)** reveals that the first received impulses (line-of-sight rays) vary smoothly across sub-bands, except at the notched sub-bands. As expected, the amplitude and number of multipath components significantly decrease at sub-bands 2 and 5 due to the band-notched characteristics of the proposed antenna in [53]. **Figure 21(b)** compares the simulated and measured power delay profiles (PDPs) for the indoor propagation scenario. The PDP is obtained by spatially averaging the squared CIRs over all receiver positions. The simulated results closely match the measured results. Discrepancies, particularly at higher excess delays, are attributed to the limited order of reflections and diffractions considered in the RT model and the limited number of sub-bands used in the sub-band-divided RT model. **Table 1** summarizes the key characteristics of several UWB antenna designs, focusing on size, bandwidth, and applied methodologies for improvement.

5. Conclusion

This chapter provides a comprehensive view of UWB antennas and their critical role in modern wireless communication systems. Section 1 introduces the chapter by providing an overview of UWB technology, defining its key characteristics, and exploring its wide range of applications in communication and sensing systems. Section 2 delves into the frequency-domain behavior of UWB antennas, addressing essential techniques for bandwidth enhancement and methods for implementing band-notched characteristics to mitigate interference. Section 3 shifts focus to the time-domain characteristics of UWB antennas, examining the impact of different antenna designs on UWB pulses like modulated Gaussian and square root raised cosine pulses, alongside techniques for measuring system fidelity factor. Finally, Section 4 explores advanced simulation techniques for evaluating UWB indoor propagation channels, comparing the benefits and challenges of methods such as FDTD, ray tracing, and sub-band divided ray tracing approaches. Together, these insights form a robust foundation for understanding and advancing UWB technology in various real-world

Author details

Seyed Ramin Emadian[1*] and Masoumeh Forgonchi Nojavan[2]

1 Independent Researcher, Tehran, Iran

2 Telecommunications Engineering Department, Islamic Azad University (IAU), Tehran, Iran

*Address all correspondence to: ramin.emadian@gmail.com

References

[1] Zhang R, Hou Y, Yang S, Huo K, Fu Z, Liang J, et al. A reconfigurable ultra-wideband high-purity multimode terahertz OAM antenna array based on graphene and vanadium dioxide. AEU-International Journal of Electronics and Communications. 2024;7:155514

[2] Xiang L, Wu F, Xia X, Jiang ZH, Yu C, Hong W. Millimeter-wave wideband dual-polarized and circularly-polarized ME-dipole reflect arrays with linearly polarized feed. IEEE Transactions on Antennas and Propagation. 2024;**72**:8022-8027

[3] Kotchapansompote P, Kumchaiseemak N, Kaemarungsi K, Suwajanakorn S. A hybrid approach to impulse-radio UWB self-localization with passive reflectors. IEEE Sensors Journal. 2024. DOI: 10.1109/JSEN.2024.3449376

[4] Nosrati L, Hoseini Semnani S, Fazel MS, Rakhshani S, Ghavami M. Improving indoor localization using mobile UWB sensor and deep reinforcement learning. IEEE Sensors Journal. 2024;**24**:32546-32553

[5] Rizvi SN, Awan WA, Choi D, Hussain N, Park SG, Kim N. A compact size antenna for extended UWB with WLAN notch band stub. Applied Sciences. 2023;**13**(7):4271

[6] Lakshmaiah YV, Roy B. A circular monopole antenna for bandwidth enhancement using cylinder slots with triple band notch characteristics. International Journal of Communication Systems. 2023;**36**(7):1-10

[7] Mayuri P, Rani ND, Subrahmanyam NB, Madhav BT. Design and analysis of a compact reconfigurable dual band notched UWB antenna. Progress in Electromagnetics Research C. 2020;**98**:141-153

[8] Kumar OP, Kumar P, Ali T. A compact dual-band notched UWB antenna for wireless applications. Micromachines. 2021;**13**(1):12

[9] Koteswara Rao Devana VN, Kusuma Kumari E, Chakradhar KS, Kadali Sharma P, Rama Devi D, Manohar Kumar C, et al. A novel foot-shaped elliptically embedded patch-ultra wide band antenna with quadruple band notch characteristics verified by characteristic mode analysis. International Journal of Communication Systems. 2022;**35**(15):e5284

[10] Alibakhshikenari M, Ali EM, Soruri M, Dalarsson M, Naser-Moghadasi M, Virdee BS, et al. A comprehensive survey on antennas on-chip based on metamaterial, metasurface, and substrate integrated waveguide principles for millimeter-waves and terahertz integrated circuits and systems. IEEE Access. 2022;**4**(10):3668-3692

[11] Elhabchi M, Srifi MN, Touahni R. A novel modified U-shaped microstrip antenna for super wide band (SWB) applications. Analog Integrated Circuits and Signal Processing. 2020;**102**:571-578

[12] Singh VK, Bangari N, Ali Z, Vyas A, Verma RK, Saxena A. Parachute shape ultra-wideband wearable antenna for remote health care monitoring. International Journal of Communication Systems. 2023;**36**(10):e5488

[13] Sediq HT, Nourinia J, Ghobadi C. A cat-shaped patch antenna for future super wideband wireless microwave applications. Wireless Personal Communications. 2022;**125**(2):1307-1333

[14] Abbas A, Hussain N, Lee J, Park SG, Kim N. Triple rectangular notch UWB antenna using EBG and SRR. IEEE Access. 2020;**25**(9):2508-2515

[15] Alibakhshi-Kenari M, Naser-Moghadasi M, Ali Sadeghzadeh R, Singh VB. Hexa-band planar antenna with asymmetric fork-shaped radiators for multiband and broadband communication applications. IET Microwaves, Antennas & Propagation. 2016;**10**(5):471-478

[16] Ali EM, Awan WA, Alzaidi MS, Alzahrani A, Elkamchouchi DH, Falcone F, et al. A shorted stub loaded UWB flexible antenna for small IoT devices. Sensors. 2023;**23**(2):748

[17] Jameel MS, Mezaal YS, Atilla DC. Miniaturized coplanar waveguide-fed UWB antenna for wireless applications. Symmetry. 2023;**15**(3):633

[18] Danjuma IM, Akinsolu MO, See CH, Abd-Alhameed RA, Liu B. Design and optimization of a slotted monopole antenna for ultra-wide band body centric imaging applications. IEEE Journal of Electromagnetics, RF and Microwaves in Medicine and Biology. 2020;**4**(2):140-147

[19] Peñafiel-Ojeda CR, Andrade CE, Baez-Egas R, Garcia-Santos V. An ultrawideband printed monopole antenna analyzed with the theory of characteristic modes. IEEE Latin America Transactions. 2022;**20**(6): 948-954

[20] Mohamadzade B, Simorangkir RB, Hashmi RM, Lalbakhsh A. A conformal ultrawideband antenna with monopole-like radiation patterns. IEEE Transactions on Antennas and Propagation. 2020;**68**(8):6383-6388

[21] Mohamadzade B, Simorangkir RB, Hashmi RM, Chao-Oger Y, Zhadobov M, Sauleau R. A conformal band-notched ultrawideband antenna with monopole-like radiation characteristics. IEEE Antennas and Wireless Propagation Letters. 2019;**19**(1):203-207

[22] Saleh S, Ismail W, Abidin IS, Bataineh MH, Al-Zoubi AS. Novel compact UWB Vivaldi nonuniform slot antenna with enhanced bandwidth. IEEE Transactions on Antennas and Propagation. 2022;**70**(8):6592-6603

[23] Lee D, Shaker G, Melek W. A broadband wrapped bowtie antenna for UWB pulsed radar applications (vol 68, pg 7803, 2020). IEEE Transactions on Antennas and Propagation. 2021;**69**(11): 8033

[24] Fiser O, Hruby V, Vrba J, Drizdal T, Tesarik J, Vrba J Jr, et al. UWB bowtie antenna for medical microwave imaging applications. IEEE transactions on Antennas and Propagation. 2022;**70**(7): 5357-5372

[25] Sadeghzadeh RA, Alibakhshi-Kenari M, Naser-Moghadasi M. UWB antenna based on SCRLH-TLs for portable wireless devices. Microwave and Optical Technology Letters. 2016;**58**(1): 69-71

[26] Alibakhshi-Kenari M, Naser-Moghadasi M, Ali Sadeghzadeh R, Singh VB. Metamaterial-based antennas for integration in UWB transceivers and portable microwave handsets. International Journal of RF and Microwave Computer-Aided Engineering. 2016;**26**(1):88-96

[27] Alibakhshikenari M, Virdee BS, Azpilicueta L, Naser-Moghadasi M, Akinsolu MO, See CH, et al. A comprehensive survey of "metamaterial transmission-line based antennas: Design, challenges, and applications". IEEE Access. 2020;**3**(8):144778-144808

[28] Alibakhshi-Kenari M, Naser-Moghadasi M, Ali Sadeghzadeh R, Singh Virdee B, Limiti E. New compact antenna based on simplified CRLH-TL for UWB wireless communication systems. International Journal of RF and Microwave Computer-Aided Engineering. 2016;**26**(3):217-225

[29] Alibakhshi KM. Design and modeling of new UWB metamaterial planar cavity antennas with shrinking of the physical size for modern transceivers. International Journal of Antennas and Propagation. 2013; **2013**(1):562538

[30] Alibakhshi-Kenari M, Naser-Moghadasi M, Sadeghzadeh RA, Virdee BS, Limiti E. Miniature CRLH-based ultra wideband antenna with gain enhancement for wireless communication applications. ICT Express. 2016;**2**(2):75-79

[31] Althuwayb AA, Alibakhshikenari M, Virdee BS, Shukla P, Limiti E. Realizing UWB antenna array with dual and wide rejection bands using metamaterial and electromagnetic bandgaps techniques. Micromachines. 2021;**12**(3):269

[32] Alibakhshikenari M, Babaeian F, Virdee BS, Aïssa S, Azpilicueta L, See CH, et al. A comprehensive survey on "Various decoupling mechanisms with focus on metamaterial and metasurface principles applicable to SAR and MIMO antenna systems". IEEE Access. 2020;**21**(8):192965-193004

[33] Alharbi AG, Rafique U, Ullah S, Khan S, Abbas SM, Ali EM, et al. Novel MIMO antenna system for ultra wideband applications. Applied Sciences. 2022;**12**(7):3684

[34] Ahmad S, Khan S, Manzoor B, Soruri M, Alibakhshikenari M, Dalarsson M, et al. A compact CPW-fed ultra-wideband multi-input-multi-output (MIMO) antenna for wireless communication networks. IEEE Access. 2022;**2**(10):25278-25289

[35] Park S, Jung KY. Novel compact UWB planar monopole antenna using a ribbon-shaped slot. IEEE Access. 2022; **13**(10):61951-61959

[36] Emadian SR, Ahmadi-Shokouh J. Modified ground circle like-slot antenna with dual band-notched characteristics for super UWB applications. The Applied Computational Electromagnetics Society Journal (ACES). 2015;**30**:436-443

[37] Emadian SR, Ahmadi-Shokouh J, Raeesi A, Askani A, Nazari J. Frequency and time domain investigation of compact UWB slot antenna with triple band notched characteristics. The Applied Computational Electromagnetics Society Journal (ACES). 2015;**30**:1348-1353

[38] Emadian SR, Ghobadi C, Nourinia J, Mirmozafari MH, Pourahmadazar J. Bandwidth enhancement of CPW-fed circle-like slot antenna with dual band-notched characteristic. IEEE Antennas and Wireless Propagation Letters. 2012; **14**(11):543-546

[39] Emadian SR, Ahmadi-Shokouh J. Very small dual band-notched rectangular slot antenna with enhanced impedance bandwidth. IEEE Transactions on Antennas and Propagation. 2015;**63**(10):4529-4534

[40] Koohestani M, Azadi-Tinat N, Skrivervik AK. Compact slit-loaded ACS-fed monopole antenna for Bluetooth and UWB systems with WLAN band-stop capability. IEEE Access. 2023;**20**(11):7540-7550

[41] Emadian SR, Ahmadi-Shokouh J. Time-domain analysis of compact dual

band-notched slot antenna in indoor environments. International Journal of Microwave and Wireless Technologies. 2018;**10**(10):1186-1195

[42] Emadian R, Mirmozafari M, Ghobadi C, Nourinia J. Bandwidth enhancement of dual band-notched circle-like slot antenna. Electronics Letters. 2012;**48**(7):356-357

[43] Emadian SR, Ghobadi C, Nourinia J. A novel compact dual band-notched slot antenna for ultrawideband applications. Microwave and Optical Technology Letters. 2012;**54**(6):1365-1368

[44] Emadian SR, Ahmadi-Shokouh J, Ghobadi C, Nourinia J. Study on frequency and impulse response of novel triple band notched UWB antenna in indoor environments. AEU-International Journal of Electronics and Communications. 2018;**1**(96):93-106

[45] Dumoulin A. A Study of Integrated UWB Antennas Optimised for Time Domain Performance. Dublin, Ireland: Dublin Institute of Technology (now part of Technological University Dublin);

[46] Quintero G, Zurcher JF, Skrivervik AK. System fidelity factor: A new method for comparing UWB antennas. IEEE Transactions on Antennas and Propagation. 2011;**59**(7): 2502-2512

[47] Emadian SR, Ahmadi-Shokouh J. Study on frequency and time domain properties of novel triple band notched UWB antenna in indoor propagation channel. International Journal of RF and Microwave Computer-Aided Engineering. 2018;**28**(9):e21428

[48] Garces E, Rodriguez-Pardo C, Casas D, Lopez-Moreno J. A survey on intrinsic images: Delving deep into lambert and beyond. International Journal of Computer Vision. 2022; **130**(3):836-868

[49] Saleh AA, Valenzuela R. A statistical model for indoor multipath propagation. IEEE Journal on Selected Areas in Communications. 1987;**5**(2):128-137

[50] Zhao Y, Hao Y, Parini C. FDTD characterization of UWB indoor radio channel including frequency dependent antenna directivities. IEEE Antennas and Wireless Propagation Letters. 2007; **30**(6):191-194

[51] Tiberi G, Bertini S, Malik WQ, Monorchio A, Edwards DJ, Manara G. Analysis of realistic ultrawideband indoor communication channels by using an efficient ray-tracing based method. IEEE Transactions on Antennas and Propagation. 2009;**57**(3):777-785

[52] Rubio L, Reig J, Fernández H, Rodrigo-Peñarrocha VM. Experimental UWB propagation channel path loss and time-dispersion characterization in a laboratory environment. International Journal of Antennas and Propagation. 2013;**2013**(1):350167

[53] Emadian SR, Ahmadi-Shokouh J, Raisi A, Ghobadi C, Nourinia J. Time-domain properties of dual-band-notched circle-like slot antenna in indoor environments. Electronics Letters. 2019; **55**(1):10-12

[54] Emadian SR, Poustindouz S. Study on extremely compact triple band-rejected antenna in indoor propagation channel for IR UWB and biomedical applications. IEEE Transactions on Antennas and Propagation. 2024. DOI: 10.1109/TAP.2024.3489872

Chapter 7

Ultra-Wideband Positioning Technology: Exploring TWR, TDoA, and PDoA Methods for Precise Ranging and Positioning

Jia Li and Fengyun Zhang

Abstract

Ultra-wideband (UWB) technology, recognized for its broad bandwidth and low-power consumption, delivers precise and reliable solutions for ranging and positioning in diverse environments. This chapter delves into the time of flight (ToF) technique, which measures the travel time of a radio signal between the transmitter and receiver. To improve accuracy despite synchronization issues, different two-way ranging (TWR) methods are discussed. For precise indoor positioning, the chapter examines methods such as the geometric analysis method, time difference of arrival (TDoA), and phase difference of arrival (PDoA). The geometric analysis method uses at least three receivers to determine the position, while TDoA requires synchronized receivers to analyze time differences. PDoA bases positioning on phase differences received by antennas. The discussion includes advanced clock synchronization strategies to enhance TDoA accuracy. These methodologies collectively bolster the effectiveness of UWB positioning systems, making them suitable for applications in industrial automation and emergency response.

Keywords: ultra-wideband (UWB), two-way ranging (TWR), time difference of arrival (TDoA), phase difference of arrival (PDoA), clock synchronization (CS)

1. Introduction

Ultra-wideband (UWB) is a wireless communication technology known for its ability to provide high-speed data transmission over short distances. UWB technology achieves high-precision positioning services and low-power communication by transmitting information over a wide frequency band. In recent years, it has been widely used in the Internet of Things (IoT), indoor positioning, and smart devices [1, 2]. The spectrum utilization mode and working principle of UWB render it highly suitable for high-precision positioning and short-range, high-speed data transmission. Due to its dispersion of signals across a wide frequency bandwidth, UWB exhibits minimal interference with other radio waves in the same frequency band, demonstrating excellent anti-interference performance. This characteristic serves as the foundation for the widespread application of UWB technology. In terms of positioning accuracy, UWB technology achieves exceptional precision through its use of high bandwidth and narrow pulse width for time measurement. This capability proves particularly advantageous in indoor environments where traditional GPS signal delivery faces significant limitations and errors. Furthermore, UWB's strong identification ability effectively enhances its positioning accuracy and reliability in complex environments.

According to records, the earliest use of UWB technology can be traced back to the development of new radar system projects in the 1960s, as noted in Ref. [3]. The concept of "ultra-wideband" was first proposed by the US Department of Defense in 1990 as part of a project involving short-pulse microwaves for data transmission and object detection in radar equipment [4]. The first UWB signal was generated by Heinrich Rudolf Hertz during his experiments in 1887 [5]. UWB technology is capable of distributing narrow pulses over a wide spectrum through the allocation of a large bandwidth while maintaining low signal energy, as noted by Zwirello [6]. However, due to a lack of relevant regulations, suitable algorithms, and matching measurement instruments, the large bandwidth property of UWB technology did not receive sufficient attention for quite some time. It was not until 2002 that the Federal Communications Commission (FCC) defined UWB systems as those with an absolute bandwidth greater than 500 MHz and (f_c) greater than 2.5 GHz, or (f_c) less than 2.5 GHz and (B_f) greater than 0.2, as noted in [7]. (B_f) can be defined as:

$$B_f = \frac{B}{f_c}, \tag{1}$$

where f_c is the frequency at which the system has the maximum power density, B denotes the bandwidth of the UWB system. UWB is defined in terms of high and low frequencies as follows:

$$B = (f_H - f_L), \tag{2-1}$$

$$f_c = \frac{f_H + f_L}{2}, \tag{2-2}$$

Combining formulas (1–3) gives the following:

$$B_f = \frac{2(f_H - f_L)}{f_H + f_L}. \tag{3}$$

As a widely adopted wireless communication technology, UWB standardization organizations and institutions play a crucial role in driving the advancement of this technology. Apart from IEEE SA, notable UWB standardization bodies include fine ranging (FiRa), the UWB Alliance, the Car Connectivity Consortium (CCC), and The open locating standard (OMLOX). Furthermore, the FCC has established specific standards for protocol and frame format after normalizing the definition of UWB signals, which can be found in Ref. [8]. In 2007, the European Electronic Communications Committee (ECC) standardized the utilization of UWB technology, as noted by Breed [9], thereby fostering its advancement. Alongside defining bandwidth, stringent regulations on the transport layer were also established. **Figure 1** illustrates a comparative analysis of effective isotropic radiated power (EIRP) regulatory measures for UWB implemented by four governing bodies (FCC, ETSI, Japan, and China). UWB radios are

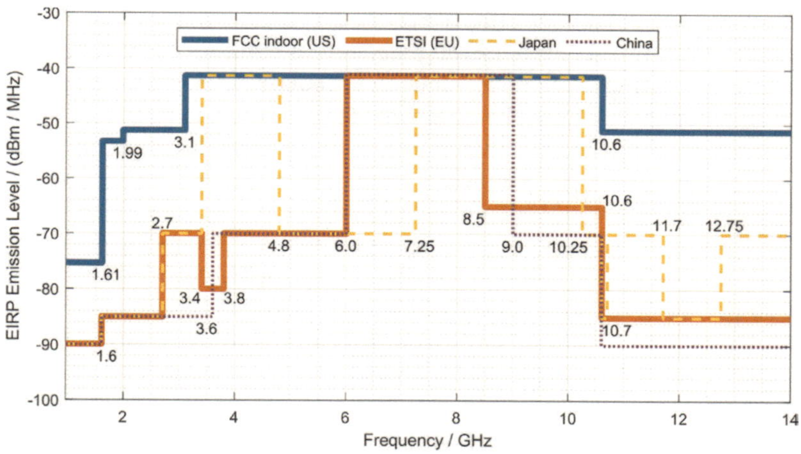

Figure 1.
Specification of EIRP for UWB spectrum in USA, EU, Japan, and China [10].

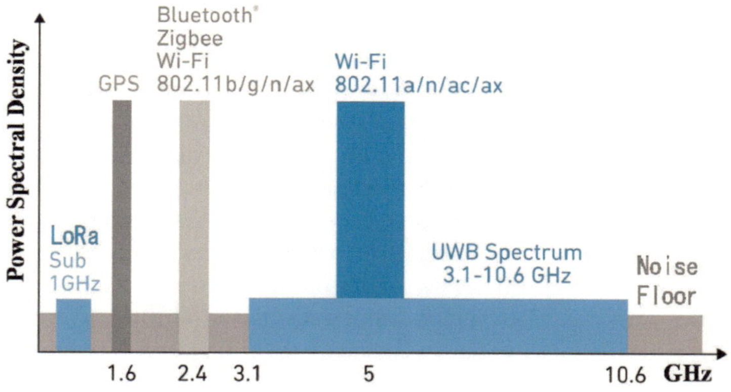

Figure 2.
Comparison of UWB with other radio communication technologies [11].

alternatively referred to as "pulse," "short pulse," "non-sinusoidal," "carrier-free," "time-domain," "ultra-wideband," "fast frequency chirp," and "monopulse" radios.

The pulsed radio communication systems and pulsed radar both utilize extremely short pulses to transmit signals in the UWB spectrum. By employing pulse position modulation (PPM) for data modulation, this communication technique is also categorized as a form of pulse modulation. Due to the low-power spectral density of UWB signals, which is comparable to background noise levels, their interception and detection pose significant challenges. However, this characteristic minimizes interference with existing narrowband radio systems. **Figure 2** illustrates the disparity in frequency band and power spectral density between UWB and other radio systems.

Antenna systems play a crucial role in radio signal-based localization methods; therefore, some recent advancements in antenna technology will be discussed. Radio-over-Fiber (RoF) technology has been explored in Ref. [12] to enhance multichannel wireless communications using modulation instability (MI). Operational flexibility and significant MI gains have been demonstrated, addressing beam squint challenges. The integration of MI holds the potential for improved performance with lower costs and space requirements. Phased array antenna (PAA) performance is enhanced using a RoF system and MI, achieving gains of 5.2 dB, 7.4 dB, and 7.8 dB in the n78, n79, and n46 bands. The design features elliptical patches and employs the true time delay (TTD) technique for improved beam control. Frequency combs are utilized to apply MI gain to all antenna elements [13]. A novel MIMO antenna array with metamaterial isolation surfaces is proposed by Hussein et al. [14], and its directivity is electronically reconfigured using PIN diode switches. Solar panels integrated into the array reduce dependence on traditional power sources, supporting energy efficiency. An AI-enabled equalizer enhances communication reliability by mitigating channel capacity degradation. A reconfigurable MIMO antenna array for 5G portable devices is presented in Ref. [15], utilizing an EBG structure for improved isolation and a Minkowski fractal for enhanced bandwidth and directivity. Impedance matching is achieved with L-shaped reflectors and LDR switches. A compact end-fire antenna array for mmWave 5G mobile terminals is proposed by Zidour et al. [16], achieving wideband operation from 24 to 34 GHz with peak gains of 7.56–8.14 dBi, and meeting 3GPP standards for coverage. A 4-element MIMO antenna system for sub-6 GHz applications is presented in Ref. [17], achieving over 20 dB isolation and ECC less than 0.01 by rotating etched areas and radiating elements, suitable for handheld devices. A feasibility study on a terahertz on-chip antenna array is presented in Ref. [18], demonstrating enhanced impedance bandwidth, gain, and radiation efficiency using metasurface and AMC technologies, with an average gain of 20.36 dBi and 37.5% efficiency in the 0.3–0.314 THz range. A modeled THz receiver using this antenna achieved a gain of 25 dB when matched to a low-noise amplifier. An innovative virtual antenna array (VAA) is proposed by Alibakhshikenari et al. [19], consuming approximately 3 dB less energy per bit than conventional MIMO arrays, thereby potentially reducing the carbon footprint of wireless systems. The VAA, utilizing substrate-integrated waveguide (SIW) and metasurface technologies, achieves a measured gain of 25 dBi and 90% radiation efficiency, with isolation better than 27 dB across the sub-5 GHz frequency range. A high-performance antenna-on-chip (AoC) is implemented on a gallium arsenide (GaAs) wafer using substrate-integrated waveguide (SIW) and metasurface (MTS) technologies in Ref. [20], achieving a bandwidth of 25 GHz from 0.445 to 0.470 THz with a return loss of -27 dB. The antenna demonstrates a gain of 4.6 dBi and radiation efficiency of 74%, suitable for terahertz applications. A metamaterial-inspired antenna array for millimeter-wave applications is proposed by Althuwayb et al. [21], designed for nonmechanical beam-steering in 5G and 6G communications. The 3×3 matrix array achieves a peak gain of 13.6 dBi and 89.54% efficiency while improving impedance bandwidth and radiation characteristics.

UWB-based indoor positioning schemes include time of flight (ToF) [22], two-way ranging (TWR) [23], time of arrival (ToA) [24], time difference of arrival (TDoA) [25], angle of arrival (AoA) [26], and phase difference of arrival (PDoA) [27]. Among these schemes, positioning methods based on TWR or ToF are constrained by the bidirectional communication mechanism between anchors and tags, requiring more communication resources to achieve centimeter-level positioning accuracy. AoA and PDoA positioning modes are limited by UWB antenna technical specifications, which impose higher requirements for line-of-sight transmission. Therefore, TDoA-based positioning, supporting higher tag densities and suitable for cascaded anchor deployment, has garnered more attention from researchers.

UWB technology offers several key benefits that enhance its performance. It provides high channel capacity for quick data transmission at maximum rates due to its wide bandwidth and ensures exceptional time resolution and centimeter-level ranging accuracy from its short pulse duration, which is critical for accurately determining signal arrival times. Operating at low transmission power (less than −41.3 dBm), UWB demonstrates strong anti-multipath interference capabilities while having minimal impact on other wireless systems. Compared to narrowband systems, UWB's short pulse duration improves robustness against multipath effects and propagation fading, facilitating easier pulse detection. Additionally, the low-frequency components of UWB signals allow for better penetration of obstacles like walls, although some signal attenuation or dispersion may occur.

2. UWB-based ranging methods

In this section, we will describe the most commonly used two-way ranging (TWR) methods [28]. In wireless communication, the distance between two transceivers can be determined using the formula $d = T_{tof} \cdot C$, where d represents the distance between two wireless transceivers, T_{tof} is the duration of signal propagation time between transceivers in free space, and c is the speed of light (2.9979·108 m/s). In general, ToF (T_{tof}) refers to the time it takes for a signal to travel between transceivers in a medium such as air or vacuum. In practice, T_{tof} can be calculated by finding the difference between two timestamps extracted from the transceiver device: T_{se} when the signal leaves the transmitter and T_{re} when it arrives at the receiver (as shown in **Figure 3**). It is important to note that both transceivers need to have synchronized clocks for accurate extraction of T_{tof}. When synchronization cannot be achieved or is not available, TWR becomes a preferred choice for measuring distance in wireless transceivers. The TWR technique utilizes time periods or intervals like signal round-trip time (T_{ro}) and reply time (T_{re}) to calculate distances between two transceivers instead of direct timestamps as depicted in **Figure 3**. This approach accounts for clock drifts that occur due to imperfections in clock oscillators even when they are perfectly tuned initially. Eliminating these small errors caused by clock drifts becomes crucial, especially when centimeter-level accuracy is required.

2.1 Single-sided two-way ranging

The single-sided two-way ranging (SS-TWR) method is considered to be the fundamental form of the TWR technique [29], as illustrated in **Figure 4**. The operation of SS-TWR involves measuring the round-trip time of a signal between device 1 and device 2 for determining their reply time.

Hence, the signal round-trip time of SS-TWR can be expressed as follows:

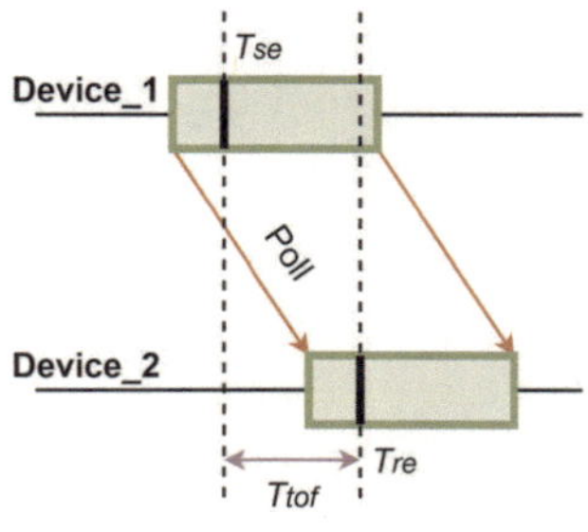

Figure 3.
Illustration of ToF measurements.

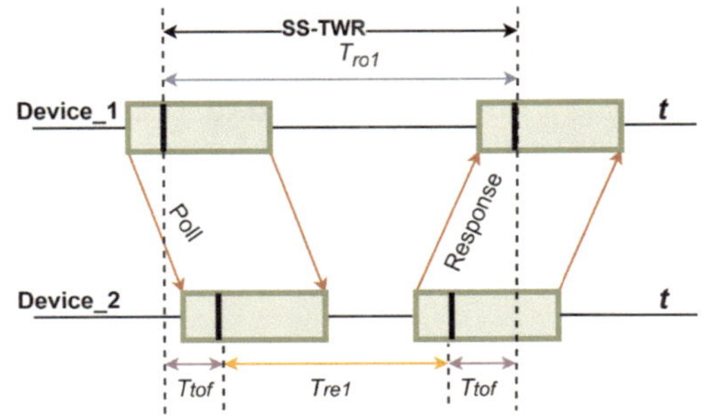

Figure 4.
Illustration of the single-sided TWR method.

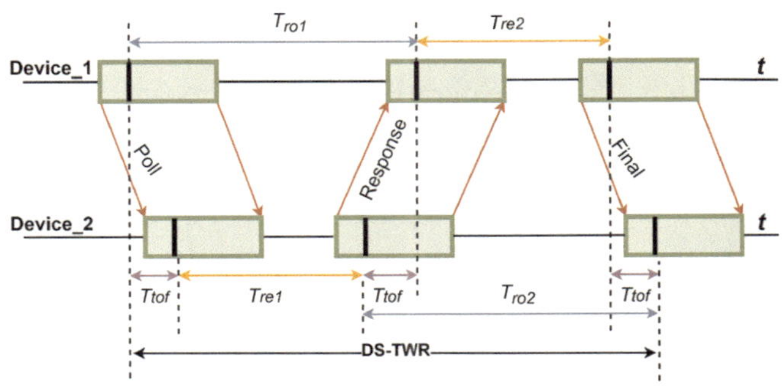

Figure 5.
Illustration of the double-sided TWR method.

$$T_{ro1} = 2T_{tof} + T_{re1} \tag{4}$$

where T_{ro1} is the round-trip time of the signal measured on device 1, and T_{re1} is the response time measured on device 2. So, the ToF for the SS-TWR method can be expressed as:

$$T_{tof} = \frac{1}{2}(T_{ro1} - T_{re1}) \tag{5}$$

2.2 Double-sided two-way ranging

In practical applications, the double-sided two-way ranging (DS-TWR) method is commonly utilized to compensate for ranging errors caused by clock drift in the SS-TWR method [30], as illustrated in **Figure 5**. The DS-TWR algorithm extends the SS-TWR algorithm by incorporating two round-trip time measurements to estimate ToF.

The two round-trip times of DS-TWR can be mathematically expressed as:

$$T_{ro1} = 2T_{tof} + T_{re1} \tag{6}$$

$$T_{ro2} = 2T_{tof} + T_{re2} \tag{7}$$

where the round-trip time of the signal measured on device 1 and device 2 is denoted as T_{ro1} and T_{ro2}, respectively. Similarly, the response times measured on device 2 and device 1 are denoted as T_{re1} and T_{re2}. By combining Eqs. (6) and (7), we could get:

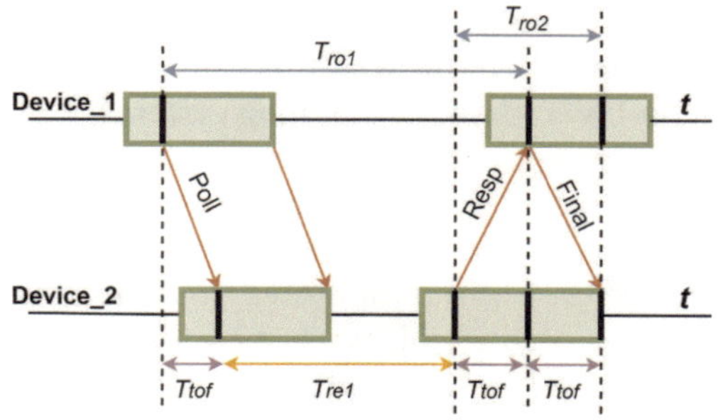

Figure 6.
Illustration of the asymmetric double-sided TWR method.

$$T_{ro1} \cdot T_{ro2} = \left(2T_{tof} + T_{re2}\right) \cdot \left(2T_{tof} + T_{re1}\right) \tag{8}$$

The T_{tof} for DS-TWR method can be obtained by simplifying Eq. (8) as follows:

$$T_{tof} = \frac{T_{ro1} \cdot T_{ro2} - T_{re1} \cdot T_{re2}}{T_{ro1} + T_{re1} + T_{ro2} + T_{re2}} \tag{9}$$

2.3 Asymmetric double-sided two-way ranging

The drawback of the DS-TWR method lies in its requirement for a minimum of three ranging messages, with two response times included in the distance calculation. Typically, the delay in response time (usually measured in microseconds) is relatively longer compared to the delay in ToF (usually measured in nanoseconds). The asymmetric double-sided two-way ranging (ADS-TWR) method has been designed to enhance the ranging efficiency of DS-TWR without compromising system performance [31], as depicted in **Figure 6**.

Consequently, the round-trip time of ADS-TWR can be expressed as follows:

$$T_{ro1} = 2T_{tof} + T_{re1} \tag{10}$$

$$T_{ro2} = 2T_{tof} \tag{11}$$

Therefore, T_{tof} of ADS-TWR can be obtained by using Eqs. (10) and (11) as follows:

$$T_{tof} = \frac{1}{4}(T_{ro1} + T_{ro2} - T_{re1}) \tag{12}$$

3. UWB-based positioning methods

The UWB-based ranging methods, namely SS-TWR, DS-TWR, and ADS-TWR, are detailed in *Section 2*. Additionally, this subsection provides an introduction to the commonly employed positioning methods in UWB systems.

3.1 Positioning method based on geometric analysis

The trilateration algorithm is the fundamental and intuitive positioning method in UWB positioning systems. This approach is derived from traditional distance measurement techniques and employs two-dimensional analytical geometry, as depicted in **Figure 7a**, to determine the position accurately. The widely adopted multilateration method in UWB-based positioning systems is illustrated in **Figure 7b**. The geometry-based positioning methods shown in **Figure 7** include triangulation and multilateration, which primarily derive from analyzing the positions of the anchors and the tag, leading to positioning methods that follow mathematical theorems. In **Figure 7a**, it is assumed that the positioning system includes three anchors arranged in a triangular configuration, located at points A1, A2, and A3, with the tag positioned within the triangular area. Based on the principle of circle intersection, the position of the tag can be estimated. Similarly, in **Figure 7b**, with four or more anchors arranged in a polygonal configuration and the tag located within the polygon, the position of the tag can be estimated based on the principle of multiple intersecting circles in a two-dimensional plane.

For the purpose of demonstrating true-range multilateration and triangulation methods, as noted in Refs. [32, 33], let us assume that the tag in 2D is positioned at coordinates $P = \left[x_t, y_t\right]^T$, while the anchors are located at positions $A_i = \left[x_i, y_i\right]^T$, where $i = 1, 2, \dots, N$ represents the known anchors. Consequently, the general spherical equation for true range can be expressed by utilizing measured distances between anchors and tags.

$$d_i^2 = (x_i - x_t)^2 + (y_i - y_t)^2 \tag{13}$$

where d_i is the measured distance (range) between the anchor A_i and tag P.

The formula derivation of multilateration for four anchors will be demonstrated in this section. It is important to note that the algorithm is not limited to only four anchors but can be extended to accommodate any number of available anchors in the system. The measured distances between the four anchors and a tag using formula (13) can be expressed as:

$$d_1^2 = (x_1 - x_t)^2 + (y_1 - y_t)^2 \tag{14}$$

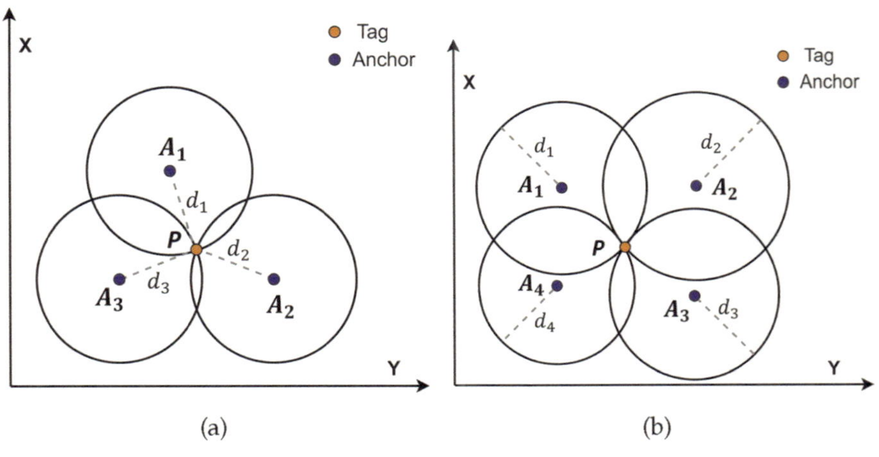

Figure 7.
Diagram of the positioning methods based on geometric analysis. (a) Triangulation method. (b) Multilateration method.

$$d_2^2 = (x_2 - x_t)^2 + (y_2 - y_t)^2 \tag{15}$$

$$d_3^2 = (x_3 - x_t)^2 + (y_3 - y_t)^2 \tag{16}$$

$$d_4^2 = (x_4 - x_t)^2 + (y_4 - y_t)^2 \tag{17}$$

Subtracting (14) from (15) and simplifying yields:

$$(x_2 - x_1) \cdot x_t + (y_2 - y_1) \cdot y_t = \frac{1}{2}\left[d_1^2 - d_2^2 + (x_2^2 + y_2^2) - (x_1^2 + y_1^2)\right] \tag{18}$$

Similarly, by subtracting (14) from (16) and (17), we can achieve the system of equations in a matrix notation as:

$$Ax = b \tag{19}$$

where,

$$A = \begin{bmatrix} x_2 - x1 & y_2 - y_1 \\ x_3 - x1 & y_3 - y_1 \\ x_4 - x1 & y_4 - y_1 \end{bmatrix}, x = \begin{bmatrix} x_t \\ y_t \end{bmatrix}$$

$$b = \frac{1}{2}\begin{bmatrix} d_1^2 - d_2^2 + (x_2^2 + y_2^2) - (x_1^2 + y_1^2) \\ d_1^2 - d_3^2 + (x_3^2 + y_3^2) - (x_1^2 + y_1^2) \\ d_1^2 - d_4^2 + (x_4^2 + y_4^2) - (x_1^2 + y_1^2) \end{bmatrix}$$

It is worth noting that the elements of matrix A in Eq. (19) are derived from the coordinates of anchor points. All elements in vector b are also known; thus, vector x is the only unknown variable in (19), representing the position of the label. In fact, if matrix A^TA is invertible, we can treat Eq. (19) as a least squares problem. Theoretically, matrix A^TA becomes singular or irreversible only when all anchor coordinates lie on a collinear line in 2D or coplanar surface in 3D. This approach can be easily verified by determining the column rank of matrix A, which represents its independent column vectors. Therefore, when satisfying the requirement for column rank verification, the estimated solution to Eq. (19) is as follows:

$$x = \left(A^TA\right)^{-1}A^Tb \tag{20}$$

The ordinary least squares method defined by Eq. (20) is typically regarded as an overdetermined system when the number of anchors in the system exceeds three in two dimensions and four in three dimensions, where x represents the estimated coordinate of the tag.

3.2 Positioning method based on time difference of arrival

Time difference of arrival (TDoA) positioning is a kind of range difference location (RDL), which uses the time of arrival (ToA) of the signal sent by the tag to the anchor to obtain the time difference between the tag and any two anchors, and multiplies the signal propagation speed to obtain the distance difference between the tag and any two anchors. Using this distance difference, the position of the target can be calculated, as noted in [34]. Because the position of the tag is above the hyperbola when using this method, it is also called hyperbola positioning.

TDoA-based positioning is a common strategy in UWB systems, and the TDoA model is shown in **Figure 8**. In TDoA-based positioning system, assume four anchors (A_1, A_2, A_3, and A_4) are arranged in a rectangular configuration. For any two anchors, they can be considered as the foci of a hyperbola. According to the geometric definition of a hyperbola, the difference in distances from any point on the hyperbola to the two foci is constant. Therefore, by using the other anchors to establish multiple hyperbolic equations, the intersection of these hyperbolas within the rectangular area will indicate the position of the tag P. According to the TDoA model, using the coordinates of the base station information A_i, or $(x_i, y_i)_{i=1}^{n}$, tags can be calculated the relative position of $P(x,y)$. In this section, a two-dimensional system is used to illustrate the TDoA model, and the algorithm to solve the TDoA problem is introduced. In the actual positioning scene, the model and algorithm can be extended to three-dimensional scene according to the actual situation, and the mentioned model and algorithm are still applicable.

We could get the timestamp τ_i when the anchor receives the signal, we could further assume that the measurement of $\hat{\tau}_i$ satisfies $\hat{\tau}_i \sim N(\tau_i, \delta_i^2)$. We set $d_i = \|P - A_i\|$ are the distances of anchors from the target. The core formula of TDoA is

$$\hat{d}_{ij} \triangleq \hat{d}_i - \hat{d}_j = c(\hat{\tau}_i - \hat{\tau}_j) := c\hat{\tau}_{ij}, \quad \forall i,j = 1, 2, \dots, n, \tag{21}$$

where c is the velocity of the electromagnetic wave. In this way, we represent range differences (RD) by the TDoA measurements, the distance $\hat{d}_{ij}$ of here satisfy $\sum_{i,j} \hat{d}_{ij} \equiv 0$.

TDoA algorithms aim to find P, such that

$$d_{ij}(P) = \hat{d}_{ij}, \qquad \forall\ i,j = 1, 2, \dots, n, \tag{22}$$

where $d_{ij}(P) = d_i(P) - d_j(P) = \|P - A_i\| - \|P - A_j\|$.

All above, the inputs of TDoA positioning algorithms are the anchors' coordinates A_i and the measured TDoA $\hat{\tau}_{ij}$, and the output P is the tag's coordinates. In practice,

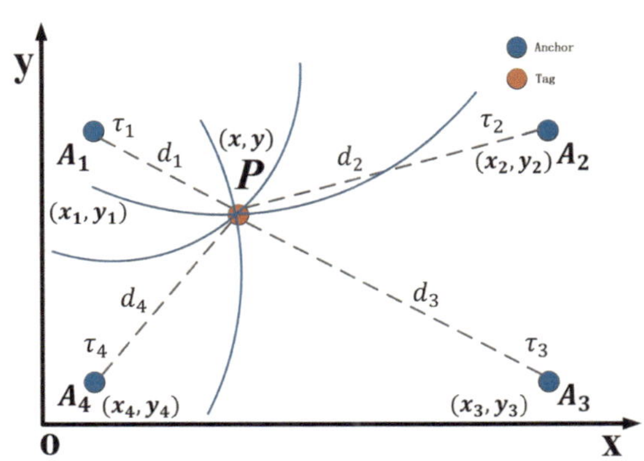

Figure 8.
Diagram of TDoA model based on hyperbola.

due to the measurement error, the equal sign is usually not satisfied in the previous formula. So the mathematical model for solving the nonlinear estimator is as follows:

$$\underset{P}{argmin}\frac{1}{2}\|r\|^2, \quad r = \left[d_{ij} - \hat{d}_{ij}\right]_{ij} \tag{23}$$

When the iterative algorithm is adopted, the Jacobi matrix **J** of the residual term r has an important effect on the error and stability of the algorithm, namely

$$\text{Error of } P = \sigma^{-1}(\mathbf{J}) \cdot \text{Error of} f. \tag{24}$$

In general, we want the minimum eigenvalue of **J** to have a lower bound, so ensuring it is not rank deficient. Notice the form $\mathbf{J}(p)$,

$$\mathbf{J}(P) = \left[\frac{x - x_i}{d_i} - \frac{x - x_j}{d_j}\ \frac{y - y_i}{d_i} - \frac{y - y_j}{d_j}\right]_{1 \le i < j \le n} \tag{25}$$

So we could get that,

$$\sigma_{\min}(\mathbf{J}(p)) = k(P(p)) \tag{26}$$

Where $P(p) = [p - A_1, p - A_2, \ldots, p - A_n]^T$. Note that, $\kappa(P(p))$ is the degree of dispersion with the center p and the coordinates $A_1, A_2, \ldots, A_n$. The TDoA localization rationale and the mathematical model for solving the nonlinear estimation problem are introduced in this section. Due to the presence of measurement errors, nonlinear filtering estimation methods, such as the extended Kalman filter (EKF), are commonly employed to calculate the tag's position. The detailed procedure of utilizing the EKF for tag position estimation in time difference of arrival (TDoA) positioning method is comprehensively presented in [35].

3.3 Positioning method based on the phase difference of arrival

The angle of arrival (AoA) measurement method, based on the phase difference of arrival (PDoA) scheme, is a technique that utilizes the phase discrepancy between signals received by each element in an antenna array to determine the direction of propagation for incident RF waves, as noted by Ge and Shen [36]. The localization method employing PDoA computes the tag's location by utilizing the measured PDoA at an anchor with a known position, along with factors such as the distance between antenna arrays and ToF from the tag to the anchor. **Figure 9a** illustrates the geometry of how AoA is calculated from PDoA, and **Figure 9b** illustrates how to calculate the position of the tag based on the signal transmission between the tag and the PDoA anchor.

The calculation of AoA can be derived from the rationale presented in **Figure 9a**, as demonstrated by the following formulas.

$$p = d\sin(\theta) \tag{27}$$

$$\phi = 2\pi\frac{p}{\lambda} = \frac{fp}{c} \tag{28}$$

$$p = \frac{\phi\lambda}{2\pi d} \tag{29}$$

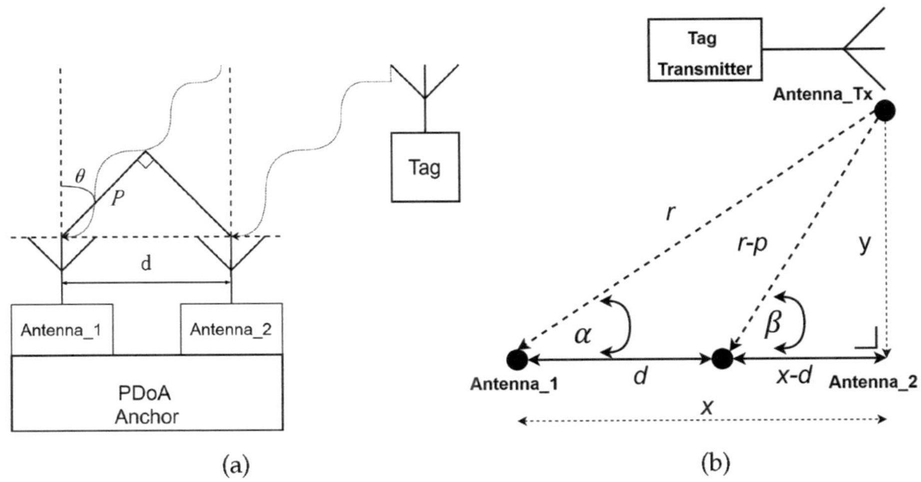

Figure 9.
Diagram of PDoA-based AoA measurement and positioning. (a) AoA measurement method. (b) Positioning method.

where $d = \lambda/2$, combining Eqs. (7) and (8),

$$\theta = \arcsin\left(\frac{\phi\lambda}{2\pi d}\right) = \arcsin\left(\frac{\phi}{\pi}\right) \tag{30}$$

According to trigonometric functions and the law of cosines in **Figure 9b**, we can derive the following equations:

$$cos\,(\alpha) = \frac{r^2 + d^2 - (r-p)^2}{2rd} \tag{31}$$

$$\frac{x}{r} = \frac{r^2 + d^2 - (r)^2 + 2rp - p^2}{2rd} \tag{32}$$

$$x = \frac{d^2 + 2rp - p^2}{2d} = \left(r - \frac{p}{2}\right)\frac{p}{d} + \frac{d}{2} \tag{33}$$

Because of,

$$x^2 + y^2 = \left(\frac{d^2 + 2rp - p^2}{2d}\right)^2 + y^2 = r^2 \tag{34}$$

So, we could get

$$y = \pm\frac{\sqrt{\left(1 - \left(\frac{p}{d}\right)^2\right)\left(4r^2 - 4rp + p^2 - d^2\right)}}{2} \tag{35}$$

In the actual positioning scenario, the distance d between antennas is smaller than the distance r between tags and anchors. Therefore, d^2 is much smaller than r^2, so it can be ignored. In this way, the simplified y can be obtained. The maximum measured error is 0.22 mm for a 6.5 GHz carrier and a receiver antenna separation of $\lambda/2$ by using this way.

$$y \approx \pm \frac{\sqrt{\left(1-\left(\frac{p}{d}\right)^2\right)\left(4r^2-4rp+p^2\right)}}{2} \tag{36}$$

$$y \approx \pm \left(r-\frac{p}{2}\right)\sqrt{1-\left(\frac{p}{d}\right)^2} \tag{37}$$

If $|x| < r$ then $y = \sqrt{r^2 - x^2}$, else $y = 0$.

4. UWB-based clock synchronization

In UWB-based positioning systems, each UWB device is equipped with an independent high-resolution timer. However, the oscillation frequency of these timers may drift over time, making it necessary to eliminate the drift effects through clock calibration. It is crucial to synchronize all the anchors, as precise timestamp information is essential for accurate location estimation. The techniques for clock synchronization encompass wired and wireless approaches. In this section, we employ wireless clock synchronization as an illustrative example to elucidate the principles and methodologies of achieving precise time alignment, as noted in Ref. [37]. The synchronization process must address three main issues: clock offset synchronization, clock drift calibration, and frequency drift elimination. The clock model can be established to relate the local time of an anchor to the reference anchor, which is described by formula [38]:

$$t_i = \theta_k + \delta_k t + \frac{1}{2}\phi_k t^2 + \omega_k(t) \tag{38}$$

where the parameter t_i denotes the local time of anchor i, k indicates the moment when the anchor receives the synchronization message. θ_k is the clock offset, δ_k denotes the clock drift, t is the reference time, $\omega_k(t)$ is the random frequency noise, and ϕ_k denotes the frequency drift.

The clock model is demonstrated in **Figure 10a**. The slope of the ideal reference clock is 1, and the clock offset is 0. In a real clock system, the slope of the clock drift may be greater than one, possibly between zero and one.

A minimum structure of TDoA-based wireless clock synchronization is illustrated in **Figure 10b**, which consists of four anchors (MA, SA1, SA2, and SA3) and one tag (T1). *MA* is a master anchor and *SA2–SA4* are slave anchors. There are five timelines, with the top one representing the tag and the bottom four belonging to the individual anchors. The dark-dashed lines represent the positioning packet (*Blinks*) sent by the tag with a sending period of 1 s, while the light-dashed lines represent the clock calibration packets (*CCPs*) sent by the master anchor at an interval of 150 ms. Once the tag sends the *Blink* or the master anchor sends the *CCP*, the slave anchors in the corresponding deployment area will receive the *Blink* or *CCP* and record the timestamp in the anchor's clock system, respectively. For the master anchor, in addition to receiving the Blink from the tag similar to a slave anchor, it is also necessary to periodically send the *CCP* and record the sending timestamp. After receiving the

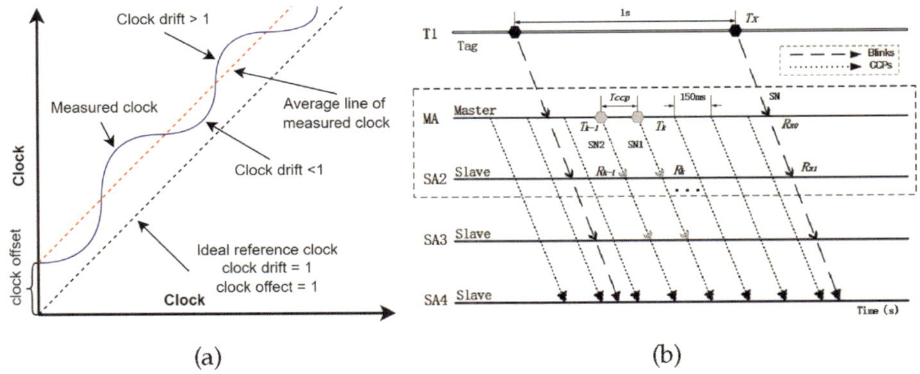

Figure 10.
Diagram of clock synchronization method. (a) Clock model (b) Structure of wireless clock synchronization.

Blink, *MA* and *SA* will record the received timestamp (R_{x0}, R_{x1}) and serial number (*SeqNum*) of this *Blink*. Meanwhile, the *MA* will record the sending timestamp (T_k, T_{k-1}) and the corresponding serial number (SeqNum1, SeqNum2) when it sends the *CCP*. When the *CCP* reaches the *SA* node, the timestamp (R_k, R_{k-1}) and the corresponding serial number will also be recorded. Based on the reference time of the master anchor, the raw *ToA* of the slave anchor can be corrected with the linear clock offset θ_k, so the relationship of the corrected ToA_i and the raw *ToA* can be expressed as $ToA_i = ToA + \theta_k$.

A three-state KF is utilized to eliminate clock offset, clock drift, and frequency drift during the synchronization period, which is proposed in [?]. So the state vector X_k can be represented as, $X_k = [ToA_i, \delta_k, \phi_k]^T$. Where $\theta_k = T_k - R_k$ and $\delta_k = \frac{T_k - T_{k-1}}{R_k - R_{k-1}}$. ϕ_k represents the minimum resolution of the crystal oscillator, such as 0.1 ppm.

$$X_k = \boldsymbol{F} X_{k-1} + W_{k-1} \tag{39}$$

where, the state transition matrix $\boldsymbol{F}$ can be expressed as:

$$\boldsymbol{F} = \begin{bmatrix} 1 & T_{ccp} & \frac{T_{ccp}^2}{2} \\ 0 & 1 & T_{ccp} \\ 0 & 0 & 1 \end{bmatrix} \tag{40}$$

where T_{ccp} denotes the interval for clock synchronization, W_k follows Gaussian distribution: $W_k \sim \mathcal{N}(0, Q)$. The noise vector obeys the zero mean Gaussian distribution, which can be expressed as a 3×3 matrix $\boldsymbol{Q}$. Based on the three-state KF algorithm, the estimating process of clock synchronization can be expressed as

$$\hat{X}^-{}_k = \boldsymbol{F} \hat{X}_{k-1} + W_{k-1} \tag{41}$$

where $\hat{X}^-{}_k$ is a prior estimated state vector, and $\hat{X}_k$ expresses a posteriori estimate. The observation value Z_k is the linear combination of $\hat{X}^-$ and $\boldsymbol{H}$, which can be defined as:

$$Z_k = \boldsymbol{H} \hat{X}_k^- + V_k \tag{42}$$

where, the observation matrix is $\boldsymbol{H} = [1\ \ 0\ \ 0]$, V_k is the observation noise, and V_k follows $V_k \sim \mathcal{N}(0, R)$. According to the variance of the measurements taken by the UWB manufacturer, $\sigma^2 = 1.5 \cdot 10^{-20} s^2$, s is the time unit, R is the covariance of the measurement, so $R = \sigma^2$.

The necessary steps of correction and prediction of the Kalman filter can be estimated below equations.

$$\boldsymbol{P^-}_{\boldsymbol{k}} = \boldsymbol{F}\boldsymbol{P}_{\boldsymbol{k-1}}\boldsymbol{F}^T + \boldsymbol{Q} \tag{43}$$

where, Eq. (6) indicates the prediction phase of the Kalman process, $\boldsymbol{P_k}$ is the estimated covariance matrix, and $\boldsymbol{P_k^-}$ is a prior estimate of $\boldsymbol{P_k}$.

$$K_k = \boldsymbol{P^-}_{\boldsymbol{k}}\boldsymbol{H}^T\left(\boldsymbol{H}\boldsymbol{P^-}_{\boldsymbol{k}}\boldsymbol{H}^T + R\right)^{-1} \tag{44}$$

where Eq. (7) represents the calculation process of Kalman gain, and K_k is the Kalman gain.

$$\hat{X}_k = \hat{X}^-{}_k + K_k\left(Z_k - \boldsymbol{H}\hat{X}^-{}_k\right) \tag{45}$$

$$\boldsymbol{P_k} = (I - K_k\boldsymbol{H})\boldsymbol{P^-}_{\boldsymbol{k}} \tag{46}$$

where Eqs. (8) and (9) represent the update phase of the Kalman process. $\boldsymbol{P_k}$ is the estimated covariance matrix, which can be recursively derived from the initial matrix $\boldsymbol{P_0}$. The covariance matrix $\boldsymbol{P_0}$ is related to the initial state vector.

$$\boldsymbol{P_0} = \begin{bmatrix} \sigma_{ToA}^2 & 0 & 0 \\ 0 & \sigma_\delta^2 & 0 \\ 0 & 0 & \sigma_\phi^2 \end{bmatrix} \tag{47}$$

where σ_{ToA}^2, σ_δ^2, and σ_ϕ^2 represent the initial variances of the state vector components.

5. Different UWB-based positioning approaches

The UWB-based positioning methods can be categorized as either clock-based or clock-free. Clock-free technology includes received signal strength (RSS), arrival angle of arrival (AoA), and phase difference of arrival (PDoA). Two-way ranging (TWR) and time difference of arrival (TDoA) are common clock-based approaches. Based on a model of radio channel propagation, UWB signals are monitored and transformed into range data. However, this strategy lacks precision. AoA employs antenna array to detect the arrived angle signal, which increases the complexity of antenna array design but has the potential to reduce the number of anchor points. The PDoA approach provides a high level of accuracy in scenes with low frequency, but the algorithm required is very complex. Both TWR and TDoA properly record the timing of signal transmission and reception based on the transmission characteristics of electromagnetic waves in the atmosphere. Various approaches based on TWR and TDoA give high precision, but tight clock synchronization is required. The pros and cons of the mentioned approaches are summarized in the **Table 1**.

Approach	Advantages	Disadvantages	Accuracy (Unit: cm)
RSS[a]	Simplicity of implementation	Low accuracy	100
TWR	Low algorithm complexity	Low system capacity	10
TDoA	High system capacity	High system capacity	10
PDoA	Low positioning accuracy with NLoS	Low positioning accuracy with NLoS	10
AoA[b]	Fewer anchors	Small positioning range	10

[a]Positioning based on RSS method is not commonly used in UWB positioning system, so this chapter will not discuss it in detail.[b]Positioning based on AoA method can be regarded as another manifestation of PDoA positioning method, so this chapter will not repeat the discussion.

Table 1.
Advantages and disadvantages of different localization approaches.

6. Conclusions

Ultra-wideband (UWB) technology proves to be an effective solution for precision ranging and localization across various environments due to its broad bandwidth and low-power consumption. This study analyzed multiple UWB-based techniques, revealing their unique advantages and challenges, particularly in wireless clock synchronization. Key ranging methods, including time of flight (ToF) and two-way ranging (TWR), illustrate the importance of accurately determining signal travel times. TWR achieves a distance measurement accuracy with an error of less than 10 cm, while dynamic positioning methods ensure an error margin of less than 30 cm.

The efficacy of UWB for indoor positioning is highlighted through methodologies such as geometric analysis, which uses at least three receivers for precise localization, as well as time difference of arrival (TDoA) and phase difference of arrival (PDoA) techniques. TDoA, leveraging synchronized algorithm, can achieve a remarkable wireless clock synchronization accuracy of up to 0.2 ns. These advancements significantly enhance UWB's positioning effectiveness, making it essential for applications in industrial automation and emergency response. Continued research will focus on improving these technologies for wider deployment and greater environmental resilience, paving the way for smarter, interconnected solutions.

Conflict of interest

The authors declare no conflict of interest.

Abbreviations

UWB	ultra-wideband
ToF	time of flight
TWR	two-way ranging
SS-TWR	single-sided two-way ranging
DS-TWR	double-sided two-way ranging

ADS-TWR	asymmetric double-sided two-way ranging
ToA	time of arrival
TDoA	time difference of arrival
PDoA	phase difference of arrival
AoA	angle of arrival
EKF	extended Kalman filter

Author details

Jia Li[1,2†] and Fengyun Zhang[2*†]

1 Chongqing College of Traditional Chinese Medicine, Chongqing, China

2 Southwest University, Chongqing, China

*Address all correspondence to: zhangfy24@swu.edu.cn

† These authors contributed equally.

References

[1] Zafari F, Gkelias A, Leung KK. A survey of indoor localization systems and technologies. IEEE Communications Surveys & Tutorials. 2019;**21**(3): 2568-2599

[2] Ridolfi M, Kaya A, Berkvens R, et al. Self-calibration and collaborative localization for UWB positioning systems: A survey and future research directions. ACM Computing Surveys (CSUR). 2021;**54**(4):1-27

[3] Barrett TW. History of ultra wideband communications and radar: Part II, UWB radars and sensors. Microwave Journal. 2001; **44**(2):22-22

[4] Frattasi S, Della RF. Mobile Positioning and Tracking: From Conventional to Cooperative Techniques. Hoboken, New Jersey: John Wiley & Sons; 2017

[5] Win MZ, Dardari D, Molisch AF, et al. History and Applications of UWB. New York: Institute of Electrical and Electronics Engineers; 2009

[6] Zwirello L. Realization Limits of Impulse-Radio UWB Indoor Localization Systems. Germany: KIT Scientific Publishing; 2014

[7] Hirt W. The European UWB radio regulatory and standards framework: Overview and implications. In: 2007 IEEE International Conference on Ultra-Wideband. Singapore: IEEE; 2007. pp. 733-738

[8] Luediger H, Zeisberg S. UWB performance assessment based on recent FCC regulation and measured radio channel characteristics. In: IST Mobile Summit. CiteSeer; 2002

[9] Breed G. A summary of FCC rules for ultra wideband communications. High Frequency Electronics. 2005;**4**(1):42-44

[10] Zandian R. Ultra-wideband based indoor localization of mobile nodes in ToA and TDoA configurations. Dissertation, Bielefeld, Universität Bielefeld, 2018, 2019

[11] Qorvo. Getting back to basics with Ultra-Wideband, Qorvo US Inc, Tech. Rep. [Online]. 2021. Available from: https://www.qorvo.com/resources/d/qorvo-getting-back-to-basics-with-ultra-wideband-uwb-white-paper

[12] Azizpour R, Zakeri H, Moradi G, et al. Multi-channel radio-over-fiber communication systems through modulation instability phenomenon. IEEE Photonics Journal. 2024;**16**(5):1-13

[13] Zakeri H, Azizpour R, Khoddami P, et al. Low-cost multiband four-port phased array antenna for sub-6 GHz 5G applications with enhanced gain methodology in radio-over-fiber systems using modulation instability. IEEE Access. 2024;**12**:117787-117799

[14] Hussein H, Abd Elmunim N, Atasoy F, et al. A novel MIMO antenna integrated with a solar panel and employing AI-equalization for 5G wireless communication networks. IEEE Access. 2024;**12**:114382-114393

[15] Elwi TA, Taher F, Virdee BS, et al. On the performance of a photonic reconfigurable electromagnetic band gap antenna array for 5G applications. IEEE Access. 2024;**12**:60849-60862

[16] Zidour A, Ayad M, Alibakhshikenari M, et al. Wideband Endfire antenna Array for 5G mmWave

mobile terminals. IEEE Access. 2024;**12**: 39926-39935

[17] Salamin MA, Zugari A, Alibakhshikenari M, et al. Compact highly isolated dual-band 4-port MIMO antenna for sub-6 GHz applications. Journal of Electromagnetic Waves and Applications. 2023;**37**(10–12):1023-1043

[18] Alibakhshikenari M, Virdee BS, Salekzamankhani S, et al. On-Chip terahertz antenna array based on amalgamation of metasurface-inspired and artificial magnetic conductor technologies for next generation of wireless electronic devices. AEU-International Journal of Electronics and Communications. 2023;**167**:154684

[19] Alibakhshikenari M, Virdee B, Mariyanayagam D, et al. Virtual antenna array for reduced energy per bit transmission at Sub-5 GHz mobile wireless communication systems. Alexandria Engineering Journal. 2023;**71**: 439-450

[20] Alibakhshikenari M, Virdee BS, Rajaguru RK, et al. High performance antenna-on-chip inspired by SIW and metasurface technologies for THz band operation. Scientific Reports. 2023; **13**(1):56

[21] Althuwayb AA, Alibakhshikenari M, Virdee BS, et al. Design technique to mitigate unwanted coupling in densely packed radiating elements of an antenna array for electronic devices and wireless communication systems operating in the millimeter-wave band. AEU-International Journal of Electronics and Communications. 2023;**159**:154464

[22] Dong F, Shen C, Zhang J, et al. A TOF and Kalman filtering joint algorithm for IEEE802. 15.4 a UWB locating. In: 2016 IEEE Information Technology, Networking, Electronic and Automation Control Conference. Chongqing, China: IEEE; 2016. pp. 948-951

[23] Jiang Y, Leung VCM. An asymmetric double sided two-way ranging for crystal offset. In: 2007 International Symposium on Signals, Systems and Electronics. Montreal, QC, Canada: IEEE; 2007. pp. 525-528

[24] Sun F, Yin H, Wang W. Finite-resolution digital receiver for UWB TOA estimation. IEEE Communications Letters. 2011;**16**(1):76-79

[25] Bocquet M, Loyez C, Benlarbi-Delai A. Using enhanced-TDOA measurement for indoor positioning. IEEE Microwave and Wireless Components Letters. 2005; **15**(10):612-614

[26] Smaoui N, Heydariaan M, Gnawail O. Single-antenna AoA estimation with UWB radios. In: 2021 IEEE Wireless Communications and Networking Conference (WCNC). Nanjing, China: IEEE; 2021. pp. 1-7

[27] Qian X. Design of embedded system based on UWB PDOA technology. International Core Journal of Engineering. 2024;**10**(6):127-145

[28] Qian M, Zhao K, Li B, et al. An overview of ultra-wideband technology and performance analysis of UWB-TWR in simulation and real environment. IPIN-WiP. 2022;**12**:1-16

[29] Figueroa A, Al-Qudsi B, Joram N, et al. Comparison of two-way ranging with FMCW and UWB radar systems. In: 2016 13th Workshop on Positioning, Navigation and Communications (WPNC). Bremen, Germany: IEEE; 2016. pp. 1-6

[30] Neirynck D, Luk E, McLaughlin M. An alternative double-sided two-way

ranging method. In: 2016 13th Workshop on Positioning, Navigation and Communications (WPNC). IEEE; 2016. pp. 1-4

[31] Navrátil V, Vejražka F. Bias and variance of asymmetric double-sided two-way ranging. Navigation. 2019; **66**(3):593-602

[32] Rahman MN, Hanuranto MTIAT, Mayasari SR. Trilateration and iterative multilateration algorithm for localization schemes on wireless sensor network. In: 2017 International Conference on Control, Electronics, Renewable Energy and Communications (ICCREC). Yogyakarta, Indonesia: IEEE; 2017. pp. 88-92

[33] Widdison E, Long DG. A review of linear multilateration techniques and applications. IEEE Access. 2024;**12**: 26251-26266

[34] Cheng Y, Zhou T. UWB indoor positioning algorithm based on TDOA technology. In: 2019 10th International Conference on Information Technology in Medicine and Education (ITME). Qingdao, China: IEEE; 2019. pp. 777-782

[35] Zhang F, Yang L, Liu Y, et al. Design and implementation of real-time localization system (RTLS) based on UWB and TDoA algorithm. Sensors. 2022;**22**(12):4353

[36] Ge F, Shen Y. Single-anchor ultra-wideband localization system using wrapped PDoA. IEEE Transactions on Mobile Computing. 2021;**21**(12): 4609-4623

[37] Zhang F, Hong S, Ding Y, et al. Ultra wideband based real-time positioning with cascaded wireless clock synchronization method. IEEE Internet of Things Journal. 2024;**11**(9): 16731-16745

[38] Zhang F, Duan S, Tan J. UWB-based wireless clock synchronization and real-time positioning in smart parking. IEEE Sensors Journal. 2023;**24**(3):3822-3832

Chapter 8

Estimation of Intra-Pulse Modulation Parameters of Wideband LPI Radar Using Cyclostationary Algorithm

Raja Kumari Chilukuri, Hari Kishore Kakarla and Subbarao Kakarla

Abstract

Ultra-wideband (UWB) radar discharges signals over a broad range of frequencies, which is from 3 to 10GHz. One of the important characteristics of UWB radar is low probability of intercept (LPI) capability. The advantages of LPI radar are large bandwidth, low power, and high resolution. It is very difficult to intercept the LPI radar signals by passive receivers since they use special types of waveforms for intra-pulse modulation. Precise estimation of parameters helps to design smart intercept receivers. Cyclostationary (CS) algorithm is good for analysis of periodic like signals. Cyclic spectral correlation density (SCD) function is an important parameter to be computed. From the bifrequency plot of SCD function, the radar parameters are measured. Five types of polyphase codes (Frank, P1-P4 codes) are analyzed. With CS algorithm alone, the parameters are extracted with an error of less than 5% for signals up to –6 dB SNR only. To analyze high noisy signals, the signals are preprocessed first using two denoising filters and then CS algorithm is applied. With this process the same accuracy is obtained for signals up to –12 dB SNR. This work is useful in the war field to protect the weapon system from the enemy.

Keywords: ultra-wideband (UWB), low probability of intercept (LPI), polyphase, cyclostationary (CS), autocorrelation function (ACF), spectral correlation density (SCD), bifrequency, contour plot, intra-pulse modulation parameters, low probability of recognition (LPR)

1. Introduction

Ultra-wideband (UWB) is a radio technology for transmitting large amounts of data with low power over a wide frequency band for short distances. In UWB, the bandwidth of the signal is larger by about 10%. If the lower and upper frequencies of the signal power spectral density are f_l and f_u, respectively, the relative bandwidth (BW) is defined as $(f_u - f_l)/(f_u + f_l)$. UWB was first developed for communication and security purposes. Radar performance depends on BW. The larger the BW, the better is the range resolution. The size of the objects that can be distinguished between each other depends on the BW. The larger the BW, the smaller is the size of the object that can be detected. However, increasing the BW decreases the maximum detection range of the radar. Hence, BW is to be selected optimally depending on the application. It means that long-range radars need smaller BWs, and high-resolution radars need large BWs [1–3].

There are two general approaches to increasing the signal bandwidth of a radar system. The first one is to shorten the pulse duration since the BW is inversely proportional to the pulse duration. This approach is the basis for impulse radar. No carrier signal is used in this type of radar. The second approach is known as the spread spectrum. Various types of signal waveforms are designed to increase the BW without decreasing the pulse duration. This type requires some type of modulation within the pulse, resulting in a large time-bandwidth product [4, 5]. A number of modulation techniques are employed to increase the BW. They are linear FM, nonlinear FM, stepped FM, and pseudo-random coding. In pseudo-random coding, the carrier signal is biphase modulated throughout the pulse by a pseudo-random binary sequence. Binary Barker sequences of length of more than 13 do not exist. But if the number of phases is increased beyond two, polyphase Barker sequences of length more than 13 do exist. Hence, polyphase sequences are preferred.

Some of the advantages of UWB radar are

- Range measurement with high resolution (1 ns =15 cm)
- Low probability of exploitation and low probability of intercept (LPI)
- Significant capacity for target detection and discrimination
- Identification of low-observable (LO) targets (stealth and low-RCS)
- Target imaging with high resolution (single and multiple targets)
- Improved ability to penetrate obstacles (soil, walls, foliage, rain)
- Extremely shortrange ability
- Reduced sensitivity to electronic warfare countermeasures

One of the important characteristics of UWB radars is LPI capability. LPI radars also work on the principles of low power and wideband as well as ultra-wideband frequencies. Various types of modulation techniques are used in LPI radar to increase the bandwidth. Two important modulation techniques generally used are

i. Frequency modulation (FM)

ii. Phase shift modulation techniques.

Most widely used techniques are phase shifting techniques. Due to the high-resolution waveform and significant SNR processing gain, these methods are used for the radar. Frank code and Barker polyphase sequences are two examples of this. Additionally used are the polyphase (P_1, P_2, P_3, and P_4) codes, and polytime codes.

In this chapter, the procedure for measuring modulation parameters of low-probability of intercept radar and the results are discussed. Polyphase (Frank, P_1-P_4) codes are used for analysis. The analysis is carried out using cyclostationary (CS) algorithm for noise-free and various noise conditions up to -15 dB SNR. Two different types of denoising filters are used on noisy data in order to enhance the signal quality. LPI radar is discussed briefly in Section 2. The importance of estimation of the LPI radar's modulation parameters is discussed in Section 3. Polyphase codes are discussed in Section 4. Cyclostationary (CS) algorithm is discussed in Section 5, and noise-free polyphase code analysis is discussed in Section 6. Denoising techniques are discussed in Section 7, and Section 8 discusses the analysis of noisy polyphase codes.

2. Low probability of intercept (LPI) radar

There are two types of radar systems-frequency modulated continuous wave (FMCW) and pulsed radar. The block diagram of FMCW radar is shown in **Figure 1**. The important components of the radar are waveform generator, transmitter, receiver, antenna, signal processing unit and data processing unit. The transmitter transmits the modulated waveform, and the received signal is processed in the signal processing unit to reduce noise and other disturbances [6]. The filtered signal is processed in the data processing unit to get the required information. The signal processing unit and the data processing are available in the personal computer block. Low sidelobe antennas are used by low probability of intercept (LPI) radars to evade detection by passive radar detecting devices. Here are a few more traits and varieties of antennas. Parabolic antenna is a common kind of radar antenna that concentrates electromagnetic waves into a narrow beam using a concave dish-shaped reflector. Applications for parabolic antennas include weather radar detectors, air traffic control

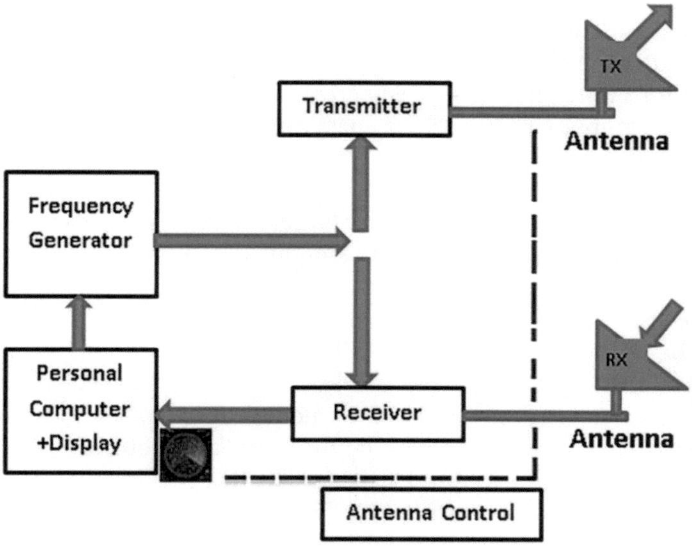

Figure 1.
Block diagram of LPI radar [6].

and military surveillance. The dipole antenna, is a balanced component, with equal but opposite voltages and currents applied at its two terminals through a balanced transmission line. The majority of antenna designs start with a dipole antenna. Phased array is an antenna composed of two or more basic antennas linked by an electrical network. Adaptive array is an antenna that modifies the weights of its constituent parts to attain the intended functionality. 5G massive multiple input multiple output (MIMO) systems require beam steering antenna arrays in order to increase system capacity and provide coverage. In actuality, antenna arrays are used in many different gadgets, including cars and radars. To increase a broadcast signal's channel capacity and resilience (signal-to-noise ratio) without consuming more spectrum, this technique is required. It enables the simultaneous direction of a common frequency in multiple directions. It has been demonstrated as multi-antenna technologies, like MIMO, reduced power consumption, overcome multipath phenomena and increase data throughput [7, 8].

Most radars in the modern Warfield, such as the ones used for reconnaissance, surveillance and target tracking, must contend with very capable and sophisticated attacks that are intended to degrade radar operations. A few instances of such danger include electronic attack (EA), electronic warfare support (ES) systems, Radar warning receivers (RWRs) and anti-radiation missiles (ARMs). As a result, radars must be unnoticed and are referred to as LPI radars. LPI radars use a large operational bandwidth, antenna side lobe reduction, low power management, frequency agility and improved scan patterns (modulations) to achieve this goal and conceal their presence. LPI radars are the name given to this class of radar, and they employ strategies "to see and not to be seen" by contemporary and sophisticated intercept receivers. Passive receivers are unable to detect and identify LPI radars with sufficient accuracy when using conventional interception techniques. The main advantages of LPI radar are its large bandwidth, low power, low cost and high resolution. Since LPI radars use specially designed waveforms, it is very difficult for the passive receivers to detect [9].

LPI radar, in general, is an essential part of contemporary military technology, offering better stealth, range and resolution, and dependability while lowering the likelihood of detection and interception by adversarial forces. The modern intercept receivers must detect and estimate the parameters of radar even when the noise/interference is high. The peak side lobe of the autocorrelation function (ACF) and the side lobe level at the matched filter's output that has been analytically integrated should be low for LPI radar. The phase or frequency of the modulated waveform should be chosen properly so that all the energy is concentrated in the main lobe of the ACF. If the peak side lobe level is low, the radar can identify weak targets while solid targets are around. Stated otherwise, if any of the side lobe levels of the ACF are high, the radar may not be able to detect another target that is nearby. These characteristics are achieved only by properly designing the modulation waveforms. Barker, nonlinear frequency modulation (NLFM), LFM, polyphase and polytime are some of the waveforms that are also used in many warfare devices [10–12].

Figure 2 shows the block diagram of LPI radar signal detection based on autocorrelation (ACF) function and Wigner-Ville distribution (WVD). The received signal from the antenna is down-converted to intermediate frequency (IF) in the RF filter unit. The IF frequency is selected such that the sampling frequency is sufficiently greater than the Nyquist rate to avoid aliasing. The IF signal is converted to analytic signal form using Hilbert Transform (HT). The output of the HT block is given to autocorrelation (AC) calculator block, where the ACF is calculated. A threshold value is selected for reducing noise. Since LPI signals are periodic, one can search for the periodic peaks of ACF. At least two peaks are required to get the radar signal. The

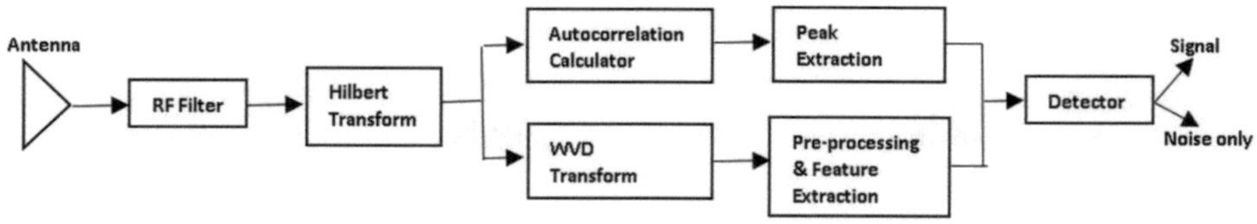

Figure 2.
Block diagram of LPI radar signal detection based on Wigner-Ville distribution and autocorrelation function [13].

number of data points between the peaks gives the code period. If the radar signal is not present, periodic peaks are not found, which indicates that only noise is present [13].

Simultaneously, the output of the HT block is also processed in the WVD block to get the pseudo-WVD (PWVD) image. The WVD image is preprocessed in the preprocessing and feature extraction block to get the binary image from which the radar signal is detected. In the detector block if any of the units return with a detection, it means that the signal is detected. With this method, the radar signal can be detected up to -7 dB SNR.

LPI radar has numerous applications across military, civilian, and scientific domains. LPI radar can be used for military surveillance and reconnaissance to find and track hostile targets while reducing the chance of being discovered and intercepted. LPI radar is frequently employed in naval warfare to find and track adversarial ships, submarines, and aircraft while reducing the chance of being discovered and intercepted. LPI radar is also utilized for border security and surveillance. LPI radar is used in scientific research for a variety of purposes, including remote sensing, studying the atmosphere, and keeping an eye on the environment. It is an essential technology in contemporary operations since it can locate and monitor targets while reducing the chance of being discovered and intercepted.

3. Importance of estimation of modulation parameters

Longer, lower energy pulses are used in LPI radars than in previous, non-LPI radars, and both LPI and LPR radars utilize complicated intra-pulse modulation methods. These complicated modulation algorithms have opened up new avenues for improving radar performance while decreasing the detectability of radar signals by third parties. While these intra-pulse modulation approaches clearly benefit radar manufacturers and users, the design of waveforms opens up new possibilities in the disciplines of pulse classification, emitter classification, and electronic intelligence. Classifiers and electronic intelligence systems rely on the radar waveform and inter-pulse characteristics (pulse width, pulse frequency, and pulse repetition interval) [14, 15].

Accurate estimation of parameters of modulation, including carrier frequency, BW, code rate, etc., and the type of modulation will provide the necessary data regarding the threat to the radar. Based on this information, the radar will be able to take immediate combat maneuver against the adversary. The same information is used to design sophisticated receivers, which can detect, intercept and measure parameters. The information is also used to develop many systems such as EA/ES systems and MIMO radar. This information is also useful to identify two radars operating at almost same frequencies.

3.1 The transmitted signal

With the receiver exciter's timing information, the phase shift key (PSK) radar's transmitter carries out the phase shifting operation. The complex signal sent over the air may be expressed as

$$S(t) = Ae^{j\left(2\pi f_c t + \varnothing_k\right)} \tag{1}$$

where $\varnothing_k$ is the time-shifted function for phase modulation, f_c is the angular frequency of the carrier. The complex signal S(t), can be represented by the in phase (I) and quadrature (Q) signals as

$$I = A\cos\left(2\pi f_c t + \varnothing_k\right) \tag{2}$$

$$Q = A\sin\left(2\pi f_c t + \varnothing_k\right) \tag{3}$$

The CW signal undergoes N_C phase shifts, with a phase shift of $\varnothing_k$ occurring every t_b seconds, for a single coding period. The BW is the reciprocal of time t_b. Pulse compression waveform of length 13 and its ACF are shown in **Figure 3**. The peak ACF value is 13.

Band width,

$$BW = \frac{1}{t_b}, \text{where} \quad t_b \text{ is the bit duration.} \tag{4}$$

$$\text{Time period, } T = \frac{1}{R_c}, \text{and } R_c \text{ is the code rate} \tag{5}$$

$$\text{Pulse compression ratio, } PCR = \frac{T}{t_b} \tag{6}$$

4. Polyphase codes

Discrete time complex sequences with variable phase k are known as polyphase sequences. The phase modification of the continuous wave carrier using a polyphase

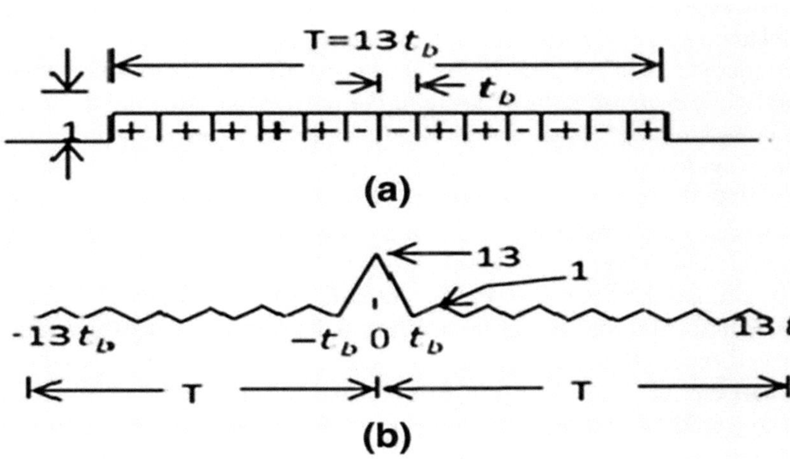

Figure 3.
(a) Pulse compression sequence of length 13. (b) Autocorrelation function [16].

sequence comprising a number of distinct phases is known as polyphase coding. In other words, more than two alphabets are used to generate the sequence elements. For example, 0^0, 90^0, 180^0, and 270^0 are the four alphabets of a Qudra phase sequence. Polyphase Barker codes of higher lengths can be obtained easily. Polyphase codes (Frank, P_1-P_4) are analyzed here. The phase codes are chosen so that the ACF of the coded waveform has the highest peak signal to side lobe (PSL) ratio [17].

FRANK CODE: Frank code is created by employing M samples and M frequency steps, and the quantity of samples in a given frequency, i, and frequency, j, together define the phase of the ith sample of the jth frequency.

$$\varnothing_k = \varnothing_{i,j} = \frac{2\pi}{M}(i-1)(j-1), i = 1, 2, \ldots, M \; and \; j = 1, 2, \ldots, \mathrm{M} \tag{7}$$

where i = 1, 2, ... , M, j = 1, 2, ... , M, and i is the number of samples and frequencies. Another way to express the Frank polyphase code is as a M x M matrix. The discrete phase values are shown in **Figure 4(a)** and **(b)** shows the phase jumps for M = 8.

P_1 CODE: The phase of the ith sample of the jth frequency is $\varnothing_{i,j}$, where i, is the number of samples in a given frequency and j, is the frequency number.

$$\varnothing_k = \varnothing_{i,j} = \frac{-\pi}{M}[M-(2j-1)][(j-1)M+(i-1)] \tag{8}$$

P_2 CODE: P_2 code is provided by

$$\varnothing_k = \varnothing_{i,j} = \frac{-\pi}{2M}[2i-1-M][2j-1-M] \tag{9}$$

P_3 CODE: P_3 code is generated theoretically from a waveform of linear frequency modulation. It provides the ith example of the P_3 code.

$$\varnothing_k = \frac{\pi}{N_c}(k-1)l^2 \tag{10}$$

P_4 CODE: P_4 signal's phase sequence is given by

$$\begin{bmatrix} 0 & 0 & 0 & \cdots & 0 \\ 0 & 1 & 2 & \cdots & (M-1) \\ 0 & 2 & 4 & \cdots & 2(M-1) \\ \vdots & \vdots & \vdots & & \vdots \\ 0 & (M-1) & 2(M-1) & \cdots & (M-1)^2 \end{bmatrix}$$

(a)

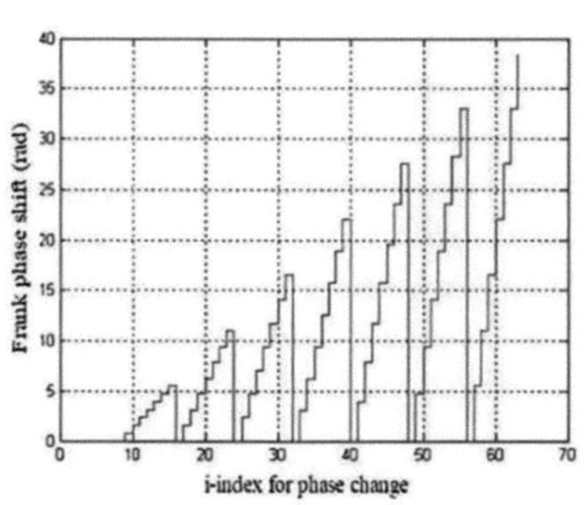

(b)

Figure 4.
(a) Phase values of Frank code for M = 8 (b) Discontinuous phase jumps for M = 8 [17].

$$\varnothing_k = \frac{\pi}{N_c}(k-1)^2 - \pi(k-1) \tag{11}$$

5. Cyclostationary (CS) algorithm

The earlier methods of signal processing fail to analyze the LPI signals. Though many time-frequency (T-F) techniques are developed, they also fail to analyze the LPI signals. The T-F methods analyze the signals simultaneously in both domains. The T-F algorithms provide time-frequency findings that make it possible to compute the LPI parameters with reasonable precision. On the other hand, the CS algorithm or the bifrequency analysis method provides some extra capabilities in LPI modulation detection and classification.

The ideal technique for the purpose of analyzing periodic like signals is the CS algorithm. All LPI radar signals are periodic because they are modulated. In CS analysis, the signals are transformed into frequency-cycle frequency or bifrequency plane. Cyclic spectral correlation density (CSCD) and cyclic autocorrelation function (CACF) are the two fundamental components of CS analysis. As an example, the autocorrelation function generates spectral lines. CS is used in many areas such as direction of arrival, estimation of parameters, signal detection, and array processing. Many parameters of LPI radar can be obtained using CACF and CSCD [18].

Let $x(t)$ be the LPI signal received by the intercept-receiver. The autocorrelation function (ACF) is given by

$$R_x(\tau) \triangleq \lim_{T \to \infty} \int_{-T/2}^{T/2} x\left(t + \frac{\tau}{2}\right) x^*\left(t - \frac{\tau}{2}\right) dt \tag{12}$$

The cyclic autocorrelation function (CACF) of $x(t)$ is given by

$$R_x^\alpha(\tau) \triangleq \lim_{T \to \infty} \frac{1}{T} \int_{-T/2}^{T/2} x\left(t + \frac{\tau}{2}\right) x^*\left(t - \frac{\tau}{2}\right) e^{-j2\pi\alpha t} dt \tag{13}$$

where α is the cycle frequency. The signal's power spectrum is calculated using Eq. (14)

$$S_x(f) = \int_{-\infty}^{\infty} R_x(\tau) e^{-j2\pi f\tau} d\tau \tag{14}$$

In the same way, the SCD function is computed using Eq. (15)

$$S_x^\alpha(f) \triangleq \int_{-\infty}^{\infty} R_x^\alpha(\tau) e^{-j2\pi f\tau} d\tau = \lim_{T \to \infty} \frac{1}{T} X_T\left(f + \frac{\alpha}{2}\right) X_T^*\left(f - \frac{\alpha}{2}\right) \tag{15}$$

$$\text{where } X(f) \triangleq \int_{-T/2}^{T/2} x(v) e^{-j2\pi f v} dv \tag{16}$$

where $X(f)$ is the Fourier transform of $x(v)$ and α represents the two dimensional plane of $S_x^\alpha(f)$ or (f, α) plane or bifrequency plane. Computation of Eqs. (13) and (15) in signal analysis is known as cyclic spectral analysis.

5.1 Estimation of spectral correlation density using direct frequency smoothing (DFS) method

Time or frequency smoothing approaches can be used to estimate the cyclic spectral correlation density (SCD). DFS method is used here as it is easy to implement. But it is computationally more time-consuming. The SCD coefficients are computed in two steps. Spectral components of the data are computed in the first step. In the second step, operations for spectral correlation are performed directly on the spectral components as shown in **Figure 5**. The basis for the DFSM method is the Fourier transform. The SCD can be computed using Eqs. (17) and (18).

$$S_x^\alpha(f) \approx S_{x_{T_W}}^\alpha(t,f)_{\Delta t} \approx \frac{1}{\Delta t} \int_{t-\left(\frac{\Delta t}{2}\right)}^{t+\left(\frac{\Delta t}{2}\right)} S_{x_{T_W}}(\mu,f)d\mu \tag{17}$$

$$\text{where } S_{x_{T_W}}(\mu,f) = \frac{1}{T_W} X_{T_W}\left(\mu,f+\frac{\alpha}{2}\right) X_{T_W}^*\left(\mu,f-\frac{\alpha}{2}\right) \tag{18}$$

The signal's total observation period is denoted by the symbol Δt, and T_W is the size of the short-time FFT window:

$$X_{T_W}(t,f) = \int_{t-\left(\frac{T_W}{2}\right)}^{t+\left(\frac{T_W}{2}\right)} x(\mu)e^{-j2\pi f\mu}d\mu \tag{19}$$

The discrete equivalent of Eq. (17) is given in Eq. (20)

$$S_{X_N}^\gamma(n,k) = \frac{1}{N}\sum_{n=0}^{N-1} X_N\left(n,k+\frac{\gamma}{2}\right) X_N^*\left(n,k-\frac{\gamma}{2}\right) \tag{20}$$

$$\text{where } X_N(n,k) = \sum_{n=0}^{N-1} w(n)x(n)e^{-\frac{j2\pi kn}{N}} \tag{21}$$

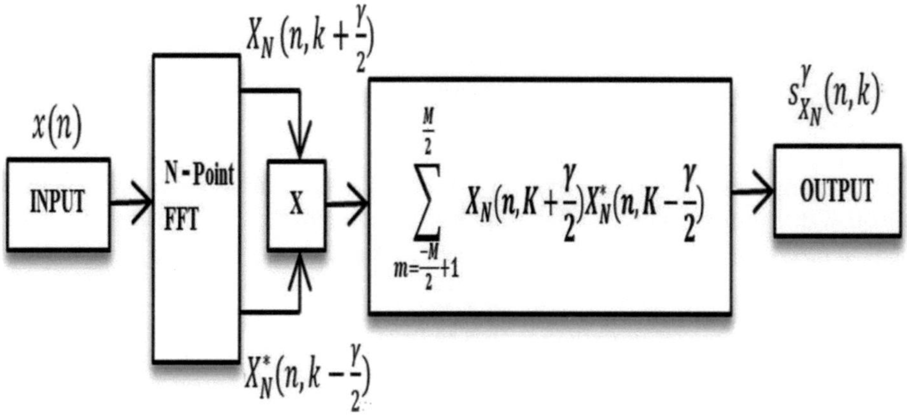

Figure 5.
Direct frequency smoothing method of computing SCD [18].

where $X_N(n,k)$ is the FFT of $x(n)$, N is the length of $x(n)$, $w(n)$ is the rectangular window and $'\gamma'$ is the discrete equivalent of cycle frequency $'\alpha'$.

The algorithm's cycle frequency resolution is stated as

$$\gamma_{RES} = \frac{f_s}{N} \tag{22}$$

and the resolution of frequency is

$$k_{RES} = \Delta k = M.\frac{f_s}{N} \tag{23}$$

where M is the Grenander's Uncertainty condition. M should be greater than 1. Here, M is taken as 2 and N = 900. f_s is the sampling frequency.

5.2 Advantages of cyclostationary algorithm

- CS algorithms can detect and analyze periodicities in a signal, even when they are buried in noise.
- It can provide information about the frequency, phase, and modulation properties of the periodic components.
- The CS algorithm is robust to noise and interference, making it useful for detecting signals in noisy environments.
- It can be applied to a wide range of signals, including communication signals, radar, and biomedical signals.

5.3 Disadvantages of cyclostationary algorithm

- Cyclostationary algorithm can be computationally intensive and requires a large amount of memory.
- It requires knowledge of the signal's underlying cyclostationary properties, which may not be available in all cases.

6. Estimation of modulation parameters using cyclostationary (CS) techniques

Estimation of parameters is very important to develop smart intercept receivers and ELINT/ES systems. Once the parameters are estimated correctly, this information is used to provide jamming and other counter actions effectively. **Figure 6** shows the block diagram for measurement of parameters using CS algorithm. Generation of polyphase codes and the complete analysis are carried out in MATLAB.

6.1 Analysis of Frank code

Frank code is generated using Eq. (7) and it is modulated with f_c=1GHz. Let the modulated signal be $x(n)$. It is also assumed that the signal is not corrupted with any

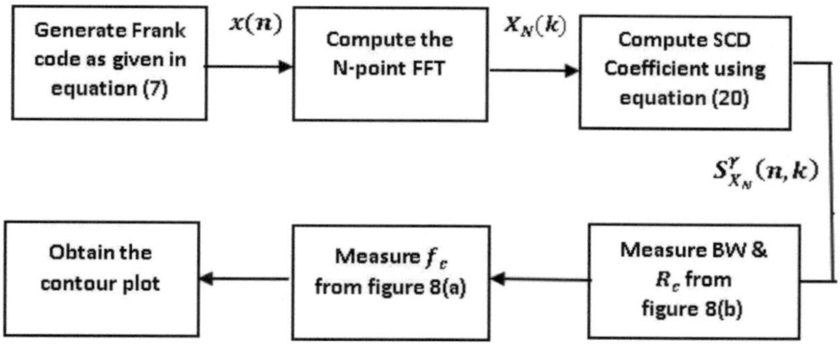

Figure 6.
Flow diagram for measurement of parameters.

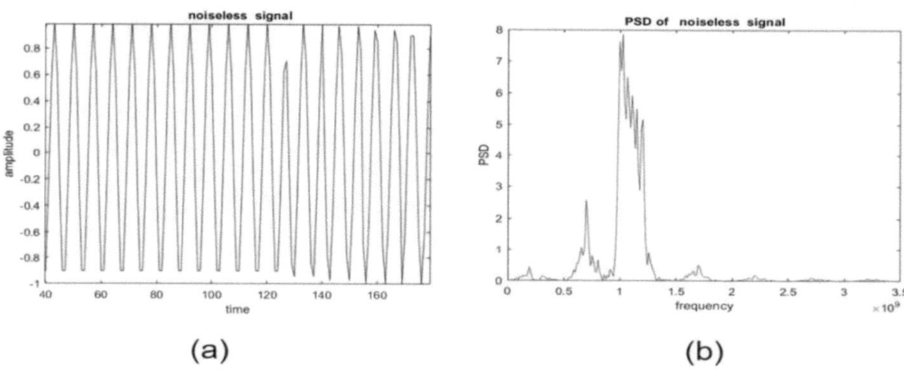

Figure 7.
(a) Time domain modulated signal of Frank code and (b) its power spectral density.

type of noise. The modulated Frank signal is shown in **Figure 7(a)**, and its power spectral density (PSD) is plotted in **Figure 7(b)**. From the figure, it is observed that most of the energy is concentrated in the main lobe which appears like an impulse. The side lobe peak is much smaller than the main lobe peak.

The spectral correlation density (SCD) coefficients, $S^{r}_{xN}(n,k)$, of the Frank code noise-free signal x(n) are calculated using Eq. (20) and its contour plot is obtained. **Figure 8(a)** shows the contour plot or the bifrequency plot of $S^{r}_{xN}(n,k)$. Y-axis represents the frequency and x-axis represents the cycle frequency; the values of $S^{r}_{xN}(n,k)$ represent the color of the figure. It may be observed that the figure is symmetric and spread over four quadrants and the shape of the patterns is like a butterfly. The distance on the x-axis from the figure's center to the center of the leftmost or rightmost pattern gives $2f_c$ where f_c is the carrier frequency used (true f_c) as shown in figure. It may be noted that at this point, the color is yellow and the $S^{r}_{xN}(n,k)$ value is maximum. Hence, $2f_c$ =1.983GHz, and hence the measured $f_c = 0.9915\ GHz$ and the true $f_c = 1.0$GHz. The difference between the two values gives the error. The percentage error is calculated using Eq. (24), and the results are tabulated in Columns 2–4 of **Table 1**. The first column shows the type of code. **Figure 8(b)** shows the zoomed part of the rightmost pattern. BW and R_c are measured from this figure. The distance

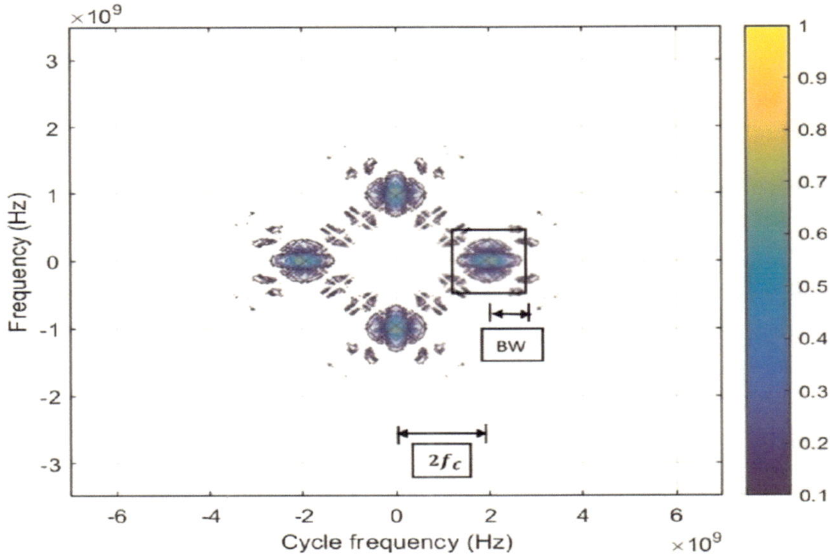

(a)

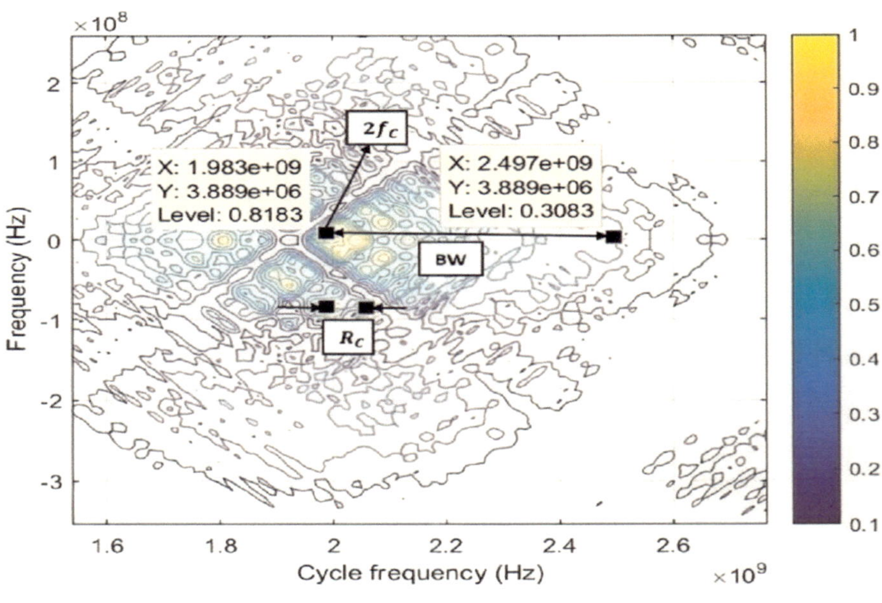

(b)

Figure 8.
Noise-free contour plot of Frank code for $f_c = 1GHz$ *when* $f_s = 7f_c$. *(a) Contour plot or bifrequency plot of Frank code. (b) Zoomed version of the selected segment of (a).*

along the x-axis from the center of the pattern to the end of the pattern gives the bandwidth. The x-coordinate of the center point is 1.983GHz and the end point is 2.497GHz, as shown in the figure. The difference between the two gives the BW, i.e., 2.497e+9-1.983e+9 = 0.514e+9. The true value of BW used is 500 MHz and measured

Type of Poly phase code (1)	f_c (MHz)			BW(MHz)			N_c		
	True value (2)	Estimated value (3)	% error (4)	True value (5)	Estimated value (6)	% error (7)	True value (8)	Estimated value (9)	% error (10)
Frank code	1000	991.5	0.85	500	514	2.8	64	64.3	0.5
	2000	1995	0.25	500	505	1.0	64	64.7	1.09
	3000	2998	0.06	500	521	4.2	64	62	3.12
P_1	1000	999.5	0.05	500	505	1.0	64	63.125	1.36
	2000	1999	0.05	500	513	2.6	64	66.125	3.3
	3000	2998	0.06	500	514	2.8	64	64.25	0.39
P_2	1000	995.5	0.45	500	506	1.2	64	63.25	1.17
	2000	1999	0.05	500	498	0.4	64	62.25	2.73
	3000	2998	0.06	500	513	2.6	64	64.125	0.19
P_3	1000	995.5	0.45	500	498	0.4	64	62.25	2.73
	2000	1999	0.05	500	489	2.2	64	61.12	4.5
	3000	2998	0.06	500	513	2.6	64	64.125	0.19
P_4	1000	999.5	0.05	500	498	0.4	64	62.25	2.73
	2000	1991	0.45	500	505	1.0	64	63.125	1.36
	3000	2986	0.46	500	524	4.8	64	65.5	2.34

Table 1.
Comparison of the estimated and true parameters of polyphase codes.

BW is 514 and 0.14 MHz is the error. The percentage error is calculated using Eq. (24) and the readings are noted in Columns 5–7 of the same table. The distance between any two adjacent locations is known as the code rate, (R_c). It is measured as 8, as shown in the figure and N_c is calculated using $N_c = \frac{BW}{R_c} = \frac{514}{8} = 64.3$. The true N_c used is 64, and the measured $N_c = 64.3$. Again, the percentage error is calculated using Eq. (24). The results are tabulated in Columns 8–10. In fact, N_c should satisfy the equation $N_c = 2^k$, where k is any positive integer. All the values are measured manually from the figures.

$$Error = \left| \frac{true\ value - measured\ value}{true\ value} \right| \tag{24}$$

It may be observed that the largest error is less than 5%. Repeating the analysis with various carrier frequencies ($f_c = 2$ GHz and $f_c = 3$ GHz) and the results are shown in rows 2 to 3 for Frank code.

6.2 Analysis of other polyphase codes

Other polyphase codes (P_1-P_4) are generated using Eqs. (8) through (11). For each of the code, the SCD coefficients are computed using Eq. (20) and the corresponding contour plots are shown in **Figure 9**. The process of measuring the parameters is

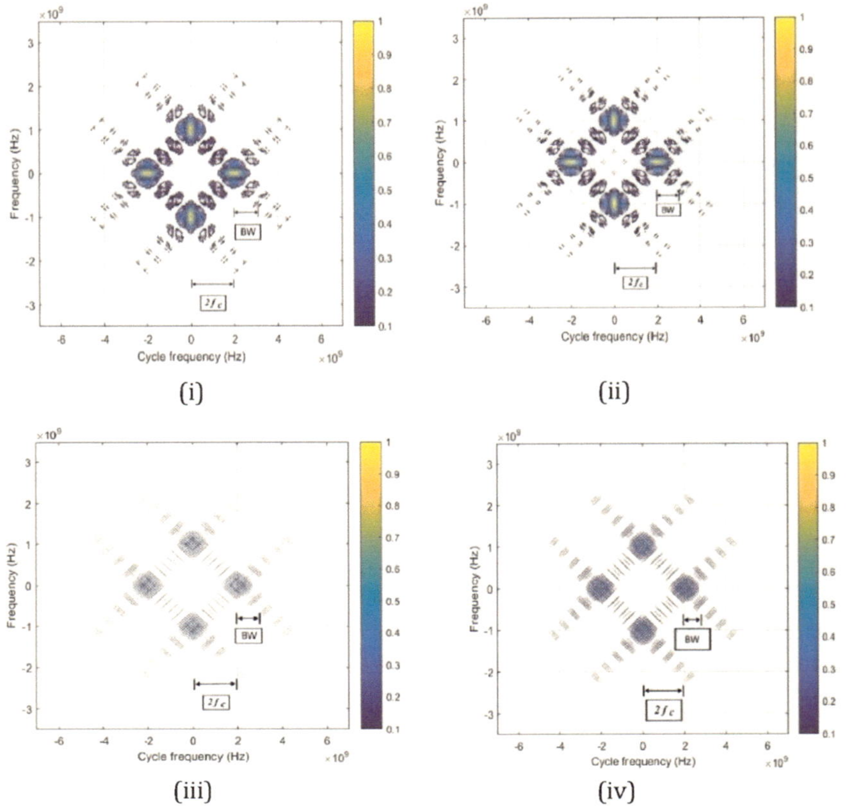

Figure 9.
Noise-free contour plots of P_1-P_4 Codes for $f_c = 1$GHz when $f_s = 7f_c$. (i) Complete bifrequency plane of P_1, (ii) complete bifrequency plane of P_2, (iii) complete bifrequency plane of P_3, and (iv) complete bifrequency plane of P_4.

described in Section 6.1. The zoomed versions of contour plot of $(P_1 - P_4)$ are not shown here. Using the same procedure, the values are measured for codes $(P_1 - P_4)$ and the results are tabulated in the table. From the table, it is clear that the accuracy achieved is more than 95%.

In all cases, the sampling frequency f_s is taken as $f_s = 7f_c$. The analysis is extended for two more sampling frequencies $f_s = 5f_c$ and $f_s = 3f_c$. But the results are not shown in the tables. From the results, it is concluded that in order to achieve an accuracy of more than 95%, $f_s \geq 5f_c$. On the other hand, if $f_s < 5f_c$, the maximum error is more than 5% and the receivers fail to take immediate counter action.

7. Denoising techniques

Noise is an inevitable part of any measurement. As the noise level increases, the accuracy in measurement decreases very badly. The accuracy of noisy signals can be improved by passing the noisy signals through denoising filters first and then, CS algorithm is applied. Generally, there are two sources of noise. They are

i. Normally, the receiver is placed far away from the command center and the median through which the signal travels introduce noise.

ii. The electronic devices produce thermal noise.

The denoising filters normally used are moving average filter, median filter, mean filter, wavelet techniques, Savitzky Golay (SG) filter, etc. SNR is used to measure the performance of the denoising filters. SG filter and median filter and are used in this work.

7.1 Savitzky Golay (SG) filter

It is based on local sequences of polynomial approximation, and it is used for data smoothing. SG filter reduces the noise level without disturbing the shape of the waveforms and the height of the peak. SG filters are mainly used in areas such as communication, spectrum estimation and biomedical engineering. The performance of SG filter depends on two key parameters: (i) length of the window and (ii) order of the filter

The parameters are to be selected optimally such that the error is minimum in the mean square sense [19–21].

7.2 Median filter

It is a filter for smoothing nonlinear data. It smoothens the data, and it is used as a preprocessing filter. It is used in several signal processing applications, such as biomedical and image processing. In this case also, the threshold values are to be selected optimally in the least mean square (LMS) sense.

7.3 Noise removal techniques of LPI signals

Precise estimation of parameters is important as it provides immediate counteractions and jamming against the enemy. Estimation of parameters of noise-free signals is discussed in Section 6. The analysis of noisy signals will be discussed here. The received signal is thought to be tainted by additive white Gaussian noise. **Figure 10**

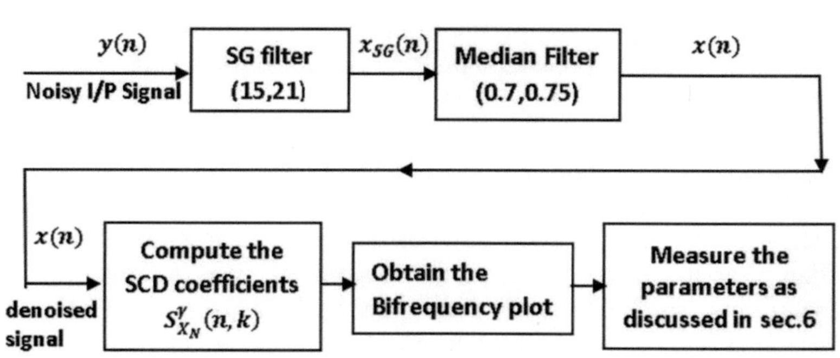

Figure 10.
Block schematic for estimating noisy waveform modulation parameters.

shows the block diagram for measuring parameters under different noise situations. Let y(n) be the signal that was noisy and was picked up by the intercept receiver, such that

$$y(n) = x_T(n) + w(n) \tag{25}$$

Where $x_T(n)$ is the transmitted signal which is free from noise, $w(n)$ is the white noise. Since $x_T(n)$ is deterministic, it may be written as

$$\varnothing(n) = 2\pi f_i(n) + \varphi(n) \text{ and } x_T(n) = A\,cos\,(\varnothing(n)) \tag{26}$$

If $f_i(n)$ is constant and $\varphi(n)$ is changed the signal is phase modulated. On the other hand, if $f_i(n)$ is changed and $\varphi(n)$ is constant, it is a frequency modulated signal.

The noisy signal, $y(n)$ is passed through the SG filter first, as shown in the block diagram. Let $x_{SG}(n)$ be the output of the SG filter, which is then passed through the median filter. The output $x(n)$ of the median filter is the filtered signal. The SNR of the denoised signal is calculated using Eq. (27):

$$\text{SNR} = -20 * \log \left[\frac{\left(\sum_{k=1}^{N} |x_T(k) - x(k)|^2 \right)^{\frac{1}{2}}}{\left(\sum_{k=1}^{N} |x_T(k)|^2 \right)^{\frac{1}{2}}} \right] \text{in dB} \tag{27}$$

After filtering, the filtered signal, $x(n)$ is nearly the same as the transmitted signal:

$$\text{that is}, x(n) \cong x_T(n) \tag{28}$$

The output of the median filter, $x(n)$ is the denoised signal. The signal $x(n)$ is analyzed using CS algorithm as explained in Section 6, and the SCD coefficients are obtained using Eq. (20). From the contour plot of $S_{x_N}^{r}(n, k)$, the parameters are measured as explained in Section 6.

7.4 Selection of denoising filter parameters

The PSD's of pure signal, $x_T(n)$, the filtered signal $x(n)$ and the noise-added signal $y(n)$ are computed using Eq. (14) for one set of values of the SG filter first. The parameters are selected optimally on the basis of minimization of error in the least mean square sense. The process is repeated for various noise levels. If the mean square error is large at any noise level, the selected set of values is not good. Choose another set of values and repeat the same process. The best set of parameters among all is selected on the basis of minimum error. The optimal values for this filter are 15 and 21. The number 15 is the order of the SG Filter and 21 is the length of the window. These values are shown in the SG filter block. In the same way, the optimal threshold values of the median filter are selected as 0.7 and 0.75. It is observed that the error is less with these arrangements. The PSDs of the denoised signal $x(n)$, the noisy signal, $y(n)$ and and the noise-free signal, $x_T(n)$ are plotted in **Figure 11** for SNR = -6 dB. From the figure, it is observed that low-frequency magnitudes are smoothed, and the high frequencies are attenuated. This combination of filters is good when compared with wavelet filter and median filter combination.

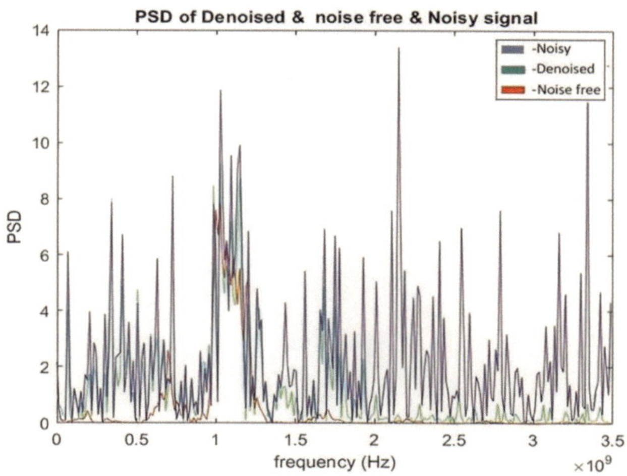

Figure 11.
PSDs of filtered, noise-free and noisy signals for Frank code for -6 dB SNR.

8. Analysis of polyphase codes under noisy conditions

The analysis of noise-free signals using CS algorithm is discussed in Section 6. In this section, analysis of noisy signals under various noise conditions is discussed.

Generate the noise-free modulated Frank code signal as given in Eq. (7) with carrier frequency f_c= 1GHz. Let this signal be $x_T(n)$. Since in this section, analysis of noisy signals is considered, white noise is added to $x_T(n)$ signal as given in Eq. (25). Let the noisy signal be $y(n)$. The noisy signal is denoised using the noise removal techniques as explained in Section 7. Let the denoised signal be $x(n)$, as shown in **Figure 10**. **Figure 11** shows PSDs of the noise-free, noisy and denoised modulated Frank code signal for -6 dB SNR. As can be seen, the spectrum is spread from 1.0GHz to 1.2GHz. It is observed that the high-frequency components are attenuated.

The SCD coefficients, $S_{x_N}^{r}(n,k)$, are computed for the denoised signal, $x(n)$, and its contour plot for -6 dB SNR is shown in **Figure 12(a)**. As mentioned in Section 6, the parameter f_c is measured from **Figure 12(a)** as $2f_c = 1991$ or $f_c = 995.5$ MHz. **Figure 12 (b)** shows the zoomed version of selected part of **Figure 12(a)**, and the parameters BW and N_c are measured from this figure as BW = 2.489–1.991 = 0.468GHz = 468 MHz and $N_c = 64.3$. The error is calculated using Eq. (24) and all the values are recorded in **Table 2**'s first row. Column 1 shows the type of polyphase code and column 2 gives the noise level. The type of the parameter is shown in column 3. The true values of the parameters are shown in column 4. Columns 5 and 6 show the results of the proposed algorithm. Column 5 shows the measured value of the parameter, and column 6 shows the error obtained. Columns 7 and 8 are same as columns 5 and 6 except for literature values. As an example, the first row shows the results of Frank code for -6 dB SNR. The true value of f_c is 1000 MHz and the measured value of the parameter is 995.5 MHz. The error in estimation is 0.45% with the proposed algorithm. Modified S-transform (MST) method is used in the literature method. The measured value of f_c with this method is 1000 MHz and the error is zero. Similarly, the other two parameters are also shown in the

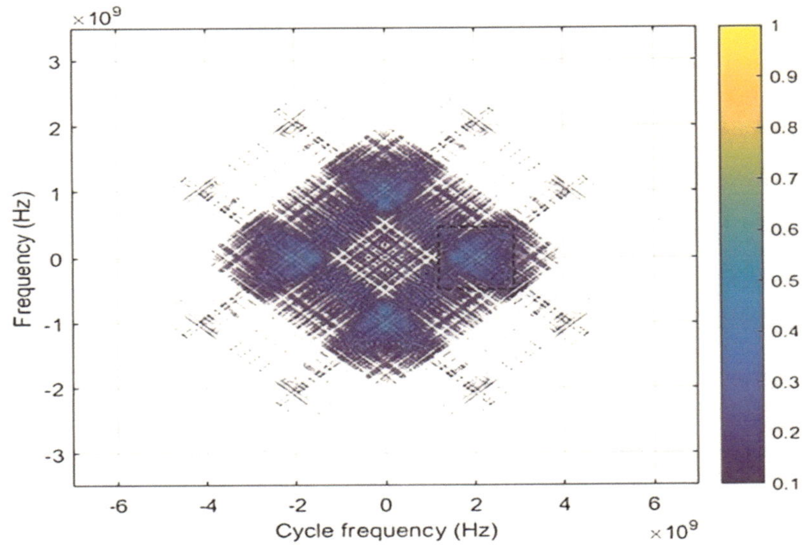

(a)

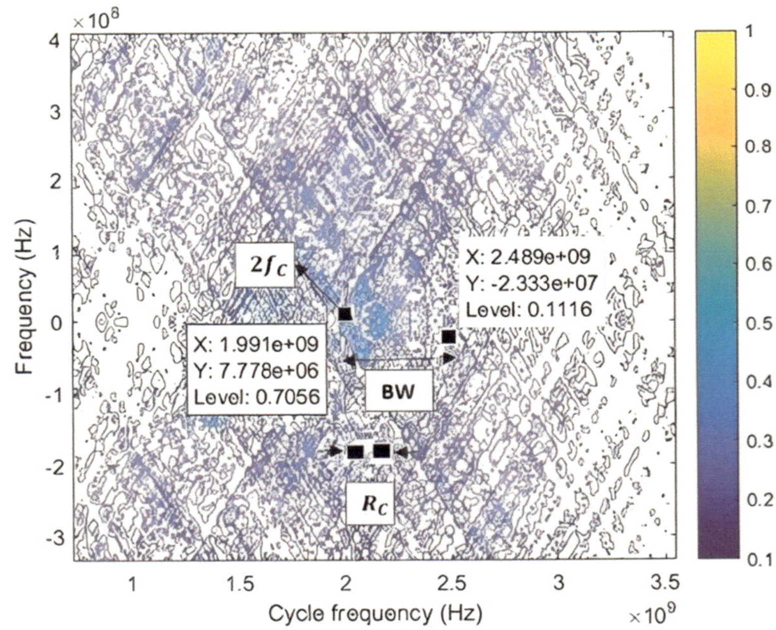

(b)

Figure 12.
Noisy contour plot of Frank Code for $f_c = 1GHz$ for -6 dB SNR. (a) Bifrequency plot and (b) closeup of the selected portion for measurement of BW and R_c.

table. Identical process is carried out for various SNRs up to -15 dB and it is found that for high noisy signals (less than -12 dB), the error in PSDs is large [22–24].

Figure 13 shows the contour plot of Frank code for -12 dB SNR. As noise level increases, the error also increases for some parameters. However, only three sets of readings are reported in the results. For high noisy signals less than -12 dB SNR, the

Type of Code (1)	SNR (dB) (2)	Parameter (3)	True value (4)	Proposed method		Literature method [21]	
				measured Value (5)	%error (6)	measured Value (7)	%error (8)
Frank code	-6	f_c(MHz)	1000	995.5	0.45	1000	0
		BW(MHz)	500	498	0.4	505	1
		N_c	64	62.25	2.73	65.2	1
	−7	f_c(MHz)	1000	991.5	0.85	1042	4.2
		BW(MHz)	500	511	2.2	472	5.6
		N_c	64	63.87	0.19	53.81	15.9
	−12	f_c(MHz)	1000	999.5	0.05	NA	NA
		BW(MHz)	500	505	1.0	NA	NA
		N_c	64	63.125	1.36	NA	NA
P_1 Code	−6	f_c(MHz)	1000	995.5	0.45	1000	0
		BW(MHz)	500	513	2.6	494	1.2
		N_c	64	64.125	0.19	63.2	1.2
	−7	f_c(MHz)	1000	999.5	0.05	1032	3.2
		BW(MHz)	500	498	0.4	514	2.8
		N_c	64	62.25	2.73	56.54	11.6
	−12	f_c(MHz)	1000	995.5	0.45	NA	NA
		BW(MHz)	500	513	2.6	NA	NA
		N_c	64	64.125	0.19	NA	NA
P_2 Code	−6	f_c(MHz)	1000	995.5	0.45	1000	0
		BW(MHz)	500	498	0.4	522	4.4
		N_c	64	62.25	2.73	66.3	3.5
	−7	f_c(MHz)	1000	999.5	0.05	1056	5.6
		BW(MHz)	500	513	2.6	556	11.2
		N_c	64	64.125	0.19	68.94	7.7
	−12	f_c(MHz)	1000	995	0.5	NA	NA
		BW(MHz)	500	498	0.4	NA	NA
		N_c	64	62.25	2.73	NA	NA
P_3 Code	−6	f_c(MHz)	1000	999.5	0.05	1000	0
		BW(MHz)	500	498	0.4	516	3.2
		N_c	64	62.25	2.73	65.8	2.8
	−7	f_c(MHz)	1000	995.5	0.45	1021	2.1
		BW(MHz)	500	507	1.4	522	4.4
		N_c	64	63.37	0.98	67.86	6.0
	−12	f_c(MHz)	1000	999.5	0.05	NA	NA
		BW(MHz)	500	490	2.0	NA	NA
		N_c	64	61.25	4.29	NA	NA

Type of Code (1)	SNR (dB) (2)	Parameter (3)	True value (4)	Proposed method		Literature method [21]	
				measured Value (5)	%error (6)	measured Value (7)	%error (8)
P_4 Code	-6	f_c(MHz)	1000	995.5	0.45	1000	0
		BW(MHz)	500	498	0.4	515	3
		N_c	64	62.25	2.73	65.92	3
	-7	f_c(MHz)	1000	991.5	0.85	1025	2.5
		BW(MHz)	500	514	2.8	495	1
		N_c	64	64.25	0.39	60.4	5.6
	-12	f_c(MHz)	1000	999.5	0.05	NA	NA
		BW(MHz)	500	505	1.0	NA	NA
		N_c	64	63.125	1.36	NA	NA

NA, not available.

Table 2.
Comparison of the results of the proposed algorithm with literature values under different noise conditions when $f_s = 7f_c$.

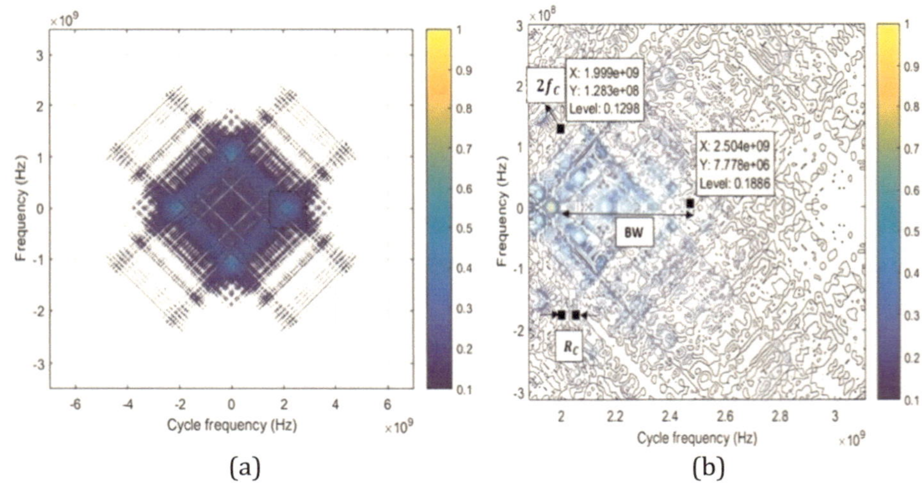

Figure 13.
Contour plot of noisy Frank Code for f_c = 1GHz for -12 dB SNR. (a) Complete bifrequency plot and (b) closeup of the selected portion.

measurement error is more than 5%. The same procedure is followed for obtaining the contour plots of denoised P_1 through P_4 codes and the plots are shown in **Figures 14-17** for -6 dB SNR. The parameters are measured from the corresponding plots for SNRs up to -15 dB and the results are shown in **Table 2**. From the table, it is clear that the error is less than 5% for all polyphase codes upto -12 dB SNR with the proposed algorithm. For high noisy signals (SNR ≤ -15 dB), the measurement error is more than 5% for some of the parameters. But the results are not shown. It indicates that CS algorithm, along with denoising filters, works well up to -12 dB SNR. But the literature values are available upto -7 dB SNR only.

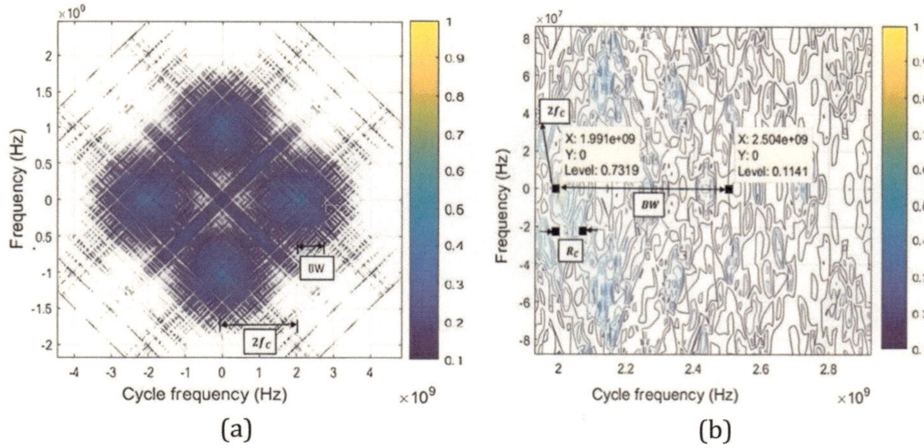

Figure 14.
Contour plot of noisy P_1 Code for $f_c = 1GHz$ for -6 dB SNR. (a) Complete bifrequency plot and (b) closeup of the selected portion.

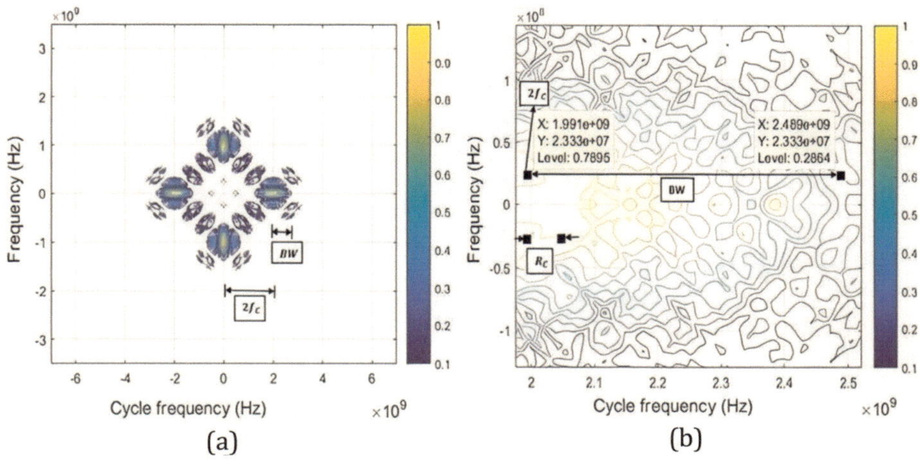

Figure 15.
Contour plot of noisy P_2 Code for $f_c = 1GHz$ for -6 dB SNR. (a) Complete bifrequency plot and (b) closeup of the selected portion.

9. Conclusions

Precise parameter measurement is crucial for designing contemporary receivers that provide immediate counteraction. This information is useful to develop an ES/EA system that could take immediate counteraction against the enemy.

It is observed that polyphase codes are periodic like waveforms. To analyze such waveforms, the most suitable and efficient algorithm is the CS algorithm. The signal is converted into frequency-cycle frequency (bifrequency) plane using this method. Applications for CS can be found in a wide range of domains, including signal recognition, arrival time estimate, array processing, and estimation of arrival of direction.

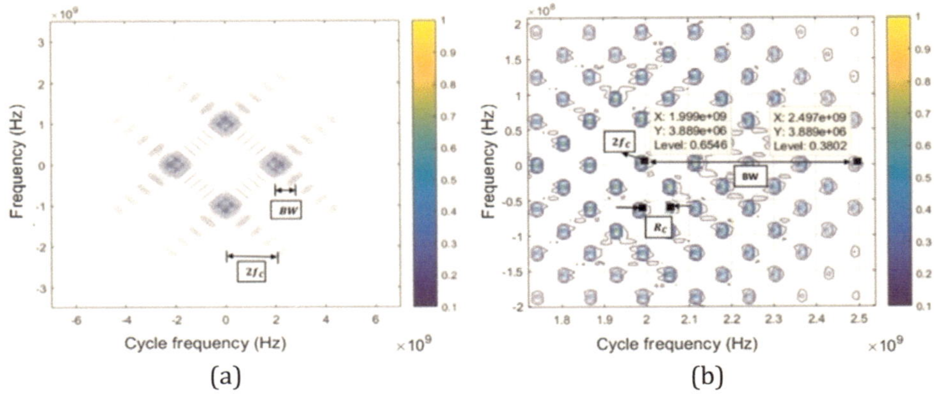

Figure 16.
Contour plot of noisy P_3 Code for f_c = 1GHz for -6 dB SNR. (a) Complete bifrequency plot and (b) closeup of the selected portion.

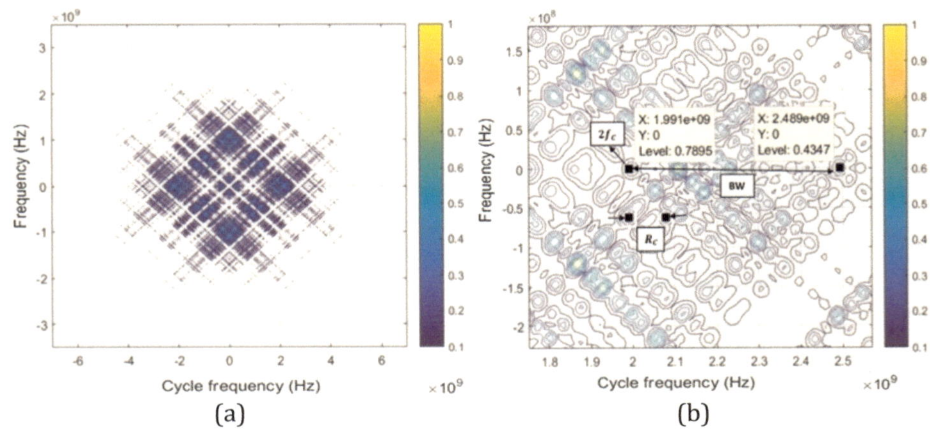

Figure 17.
Contour plot of noisy P_4 Code for f_c = 1GHz for -6 dB SNR. (a) Complete bifrequency plot and (b) closeup of the selected portion.

Five types of polyphase codes (Frank, $P_1 - P_4$) are analyzed under various noise conditions and estimated three modulation parameters (carrier frequency, bandwidth, and code rate). The SCD coefficients of the signal are calculated, and the parameters are measured manually from the contour plot of the SCD function. The simulation work is carried out under various noise conditions (zero noise and noisy environment up to -15 dB SNR). With an accuracy of greater than 95%, the CS algorithm by itself can analyze signals up to -6 dB SNR. However, the estimation error grows with increasing noise intensity. Hence, the use of denoising filters is essential to reduce the noise. The analysis of noisy signals is carried out in two phases. In the first phase, the noisy signal is passed through two denoising filters connected in cascade. After denoising, the SCD coefficients of the denoised signal are computed, and the contour plot is obtained. From the contour plot, all the parameters are measured. With these filters in cascade, we are able to analyze the signals up to -12 dB SNR with a maximum of 5% error. For highly noisy signals less than -12 dB SNR, the error in estimation is more than 5% for some of the parameters. It is also observed that the sampling frequency should satisfy the relation, $f_s \geq 5f_c$. This work is useful in the battlefield to protect the weapon system from the enemy.

Acknowledgements

Sincere gratitude is extended by the authors to the administration and principal of VNRVJIET, Hyderabad, for providing the resources needed to complete this work.

Funding

External funding was not obtained for this chapter.

Conflict of interest

There is no conflict of interest declared by the authors.

Author details

Raja Kumari Chilukuri[1]*, Hari Kishore Kakarla[2] and Subbarao Kakarla[3]

1 Department of Electronics and Communication Engineering, VNR Vignana Jyothi Institute of Engineering and Technology, Hyderabad, India

2 Department of Electronics and Communication Engineering, Koneru Lakshmaiah Education Foundation, Guntur, Andhra Pradesh, India

3 Department of Electronics and Communication Engineering (Retd.), Osmania University, Hyderabad, India

*Address all correspondence to: chrajakumari@gmail.com

References

[1] Cheraghinia M, Shahid A, StijnLuchieG-JG, Caytan O, Fontaine J, VanHerbruggen B, et al. A comprehensiveoverview on UWB radar: Applications, standards, signal processing techniques, datasets, radio chips, Trends and Future Research Directions. 2024:1-37. arXiv: 2402.05649v1[eess.SP]. DOI:10.48550/arXiv.2402.05649

[2] Džunda M, Dzurovčin P, Kaľavský P, Korba P, Cséfalvay Z, Hovanec M. The UWB radar application in the aviation security systems. Applied Sciences. 2021; **11**(10):4556.1-17. DOI: 10.3390/app11104556

[3] Immoreev IY. Practical Application of Ultra-Wideband Radars. In: 2006 3rd International Conference on Ultrawideband and Ultrashort Impulse Signals, Sevastopol, Ukraine. 2006. pp. 44-49. DOI 10.1109/UWBUS.2006.307156

[4] Paulson CN, Chang JT, Romero CE, Watson J, Pearce FJ, Levin N. Ultra-wideband radar methods and techniques of medical sensing and imaging. In: Proc. SPIE 6007, Smart Medical and Biomedical Sensor Technology III. Boston, MA, United States. 11 November 2005. p. 60070L. DOI: 10.1117/12.630004

[5] Pace PE. Detecting and ClassifyingLow Probability of Intercept Radars. 2nd ed. Artech House, U.K.; 2009. 857 p ISBN: 9781596932340

[6] Sulistyaningsih, Saputera YP, Wahab M. Radar signal processing development for low probability of intercept radar system. In: 2016 IEEE Region 10 Conference (TENCON), Singapore. 2016. pp. 946-950. DOI: 10.1109/TENCON.2016.7848144

[7] Althuwayb AA, Alibakhshikenari M, Virdee BS, Benetatos H, Rashid N, Kaaniche K, et al. Design technique to mitigate unwanted coupling in densely packed radiating elements of an antenna array for electronic devices and wireless communication systems operating in the millimeter-wave band. AEU – International Journal of Electronics and Communications. 2023;**159**:154464. ISSN: 1434-8411, DOI: 10.1016/j.aeue.2022.154464

[8] Althuwayb AA, Alibakhshikenari M, Virdee BS, Rashid N, Kaaniche K, Atitallah AB, et al. Meta surface-Inspired Flexible Wearable MIMO Antenna Array for Wireless Body Area Network Applications and Biomedical Telemetry Devices. IEEE Access. 2023;**11**:1039-1056

[9] Pace P. LPI Radar Class Notes, EC Network Centric Radar Electronic Warfare Techniques and Systems forInternational Students. Monterey, California: Naval Postgraduate School; 2002

[10] Liu X, Zhang T, Shi Q, Yu X, Cui G, Kong L. LPI radar waveform design with desired cyclic Spectrum and pulse compression properties. IEEE Transactions on Vehicular Technology. May 2023;**72**(5):6789-6793. DOI: 10.1109/TVT.2022.3233446

[11] Kishore TR, Rao KD. Automatic intrapulse modulation classification of advanced LPI radar waveforms. IEEE Transactions on Aerospace and Electronic Systems. 2017;**53**(2):901-914

[12] Singh AK, Rao KS. Digital receiver-based electronic intelligence system configuration for the detection and identification of intrapulse modulated radar signals. Defence Science Journal. 2014;**64**(2):152

[13] Ünal L, Pakfiliz AG. LPI radar signal detection based on autocorrelation function and Wigner-Ville distribution. Review of Computer Engineering Studies. 2022;**9**(4):125-135. DOI: 10.18280/rces.090401

[14] Das J, Munnavvar Hussain S, Pasha IA. Generation of poly-phase frequency-hopped spread Spectrum signal for LPI radar target detection. In: Communication, Software and Networks: Proceedings of INDIA. Singapore: Springer Nature Singapore; 2022. pp. 213-224

[15] Shi C, Ding L, Wang F, Salous S, Zhou J. Low probability of intercept-based collaborative power and bandwidth allocation strategy for multi-target tracking in distributed radar network system. IEEE Sensors Journal. 2020;**20**(12):6367-6377

[16] Anjaneyulu L, Sarma NVSN, Murthy NS. Identification of LPI radarsignals by higher order spectra andneural network techniques. InternationalJournal of Information and Communication Technology. 2009;**2**:142-155

[17] Kumari CR, Kakarla HK, Subbarao K. Estimation of intra-pulse modulation parameters of LPI radar under noisy conditions. International Journal of Microwave and Wireless Technologies. 2022;**14**(9): 1177-1194

[18] Chilukuri RK, Kakarla HK, Subbarao K. Estimation of modulation parameters of LPI radar using cyclostationary method. Sensing and Imaging. 2020;**21**:1-20

[19] Chen K, Chen S, Zhang S, et al. Automatic modulation recognition of radar signals based on histogram of oriented gradient via improved principal component analysis. SIViP 17. 2023: 3053-3061. DOI: 10.1007/s11760-023-02526-x

[20] Lu X, Xu Z, Ren H, Yi W. LPI-based Resource Allocation Strategy for Target Tracking in the Moving Airborne Radar Network. In: 2022 IEEE Radar Conference (RadarConf22). New York City, NY, USA. 2022. pp. 1-6. DOI: 10.1109/RadarConf2248738.2022.9764195

[21] Shyamsunder M, Subbarao K. Time Frequency Analysis of LPI radar signals usingModified S transform. International Journal of Electronics Engineering Research. 2017;**9**(8):1267-1283. ISSN: 0975-6450

[22] Alrubeaan T, Albagami K, Ragheb A, Aldosari S, Altamimi M, Alshebeili S. An investigation of LPI radar waveforms classification in RoF channels. IEEE Access. 2019;**7**:124844-124853

[23] Chilukuri RK. Analysis of low probability of intercept radar signals [thesis]. India: KL University; 2024

[24] Savci K, Stove AG, De Palo F, Erdogan AY, Galati G, Lukin KA, et al. Noise radar—Overview and recent developments. IEEE Aerospace and Electronic Systems Magazine. 2020; **35**(9):8-20

Chapter 9

Enhancing Energy Efficiency in Wireless Sensor Networks through Advanced Algorithms for Wideband Spectrum Utilization

Asma Bagheri

Abstract

This chapter introduces novel frameworks and analyses for enhancing energy efficiency in wireless sensor networks (WSNs). The research focuses on the development of advanced algorithms that enable dynamically allocation for cognitive radio sensors for wideband spectrum utilization and identify free channels, in order to maximize the network lifetime while optimizing spectrum usage. We present two scenarios for the network lifetime maximization, using both the linear optimization and also game theory. By integrating cognitive radio capabilities into sensor allocation strategies, this approach aims to significantly improve the overall performance and sustainability of WSNs over wideband sensing. Hence, the chapter provides insightful technical details and potential implications of this innovative methodology, shedding light on its potential to revolutionize energy-efficient and spectrum-aware WSN deployments. Additionally, it emphasizes that the proposed solutions and algorithms in this chapter can be easily extended and applied to other sensor networks with diverse functionalities. Finally, simulations are presented to demonstrate the ability of the algorithms, assuming the systems using IEEE802.15.4/Zigbee and IEEE802.11af.

Keywords: cognitive radio sensor network, dynamic sensor allocation, energy efficiency, sustainable deployment, wireless sensor network, wideband spectrum utilization

1. Introduction

Wireless sensor networks (WSNs) offer numerous advantages, including low cost, compact size, multifunctionality, self-organization, and routing capabilities facilitated by WSN protocols. However, these benefits are counterbalanced by several challenges that impede their widespread application. Key among these challenges are issues related to energy efficiency. One significant challenge in WSNs is the limitation imposed by low battery capacity, which directly impacts the network's lifetime and operational capabilities. The finite energy resources available to sensor nodes necessitate efficient energy management strategies to prolong network longevity and maintain uninterrupted operation. Reducing number of sensing nodes is one of the main directions of possible energy conservation methods, which can be performed through node selection, censoring, or voting schemes [1]. These approaches aim to optimize energy consumption by strategically activating sensors, only, when necessary, thereby maximizing the network's efficiency.

One of the efficient uses of WSNs is using the tiny, low-power, and low-cost frequency sensors for spectrum sensing (SS) [2], that is, wireless cognitive sensor network (WCSNs). SS is a functionality of cognitive systems, which is done to detect the spectrum holes, that is, the frequency channels that are not utilized by the primary users (PUs) [3]. In a cognitive radio network (CRN), a secondary user (SU) cannot sense and transmit simultaneously if it does both sensing and transmission on detected idle channels, since it needs to parallel receiver and transmitter circuits so that it reduces the opportunistic access efficiency. In addition, one SU might fail to sense all the sensing area due to improper positions. So, WCSNs are good solutions. However, the sensors might not be able to reliably detect weak PU signals due to fading or shadowing effects; cooperative spectrum sensing (CSS) schemes are proposed based on the combination of the detection results from spatially distributed multiple sensors [4]. But this scheme highly increases the energy consumption and severely limits the WCSN lifetime.

In this chapter, we specifically focus on WCSNs where sensors are employed for spectrum sensing. Then, we offer various strategies to reduce the number of sensors while enhancing energy efficiency (EE) and network lifetime, ensuring optimal performance of CSS. Key approach is intelligent sensor selection aimed at optimizing energy consumption while preserving the quality of CSS. In any case, the solutions presented here are easily generalizable and extendable to other WSNs with different sensing tasks. These strategies for optimizing EE and network performance through sensor election, whether cooperative or noncooperative decision-making, can be adapted to various WSN applications. The overarching goal remains consistent: sensor selection to maximize EE while maintaining the quality of sensing tasks.

The next subsection describes the general WCSN model. Then, the proposed solutions for EE and the network lifetime extension are presented in Sections 2 and 3.

1.1 System model

A WCSN comprising N sensors uniformly distributed across a rectangular region, with a Fusion Center (FC) situated at the origin is assumed in this chapter. The spectrum consists of M identical bandwidth channels utilized by M PUs, where each PU operates exclusively on one channel. Assumptions include the FC knows the sensor and PU positions, as well as PU-sensor channel coefficients, facilitating the computation of instantaneous Signal-to-Noise Ratios (SNRs) at the sensors. These assumptions do not impair the functionality of the proposed algorithms, which are designed for sensor selection based on either average or estimated SNR criteria. **Figure 1** illustrates a representative WCSN, and **Table 1** provides definitions of notations used throughout this chapter.

The sensors are equipped with low-complexity circuits due to their limited size and cost-effective design constraints, restricting them from simultaneous multichannel sensing. However, through collaboration and clustering within WCSNs, concurrent sensing across multiple channels becomes feasible. Each sensor is equipped with a receiver circuit comprising a synthesizer, narrowband filter, and energy detector (ED), as illustrated in **Figure 2**. The synthesizer is controlled by commands from the FC to tune into the center frequency of a specified channel. The ED, chosen for its

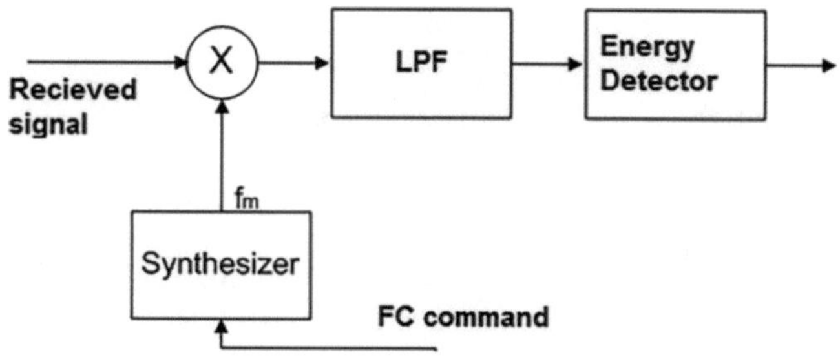

Figure 1.
The receiver circuit of every cognitive sensor [5, 6].

N	Number of sensors	M	Number of channels/PUs
m, m'	Number of a specific channel.	J_m	Selected nodes set for sensing m-th channel
f_s	Nyquist sampling	δ	Sensing time
E_{t-elec}	Energy for the electronic circuits	e_{amp}	Amplifying coefficient
EC_n	Consumed energy for sensor n	E_s/E_{tn}	Sensing/transmission energy
P_t	PU signal power	s_m	PU signal with variance σ_s^2 over channel m
$h_{m,n}$	PU-sensor channel gains	$v_{m,n}$	Noise of PU-sensor channel$\sim \mathcal{N}(0, \sigma_V{}^2)$
$Z_{n,m}$	Shadowing with variance σ_z^2	$\widetilde{h_{m,n}}$	Rayleigh fading over PU-sensor channel
$d_{m,n}$	m-th PU to n-th sensor distance	$d_{0,n}$	n-th sensor to FC distance
E_n	The n-th sensor remaining energy	$\mathcal{L}$	Live sensors ratio in lifetime moment of WCSN
$H_{0,m}/H_{1,m}$	State off/on for m-th PU	β_m/α_m	GDP/GFP limit for channel m
$P_{d_{m,n}}/P_{f_{m,n}}$	DP/FP of n-th sensor about m-th PU	P_{d_m}/P_{f_m}	GDP/GFP for the m-th channel
Λ	The PU signals wavelength	γ / γ_m	Detector threshold for every channel

Table 1.
The description of parameters.

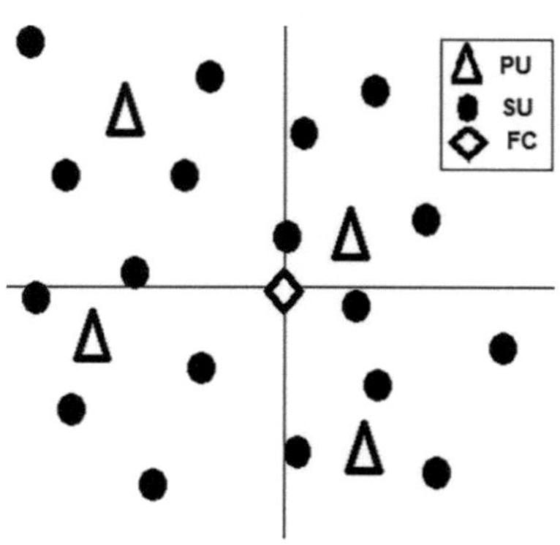

Figure 2.
A sample of system model [5, 6].

simplicity and cost-efficiency, measures the signal energy within each channel. The test statistic $T_{m,n} = \sum_{k=1}^{K=\delta f_s} |x_{m,n}(k)|^2$ serves as the detector's decision metric, where $x_{m,n}(k)$ denotes the k-th sample of the received signal on the m-th channel observed by the n-th node. Here, K is defined as $K = \delta f_s$ [7]. Each sensor compares the calculated energy against a predetermined threshold, generating a binary decision bit. A decision bit of zero indicates energy levels below the threshold, while a bit of one signifies levels above it. Thus, two hypotheses are formulated for each channel: $H_{0,m}$, indicating non-utilization by the primary system and $H_{1,m}$, indicating PU transmission on the m-th channel. So, we have [8]:

$$\begin{cases} H_{1,m};\, x_{m,n}(k) = h_{m,n}.s_m(k) + v_{m,n}(k) \\ H_{0,m};\, x_{m,n}(k) = v_{m,n}(k) \end{cases} \tag{1}$$

$$\begin{cases} H_{1,m};\, T_{m,n} \geq \gamma;\, T_{m,n} \sim \mathcal{N}(0, {\sigma_{1,m,n}}^2) \\ H_{0,m};\, T_{m,n} \leq \gamma;\, T_{m,n} \sim \mathcal{N}(0, {\sigma_v}^2) \end{cases} \tag{2}$$

The noise on the m-th channel, denoted as $v_{m,n}$, is assumed as an independent and identically distributed (i.i.d.) Gaussian distribution. The k-th sample of the primary signal transmitted on the m-th channel $s_m(k)$ is modeled as an i.i.d. Gaussian process with zero mean and variance σ_S^2. Furthermore, s_m and $v_{m,n}$ are considered independent. The channel gain $h_{m,n}$ between the m-th PU and the n-th sensor incorporates path loss, Rayleigh fading, and shadowing effects, and is modeled as [9]:

$$h_{m,n} = 10^{\frac{20 \log\left(\frac{\Lambda}{4\pi d_{m,n}}\right) + Z}{20}} \widetilde{h_{m,n}} \tag{3}$$

where $\widetilde{h_{k,n}}$ follows a standard complex Gaussian random process (Rayleigh fading), Z is a Gaussian random variable (in dB) with zero mean and variance σ_z^2, and $\left(\frac{\Lambda}{4\pi d_{m,n}}\right)^2$ accounts for the free-space path loss component, where Λ is the wavelength and $d_{m,n}$ is the distance between the PU using the m-th channel and the n-th sensor. Under $H_{1,m}$, the SNR for the n-th sensor measuring the m-th PU's signal is [8]:

$$SNR_{m,n} = \frac{p_t\, |h_{m,n}|^2 \sigma_{s_{m,n}}^2}{\sigma_v^2} \tag{4}$$

Two metrics for spectrum sensing quality are false alarm probability (FP) and detection probability (DP). These metrics are respectively defined as [8]:

$$P_{f_{m,n}} = P(H_{1,m}|H_{0,m}) = Q\left(\left(\frac{\gamma}{\sigma_v^2} - 1\right)\sqrt{\delta f_s}\right) \tag{5}$$

$$P_{d_{m,n}} = P(H_{1,m}|H_{1,m}) = Q\left(\left(\frac{\gamma}{\sigma_v^2} - SNR_{m,n} - 1\right)\sqrt{\frac{\delta f_s}{2SNR_{m,n} + 1}}\right) \tag{6}$$

where Q(.) denotes the complementary distribution function.

Due to channel fading, shadowing, or limited sensing range, sensors may fail to detect a PU correctly. Thus, to enhance sensing performance, CSS is proposed [4]. However, participation of all sensors in CSS can lead to high-energy consumption and increase the global FP (GFP) without significantly enhancing global DP (GDP) [3]. The coefficient $\varphi_{m,n}$ is introduced to designate selected sensors: $\varphi_{m,n} = 0$ if the n-th sensor does not sense the m-th channel, and $\varphi_{m,n} = 1$ if it does. Additionally, it is assumed that each sensor can sense at most one channel per sensing period due to limited Analog-to-Digital Converters (ADCs). Thus, for the m-th channel, J_m denotes a set of cooperating sensors, under the following conditions:

$$\begin{cases} J_m \cap J_{m'} = \varnothing \\ J_m \cup J_{m'} \subset \{1, \ldots, N\} \end{cases} \quad \forall m, m' \in \{1, .., M\}, m \neq m' \tag{7}$$

Then, the selected nodes transmit their decision bits to an FC, which makes the final decision based on the logical OR rule [10]. According to this rule, if at least one sensor detects PU transmission on the m-th channel, the final decision indicates channel occupancy. The GDP and GFP for the m-th channel are calculated respectively [5, 8]:

$$P_{d_m} = 1 - \prod_{n \in J_m} \left(1 - \varphi_{m,n} P_{d_{m,n}}\right) \tag{8}$$

$$P_{f_m} = 1 - \prod_{n \in J_m} \left(1 - \varphi_{m,n} P_{f_{m,n}}\right) \tag{9}$$

We want to maximize the lifetime of the multichannel CSS in WCSN under the constraints on the GDP and GFP. As mentioned before, because of battery size and weight limitations of sensors, power scarcity is an important issue that imposes time and efficiency limits on the operation network lifetime. Therefore, energy utilization of sensors has a great impact on extending the network lifetime. We use the following definition of lifetime because when only some of the nodes have enough energy, the duty of spectrum sensing can be performed:

Definition 1. We define the lifetime of a WCSN as the time that the number of live sensors drops under $\mathcal{L}.N$ where $0< \mathcal{L} \leq 1$, [11]. ■

To formulate the lifetime extension problem, first we calculate the consumed and remaining energy of sensors. It is assumed that the sensing energy of sensor, that is, E_s constant, is the same for all sensors. The energy for transmitting the n-th sensor decision bit to the FC is E_{tn}, which we calculate it as [5, 8]:

$$E_{tn} = E_{t-elec} + e_{amp}.{d_{0,n}}^2 \tag{10}$$

in which E_{t-elec} is the energy used for the electronic circuits of the transmitter, e_{amp} is the amplifying coefficient, and $d_{0,n}$ is the sensor-FC distance. Therefore, the consumed energy of the sensor for sensing every channel is calculated as [5, 8]:

$$EC_n = E_s + E_{tn} \tag{11}$$

The following sections present two methods for extending lifetime of WCSNs based on sensor selection.

2. Sensor selection methods for lifetime maximization of WCSNs

In this section, two lifetime maximization schemes are presented based on sensor selection. These schemes select sensors for sensing different channels in different sensing durations, under some detection quality constraints. The reason behind such constraints is that the purpose of a WCSN is sensing the spectrum activities. Therefore, the GDP and GFP are limiting the sensor selection problem. A high GDP must be achieved to reduce the amount of interference with the PUs. Also, when a false alarm occurs, it is decided that a PU is transmitting, while the channel is free; therefore, a high GFP is highly undesirable. In an ideal CSS, it should be GDP = 1 and GFP = 0 for every channel, but such perfect CSS is impossible in actual systems because, from [8, 12], it is concluded that increasing the GDP leads to increase the GFP. Therefore, we define an upper-bound for GFP and a lower-bound for GDP denoted by α and β, respectively.[1] Now, the problem is presented as:

$$\text{Problem 1} \qquad \max_{J_m} \text{lifetime} \qquad \forall m \in \{0, \dots, M\} \tag{12}$$

subject to;

$$P_{f_m} \leq \alpha \qquad \forall m \in \{1, \dots, M\} \tag{13}$$

$$P_{d_m} \geq \beta \qquad \forall m \in \{1, \dots, M\} \tag{14}$$

$$J_m \cap J_{m'} = \varnothing, \ \bigcup_{m=0}^{M} J_m = \{1, \dots, N\} \qquad \forall m, m' \in \{0, \dots, M\}, m \neq m' \tag{15}$$

The last constraint states that one sensor cannot sense more than one channel in a sensing duration. The "max-min" method is a good solution for optimizing the network lifetime [13]. In this method, the minimum remaining energy of sensors is maximized. Thus, the sensors that have the lower remaining energy are not being selected for sensing. It leads to the remaining energy level of sensors keep balanced and consequently to extend the network lifetime. The remaining energy of sensors and the minimum of remaining energy of sensors are denoted by E_n and E_{th}, respectively. Now, the problem is formulated as:

$$\text{Problem 2} \qquad \max_{J_m} \left\{ \min_{n} \{E_n\} = E_{th} \right\} \qquad \forall m, n \tag{16}$$

subject to;

$$P_{f_m} \leq \alpha \qquad \forall m \in \{1, \dots, M\} \tag{17}$$

$$P_{d_m} \geq \beta \qquad \forall m \in \{1, \dots, M\} \tag{18}$$

$$J_m \cap J_{m'} = \varnothing, \ \bigcup_{m=0}^{M} J_m = \{1, \dots, N\} \qquad \forall m, m' \in \{0, \dots, M\}, m \neq m' \tag{19}$$

The problem is classified as NP-complete, indicating that achieving the optimal solution necessitates an exhaustive search with prohibitively high computational

[1] Other constraints could be defined for different WSNs with other duties, similarly. These constraints describe the sensing quality of sensors.

complexity. In the following, we present two alternative methods to solve this challenge by providing reduced complexity solution, albeit at the cost of optimality. These methods are simulated to assess their efficacy in addressing the problem.

2.1 Sensor selection using convex method

First approach employs convex optimization techniques to derive a solution for prioritizing sensor selection. Subsequently, an algorithm is formulated to perform the solution. To tackle the complexity of the lifetime maximization problem due to the integer assignment parameter for the sensors, here the parameter is treated as continuous, representing the priority levels of the sensors. Upon resolving the optimization problem, the continuous parameter is mapped to the real integer domain. In this context, the optimization problem can be formulated as follows:

$$\text{Problem 3} \qquad \max_{\varphi_{m,n}} \qquad E_{th} \tag{20}$$

subject to;

$$\left(E_{th} - \varphi_{m,n}\left(E_{0,n} - \sum_{m=1}^{M} \varphi_{m,n} E_{cn}\right)\right) \le 0 \qquad \forall n \in \{1, \ldots, N\} \tag{21}$$

$$\beta_m - \left(1 - \prod_{n \in J_m} \left(1 - \varphi_{m,n} P_{d_{m,n}}\right)\right) \le 0 \qquad \forall m \in \{1, \ldots, M\} \tag{22}$$

$$\sum_{m=1}^{M} \varphi_{m,n} - 1 \le 0 \qquad \forall n \in \{1, \ldots, N\} \tag{23}$$

$$\sum_{n=1}^{N} \varphi_{m,n} - J_{m_{max}} \le 0 \qquad \forall m \in \{1, \ldots, M\} \tag{24}$$

$$\varphi_{m,n} \ge 0 \qquad \forall m, n \tag{25}$$

Although the non-convex nature of the second constraint, a convex approach can be employed to find a sub-optimal yet efficient solution. This method is particularly valuable for handling non-convex problems and has been extensively utilized in prior research [14]. Additionally, the algorithm verifies constraint (25) to ensure that the variables $\varphi_{m,n}$ remain within specified bounds, potentially allowing for its removal. By introducing Lagrange multipliers ρ_n, λ_m, η_n, and ξ_m corresponding to constraints (21)–(24), respectively, we construct the Lagrange function as [14]:

$$\begin{aligned} L(\varphi_{m,n}, \rho_n, \lambda_m, \eta_n, \xi_m) = E_{th} &- \sum_n \rho_n \left(E_{th} - \varphi_{m,n}\left(E_{0,n} - \sum_{m=1}^{M} \varphi_{m,n} E_{cn}\right)\right) \\ &- \sum_m \lambda_m \left(\beta_m - 1 + \prod_{n \in J_m} \left(1 - \varphi_{m,n} P_{d_{m,n}}\right)\right) \\ &- \sum_n \eta_n \left(\sum_{m=1}^{M} \varphi_{m,n} - 1\right) - \sum_m \xi_m \left(\sum_{n=1}^{N} \varphi_{m,n} - J_{m_{max}}\right) \end{aligned} \tag{26}$$

The optimal value of $\varphi_{m,n}{}^{*}$ is found by differentiating L with respect to $\varphi_{m,n}$ [14]:

$$\frac{\partial L}{\partial \varphi_{m,n}} = \rho_n E_{0,n} - 2\rho_n \varphi_{m,n} E_{cn} + \lambda_m P_{d_{m,n}} \prod_{\substack{\ell \in J_m \\ \ell \neq n}} \left(1 - \varphi_{m,\ell} P_{d_{m,\ell}}\right) - \eta_n - \xi_m = 0 \quad (27)$$

Now, the sensor priority to detect the m-th channel is calculated as:

$$\varphi_{m,n} = \frac{-\xi_m - \eta_n + \rho_n E_{0,n} + \lambda_m P_{d_{m,n}} \prod_{\substack{\ell \in J_m \\ \ell \neq n}} \left(1 - \varphi_{m,\ell} P_{d_{m,\ell}}\right)}{2\rho_n E_{cn}} \quad (28)$$

To obtain the optimum values of Lagrange multipliers in Eq. (28), the complimentary slackness conditions are analyzed:

$$\text{condition 1}: \rho_n{}^{*}\left(E_{th} - \varphi_{m,n}\left(E_{0,n} - \sum_{m=1}^{M} \varphi_{m,n} E_{cn}\right)\right) = 0$$

$$\begin{cases} \rho_n{}^{*} > 0, E_{th} = \varphi_{m,n}\left(E_{0,n} - \sum_{m=1}^{M} \varphi_{m,n} E_{cn}\right) \\ \rho_n{}^{*} = 0, E_{th} < \varphi_{m,n}\left(E_{0,n} - \sum_{m=1}^{M} \varphi_{m,n} E_{cn}\right) \end{cases} \quad (29)$$

$$\text{condition 2}: \lambda_m{}^{*}\left(\beta_m - 1 + \prod_{n \in J_m} \left(1 - \varphi_{m,n} P_{d_{m,n}}\right)\right) = 0$$

$$\begin{cases} \lambda_m{}^{*} > 0, \beta_m = 1 - \prod_{n \in J_m} \left(1 - \varphi_{m,n} P_{d_{m,n}}\right) \\ \lambda_m{}^{*} = 0, \beta_m < 1 - \prod_{n \in J_m} \left(1 - \varphi_{m,n} P_{d_{m,n}}\right) \end{cases} \quad (30)$$

$$\text{condition 3}: \eta_n^{*}\left(\sum_{m=1}^{M} \varphi_{m,n} - 1\right) = 0 \begin{cases} \eta_n^{*} > 0 \sum_{m=1}^{M} \varphi_{m,n} = 1 \\ \eta_n^{*} = 0 \sum_{m=1}^{M} \varphi_{m,n} < 1 \end{cases} \quad (31)$$

$$\text{condition 4}: \xi_m{}^{*}\left(\sum_{n=1}^{N} \varphi_{m,n} - J_{m_{max}}\right) = 0 \begin{cases} \xi_m{}^{*} > 0 \sum_{m=1}^{M} \varphi_{m,n} = J_{m\_max} \\ \xi_m{}^{*} = 0 \sum_{m=1}^{M} \varphi_{m,n} < J_{m\_max} \end{cases} \quad (32)$$

The first and second conditions must be satisfied for all sensors and channels, respectively. Condition (31) ensures the selection of sensors for sensing channels, while condition (31) guarantees sensors remain unselected across all sensing durations. As selected sensors are excluded from subsequent channel allocations, η_n from

Eq. (28) becomes redundant. The fourth condition applies uniformly across all channels, allowing for the removal of ξ_m and a focus solely on the count of selected nodes. Consequently, the final calculation of sensor priorities is as follows:

$$\text{priority}_m(n) \triangleq \frac{\rho_n(E_{0,n} - E_{th}) + \lambda_m P_{d_{m,n}}}{2\rho_n EC_n} \quad (33)$$

It is observed that sensor priority for channel sensing is inversely proportional to its energy requirement for participating in the CSS and directly correlated with its residual energy and DP for channel sensing. Hence, a higher value of the priority function $\text{priority}_m(.)$ signifies greater priority for sensing the m-th channel. Determining the optimal values of Lagrange multipliers in Eq. (33) presents a multidimensional problem necessitating numerical methods. We employ the Subgradient method, known for its simplicity and rapid convergence, to address this challenge. The parameters ρ_n and λ_m are updated iteratively with step sizes $\ell 1(i) = \frac{C1}{i}$ and $\ell 2(i) = \frac{C2}{i}$ [14]:

$$\rho_n(i) = \rho_n(i-1) + \ell 1(i)(E_{th} - E_{0,n} + E_{m,n}) \quad (34)$$

$$\lambda_m(i) = \lambda_m(i-1) + \ell 2(i)(\beta_m - P_{d_m}) \quad (35)$$

Now, sensor prioritization for channel sensing is determined by its inverse energy requirement for CSS participation and direct correlation with residual energy and channel-specific DP. Consequently, higher values of $\text{priority}_m(.)$ indicate increased preference for sensing the m-th channel. Optimal Lagrange multiplier values in Eq. (33) are determined through a multidimensional numerical approach, specifically employing the Subgradient method for its efficient convergence and simplicity [14]. Iteratively, parameters ρ_n and λ_m are updated with step sizes $\ell 1(i) = \frac{C1}{i}$ and $\ell 2(i) = \frac{C2}{i}$, respectively, where "i" denotes the iteration number and C1, C2 are constants [14].

The Subgradient algorithm continues until changes in parameters fall below a small value ϵ, defined as:

$$\max(|\lambda_m(i) - \lambda_m(i-1)|, |\rho_n(i) - \rho_n(i-1)|) \geq \epsilon \quad (36)$$

Sensor selection for each channel is based on expression in Eq. (33). Sensors are assigned based on their distance from the center of the region, with farther PUs given higher priority due to likely fewer sensors in proximity. Simultaneous selection for all channels begins with initial selection of one sensor per channel. Constraints on GDP are checked sequentially; if unsatisfied, additional sensors are selected per channel, up to a maximum of J_{m_max} sensors. Detailed pseudo-code for the convex-based algorithm is provided in **Table 2**. In the subsequent section, we explore a game theory-based approach to extend the lifetime of WCSNs by solving the sensor selection problem.

2.2 Sensor selection using game theory

Game Theory (GT) serves as a robust method for modeling interactions and cooperation among players, particularly in multiuser environments [15, 16]. In the context of Wireless Sensor Networks (WSNs), coalition formation games (CFGs) are pivotal for fostering cooperation among sensors to maximize utility [17, 18]. CFGs

Step1:for m=1:M, Compute J_{m_max} , Compute the m-th PU distance **end**
[PU_pri, PU_dis]=Sort distances descending
it=0 % lifetime iteration
While (Set2 < (1- $\mathcal{L}$)N)% lifetime loop
$\lambda_m = \frac{\lambda_{m\min}+\lambda_{m\max}}{2}$ & $\rho_n = \frac{\rho_{n\min}+\rho_{n\max}}{2}$ & Set1=zeros(1,N) , i=0% Subgradient iteration
Step2: While stop>=ϵ%Start Subgradient loop
  for j=1: J_{m_max}
    for ii=1:M m= PU_pri(ii) & C1(m)=0 & C2(m,j)=0
      for n=1:N
        if (Set1=0 & $E_{0,n}$>=$E_{m,n}$)
          Compute priority of sensors according (33) & C1(m)= C1(m)+1
        else $\text{priority}_{m}(n) = 0$
        end
      end
    [S_pri, S_ind]=Sort sensor priorities descending & $P_{d_m} = 1$
  if $(\text{find}(C2(m,\ :)) < J_{m\_max})$&&$P_{d_m} < \beta_m$
ns= the selected sensor from S_ind & Set1(ns)=1 , C2(m,j)= S_ind(ns)
  $P_{d_m} = 1 - (P_{d_m} * (1 - P_{d_{m,ns}}))$
  end
 end
end
Step3: Update $E_{0,n}$ and E_{th} & Calculate λ_m and ρ_n according (34) and (35)
$\lambda_m = \max(\min(\lambda_{m\max}, \lambda_m), \lambda_{m\min})$; & $\rho_n = \max(\min(\rho_{n\max}, \rho_n), \rho_{n\min})$; Compute the Subgradient stop value, i=i+1;
 end% for the Subgradient loop Update Set2 & it=it+1
end % for lifetime loop

Set1: Saves all the selected sensors such that the sensors not be the same.
C1(m): Number of sensors with enough remaining energy for sensing the m-th channel.
C2(m,:): Selected sensors for sensing the m-th channel.
Set2: Off-sensor number

Table 2.
Convex-based algorithm.

excel in encouraging players (sensors in this case) to collaborate effectively towards achieving optimal outcomes.

Given the effectiveness of CFGs in cooperative problem-solving, this approach is explored for sensor selection in WSNs. An algorithm based on CFG is introduced to maximize the lifetime of WSNs engaged in Cooperative Spectrum Sensing (CSS) while considering sensing quality constraints and the limited capabilities of sensors in multichannel environments. In this section, sensors are conceptualized as players in a CFG, where they decide, based on their rewards, whether to join a coalition to collectively sense a channel or abstain from CSS participation. The game is formally defined as [5]:

$$\mathcal{G} = \left\{\mathcal{N}, \{\mathcal{A}_n\}_{n\in\mathcal{N}}, \{u(J_m)\}_{J_m\subset\mathcal{N}}\right\} \tag{37}$$

Here, $\mathcal{N} = \{1, \ldots, N\}$ denotes the sensors as game players. $\mathcal{A}_n = \{a_{n0}, a_{n1}, \ldots, a_{nM}\} = \{0, 1, \ldots, M\}$ denotes the available actions for player n, where $a_{n0} = 0$ signifies not sensing any channel, and $a_{nm} = m$ denotes participation in CSS for the m-th channel. An action profile $(a_1, \ldots, a_N)$ shows the selected actions by the players, when $a_n \in \mathcal{A}_n$.

Each CFG includes a utility function that quantifies the value of potential coalitions. The utility $u(J_m)$ of coalition J_m in sensing channel m is pivotal in this game. The goal is to select sensors for sensing M channels while satisfying detection quality constraints (17) and (18). To minimize energy consumption and Green Function Penalty (GFP), some sensors will be activated while others will power down their sensing modules during designated sensing durations. Consequently, the number of coalitions equals M + 1: M coalitions for sensing M channels and one coalition where sensors sense none of the channels (coalition zero). Moreover, as per constraint (19), these coalitions are mutually exclusive.

Thus, $u(J_m)$ must be an increasing function of the Detection Gain (GDP) and Energy Threshold E_{th} within coalition (J_m) and a decreasing function of GFP and energy consumption associated with sensors in the coalition. It is assumed that sensors opting out of channel sensing (coalition zero) receive zero utility, as they neither gain nor consume resources. A suitable utility function for coalition (J_m) in cooperative channel sensing is [5, 6]:

$$u(J_m) = \begin{cases} \text{if } P_{d_m}(J_m) \geq \beta, P_{f_m}(J_m) \leq \alpha \begin{cases} 1 + E_{th}; \text{if } E_{th} > E_{th,0} \\ \dfrac{\max(E) - E(J_m)}{\max(E) - \min(E)}; \text{if } E_{th} = E_{th,0} \\ \vartheta \dfrac{\max(E) - E(J_m)}{\max(E) - \min(E)}; \text{if } E_{th} < E_{th,0} \end{cases} \\ , \\ \text{if } P_{d_m}(J_m) < \beta \text{ or } P_{f_m}(J_m) > \alpha \begin{cases} \dfrac{-\min(E)}{\max(E)} \dfrac{E_0 - E_{th}}{E_0}; \text{if } E_{th} > E_{th,0} \\ \dfrac{-E(J_m)}{\max(E)}; \text{if } E_{th} = E_{th,0} \\ -1 - \dfrac{E_0 - E_{th}}{E_0} - E(J_m); \text{if } E_{th} < E_{th,0} \end{cases} \end{cases} \tag{38}$$

where $E(J_m)$ denotes the sum of energy consumption of sensors in coalition J_m; $\min(E)$ denotes the minimum energy consumption for sensing a channel, equivalent to the energy consumption of a single node with the lowest energy; and $\max(E)$ denotes the maximum energy consumption assuming all sensors are active. The coefficient $\vartheta = \frac{0.5 \min(E)}{\max(E) - \min(E)}$ facilitates sensor selection into coalitions. A sensor can join a coalition in two ways: by adding to it while all other nodes remain unchanged, or by replacing another node n_0 in the coalition, causing n_0 to move to coalition zero. Both actions yield a reward or penalty based on the Shapely value [19]. The reward $Re_n(J_m)$ for the first scenario is calculated as:

$$Re_n(J_m) = u(J_m + \{n\}) - u(J_m) \tag{39}$$

where $u(J_m)$ is the coalition utility before adding sensor n, and $u(J_m + \{n\})$ is the utility after adding n. In the second scenario, the reward $Re_n(J_m)$ is:

$$Re_n(J_m) = u(J_m + \{n\} - \{n_0\}) - u(J_m) \tag{40}$$

While (live sensors number > = 0.25*N)% lifetime loop
Step1: live sensors randomly select channels.
Step2: while (at least one of the latest N players have changed their decision) % This is the stop criteria.
1. a live sensor n is selected randomly
2. The sensor n calculates its reward from adding to any coalition, from (39).
3. The sensor n calculates its reward from replacing any of the sensors in the current coalitions, from (40), from coalition1 to M.
4. The sensor n compares the rewards and decides to sense or not, and if it senses, it joins which of the coalitions (in this way, it selects the highest reward).
5. Update the $E_{th,0}$.

end
Step3: Update the remaining energy of sensors, live sensors, and life numbers.
end % for lifetime loop

Table 3.
The GT-based algorithm.

where $u(J_m + \{n\} - \{n_0\})$ denotes the utility when sensor n_0 is replaced by sensor n_0 Positive rewards indicate an increase in coalition value, while negative rewards signify a decrease.

An algorithm is devised for the coalition formation game (CFG) aimed at extending the lifetime of Cooperative Spectrum Sensing (CSS) in WSNs. Details and proof of the algorithm's convergence to a Nash equilibrium (NE) without cycling are provided in [5, 6, 17]. An action profile $a_1, \ldots, a_N$ is a strategy NE if no player can unilaterally increase its utility by deviating (**Table 3**) [20, 21].

2.3 Simulation results of the sensor selection algorithms

This section showcases simulation results of algorithms designed for sensor selection in WCSNs using MATLAB. The Monte-Carlo method is employed across 1000 iterations. Network lifetime is defined as the number of iterations where more than 25% of sensors remain operational, multiplied by the fixed duration between sensor selections (for uniform plotting purposes).

Key parameters include a square region with length L, an FC located centrally, N sensors $(5 \leq N \leq 40)$, and uniformly distributed PUs within the region. PUs transmit at 20 mW, and cognitive sensors utilize IEEE 802.15.4/Zigbee protocols. Additional simulation parameters are detailed in **Table 4**.

We compare the average results of the convex-based with the GT-based algorithms meanwhile comparing their results with the random sensor selection algorithm and also the detection-based sensor selection algorithm (we call it as DBSS) [22]. The reasons behind these comparisons are: The random algorithm has the lowest complexity order; the detection-based algorithm finds acceptable solution, if there is,

$f_c = 2.45$ GHZ	$E_{0,n} = 0.2$ mJ
$e_{amp} = 40.4$ pJ/m^2	$E_s = 190$ nJ
$E_{t-elec} = 80$ nJ	$\sigma_z^2 = 3$ db
$\alpha = 0.1$	$\sigma_0^2 = 10^{-11}$ W
$\beta = 0.9$	L = 200 m

Table 4.
The values of simulation parameters.

meanwhile the complexity order of the algorithm is low [23]. In [5, 6, 22], more cases, such as IEEE 802.11af system (operating the TV white spaces that will be used for many Internet of Things (IoT) applications [24–26]), are assumed for simulations to evaluate these algorithms.

In **Figure 3**, the network lifetime is plotted versus the total number of sensors. It shows that the GT algorithm extends the WCSN lifetime more than the others. In a lower number of sensors, it provides (about 60%) more iterations than the convex algorithm (for N = 10, the lifetime of the GT and the convex algorithms are about 800 and 500 iterations, respectively). When the total number of sensors increases, the network lifetime of all the algorithms increases, because the number of sensors with high DP increases; the number of lifetime iterations for the convex and GT-based algorithms is closer because the difference between consumption energy of sensors decreases and also the ratio of sensors number to the channel number decreases; the number of lifetime iterations for the GT algorithm is much higher than the random and the DBSS algorithms because the algorithms ignore remaining energy and consumption energy of sensors.

For the purpose of maximizing lifetime, the minimum remaining energy of nodes was optimized. **Figure 4** illustrates the average $\mathbf{E_{th}}$ over lifetime iterations. Both GT and convex algorithms maintain high $\mathbf{E_{th}}$ values across similar iteration numbers due to their consideration of remaining node energy in sensor selection. Conversely, DBSS exhibits the lowest $\mathbf{E_{th}}$ as it prioritizes sensors with high detection probability (DP), leading to rapid depletion of $\mathbf{E_{th}}$. Comparing GT and convex algorithms, GT selects sensors ensuring higher $\mathbf{E_{th}}$ than the convex algorithm, which prioritizes channel assignment efficiency over energy optimization [22]. Given that WCSN aims to monitor PU spectral activities, the selected algorithm must extend lifetime while ensuring acceptable Detection Probability (DP) and Green Function Penalty (GFP).

To evaluate this, a success rate metric assesses algorithm performance in meeting GDP and GFP constraints. **Figure 5** compares success rates of GT-based and convex-based algorithms with random selection and exhaustive search, which evaluates all possible solutions to select the optimal one (if acceptable). Exhaustive search, due to its complexity, is plotted for a limited number of sensors. Success rate is defined as the

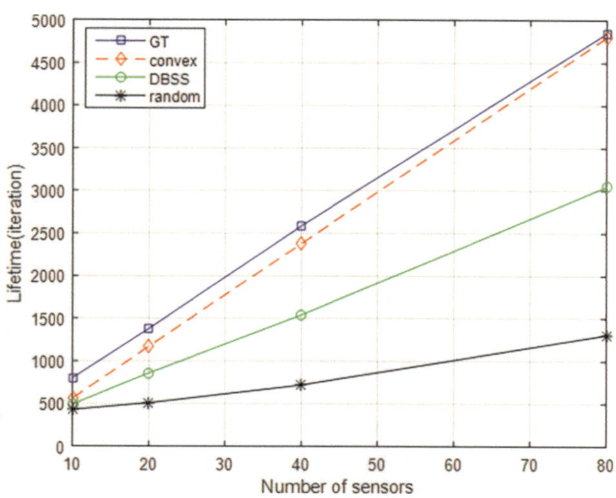

Figure 3.
The average lifetime versus the total number of sensors (M = 4) [5].

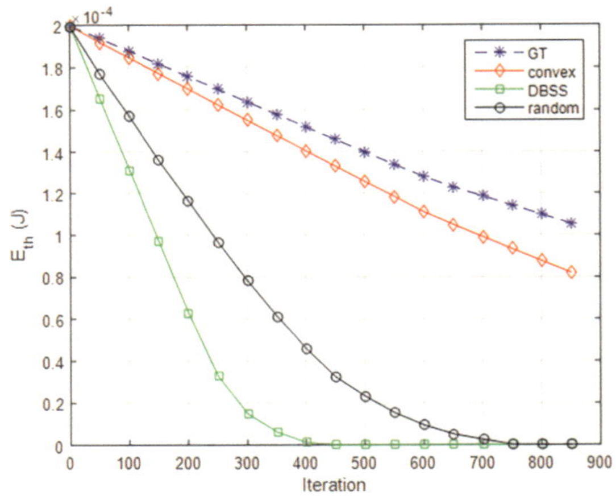

Figure 4.
The average minimum remaining energy of sensors during lifetime iterations (N = 40, M = 4) [5].

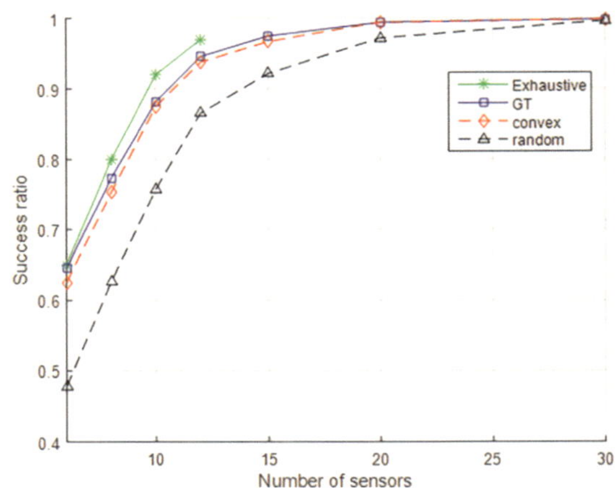

Figure 5.
The success rate versus the number of sensors (M = 2).

ratio of iterations where an algorithm finds an acceptable solution to the total iterations provided. Both GT and convex algorithms demonstrate success rates close to the optimal solution, improving with increasing sensor count due to higher DP selections.

In **Figure 6**, the average energy consumption is about 60% less than the random algorithm (for N = 10, the energy consumption of the random algorithm is approximately 1200, and the energy consumption of the GT and the convex algorithms are about 600). Average energy consumption of sensors over a network lifetime iteration. The random algorithm leads to increased average energy consumption as the number of nodes grows, due to the need to satisfy GDP constraints. For instance, with 10 nodes, the random algorithm's energy consumption is approximately 1200, significantly higher than the GT and convex algorithms, which consume about 600 each. As the number of nodes increases, from 10 to 20, other algorithms show increased average energy consumption as more sensors are selected to meet GDP constraints. However, at higher node counts, average energy consumption decreases as more sensors with higher DP become available.

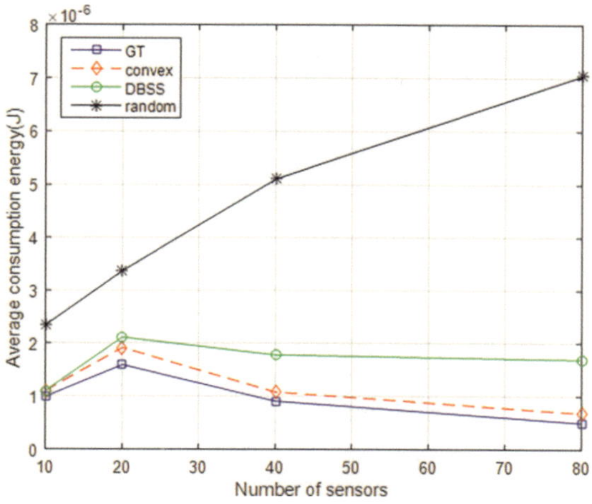

Figure 6.
The average energy consumption at different number of sensors (M = 4) [5].

Nevertheless, the GT and convex algorithms consistently offer the longest network lifetime and highest success ratio at the lowest average energy consumption, as they select sensors based on consumption, remaining energies, and DP.

Simulations affirm the efficiency of GT and convex methods in sensor selection for WCSNs, optimizing network lifetime while maintaining acceptable detection quality. It is important to note that exhaustive search, while optimal, has a computational complexity of O $((\mathbf{M}+\mathbf{1})^{\mathrm{N}})$, making it impractical for frequent sensing periods. In contrast, random and DBSS algorithms operate at O(1) complexity but suffer from shorter lifetimes. The GT algorithm operates at O(2(N-1)), significantly faster than exhaustive search, while the convex algorithm runs at O(N.M) complexity [22].

3. Joint sensor and dynamic threshold selection

In this subsection, Joint Sensor and energy detector threshold selection is utilized to maximize lifetime and Energy Efficiency (EE) in WCSNs while ensuring desired detection performance. The energy detector, favored for its simplicity, low computational complexity, and minimal energy consumption, operates using noise variance without prior knowledge of primary signals [27]. However, a fixed threshold in the energy detector can be sensitive to noise, leading to interference with the primary system and reduced secondary system throughput, particularly in low Signal-to-Noise Ratio (SNR) scenarios [28]. Dynamic threshold selection improves detector performance by adjusting thresholds according to SNR variations [29].

Concurrently, sensor energy consumption increases with detector thresholds [30]. Lowering the threshold enhances detection probability as well as increases false alarms, potentially reducing the number of required sensing nodes and overall energy consumption. Therefore, finding an optimal threshold that satisfies Detection Probability (GPD) and Green Function Penalty (GPF) constraints is crucial.

In [30], an optimal fixed threshold was determined, but for lifetime maximization, where the number of active sensors fluctuates, dynamic threshold optimization is proposed. This involves dynamically determining γ_{opt} (or $\gamma_{m,opt}$ for each channel) throughout the network lifetime under two scenarios:

- First scenario: known instantaneous SNR of each node.
- Second scenario: practical scenario where average SNR per node over the lifetime duration is used due to the complexity of obtaining instantaneous SNR.

The objective in this section is to maximize lifetime by selecting optimal cooperative nodes for channel sensing and dynamically adjusting sensor energy detector thresholds to meet GPD and GPF constraints. Similar to Section 2, the approach maximizes the minimum remaining energy of sensors, prioritizing nodes with higher remaining energy and lower energy consumption while utilizing dynamic optimal thresholds.

$$\max_{\varphi_{m,n},\gamma_m} \{E_{th}\} \tag{41}$$

subject to;

$$E_n \geq \varphi_{m,n}.E_{th} \qquad \forall m, n \tag{42}$$

$$P_{f_m} - \alpha_m \leq 0 \qquad \forall m \in \{1, \ldots, M\} \tag{43}$$

$$\beta_m - P_{d_m} \leq 0 \qquad \forall m \in \{1, \ldots, M\} \tag{44}$$

$$\sum_{m=1}^{M} \varphi_{m,n} - 1 \leq 0 \qquad \forall n \in \{1, \ldots, N\} \tag{45}$$

$$\varphi_{m,n} \in \{0, 1\} \qquad \forall m, n \tag{46}$$

The first constraint ensures that the minimum remaining energy of all sensors is $\mathbf{E_{th}}$, with at least one sensor meeting this requirement. The second constraint stipulates that the false alarm probability for the m-th channel should be below a desired parameter $\boldsymbol{\alpha_m}$. The third constraint requires the detection probability for the m-th channel to exceed a desired parameter $\boldsymbol{\beta_m}$. Both constraints 44 and 45 must be satisfied across all channels.to jointly determine the optimal thresholds and cooperative sensors for each channel dynamically, where energy detector threshold levels are independent for sensing different channels, the Bisection algorithm is employed across all methods.

3.1 The first and second algorithms for sensor and threshold selection (WLMTSS1,2)

An initial feasible set with upper and lower bounds is established for each γ_m. In each network lifetime iteration, initial thresholds are set midway within the feasible sets. The maximum number of sensing nodes per channel is determined using Eq. (20). Subsequently, the detection probabilities of all sensors across channels are computed based on average SNR. Using the priority function in Eq. (33), cooperative sensors for each channel are selected from the active sensor set until meeting the detection probability GPD constraints, up to a maximum of J_{m_max} sensors per channel. The optimal values of λ_m and $\rho_{m,n}$ in Eq. (33) are calculated using the subgradient

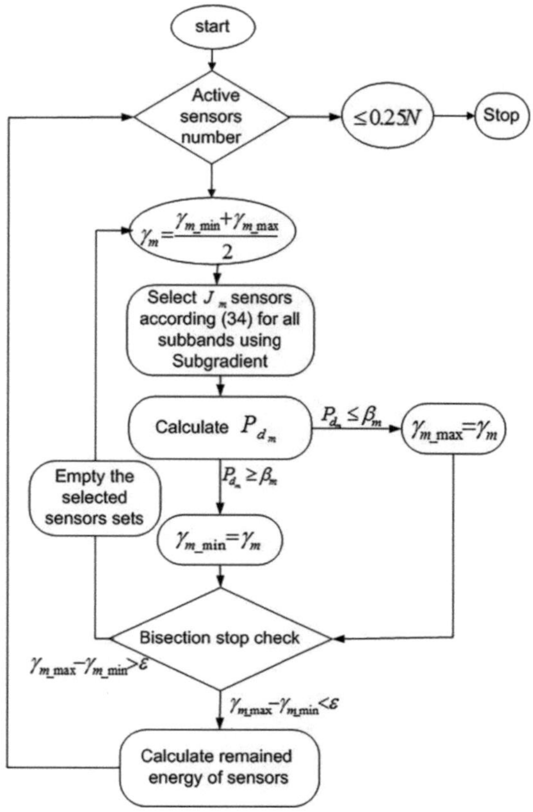

Figure 7.
The block diagram of WLMTSS1,2.

method. If convergence to an acceptable solution satisfying the global probability constraint for channel m is achieved, the threshold level for that channel is decreased (setting γ_m to the upper bound of the feasible set). If the global probability constraint is not met, γ_m is increased (setting it to the lower bound). This iterative process continues until the termination criterion of the bisection algorithm is met. Subsequently, sensor remaining energies are updated and the number of inactive sensors is determined. This algorithm is termed the first multiband lifetime maximization by threshold and Sensor Selection (WLMTSS1), as depicted in **Figure** 7 for $1 \leq m \leq M$. The interior loop iterates M times.

Given the challenge of obtaining instantaneous SNR for each node throughout the network lifetime, the second algorithm, WLMTSS2, employs average SNR for each node across the lifetime duration while maintaining identical procedures as WLMTSS1.

3.2 The third algorithm for sensor and threshold selection (WLMTSS3)

An initial feasible set with upper and lower bounds for each γ_m is established. During each network lifetime iteration, initial thresholds are set midway within these feasible ranges. The maximum number of sensors J_{m_max} is determined for each

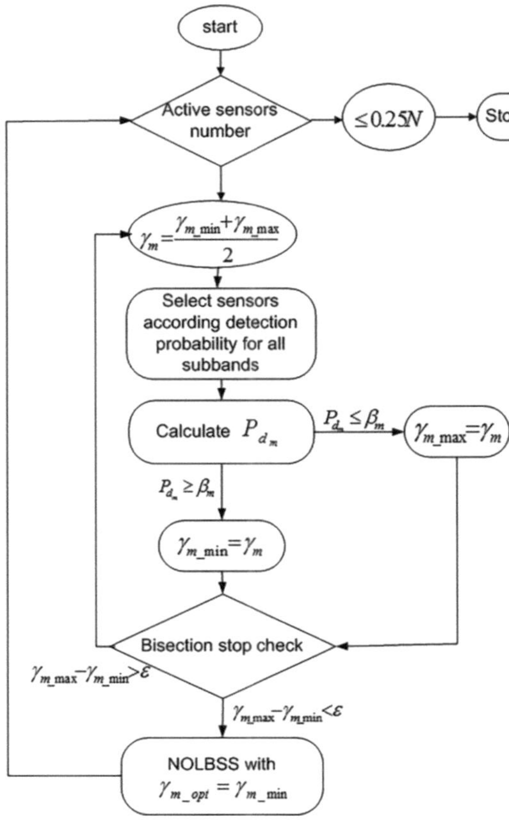

Figure 8.
The block diagram of WLMTSS3.

channel using Eq. (20). Subsequently, based on average SNR, the detection probabilities of all sensors across all channels are computed. Sensors are then selected for each channel m based on their detection probabilities, aiming to meet the Detection Probability (GPD) constraint.

Using a priority function (Eq. (33)), J_{m_max} sensors with the highest detection probabilities are chosen from the set of active sensors for each channel m. If the global probability constraint is satisfied for channel m, the threshold γ_m is decreased (set to the upper bound of its feasible set). Conversely, if the global probability constraint is not met, γ_m is increased (set to the lower bound). This iterative process continues until the termination criterion of the Bisection algorithm is met. Once the minimum values of γ_m for all channels are determined, these thresholds are chosen as the optimal detection thresholds. Cooperative sensors are then selected using the WOLBSS algorithm. These procedures are repeated throughout the network lifetime until the number of inactive sensors reaches 75% of the initial number. This algorithm is referred to as the Third Multi-Band Lifetime Maximization by Threshold and Sensor Selection (WLMTSS3), as illustrated in **Figure 8** for $\mathbf{1 \leq m \leq M}$. The inner loop iterates M times.

3.3 Simulation results of the sensor and dynamic threshold selection algorithms

This section presents the simulation results of the WLMTSS1, WLMTSS2, and WLMTSS3 algorithms for sensor and threshold selection in a WCSN, implemented

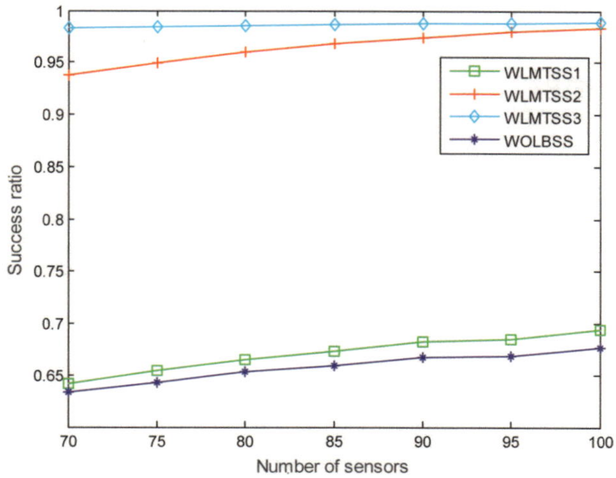

Figure 9.
The success ratio of the algorithms versus the number of sensors.

using MATLAB. The results are compared with the WOLBSS algorithm from Section 3, which uses a fixed threshold level for sensor selection. Two notable applications, IEEE802.15.4/Zigbee and IEEE802.11af, are evaluated in a square region with length L, centered on a Fusion Center (FC), with N sensors ($70 \leq N \leq 100$), and M PUs ($M = 8$) uniformly distributed. The wideband spectrum is divided into $M = 8$ equal channels, each with PU power of 20 mW. Other simulation parameters match those used in Section 2.3.

Given that WCSN aims to opportunistically use frequency bands, a suitable algorithm extends network lifetime by minimizing interference probability and maximizing detection accuracy. Simulation results ensure compliance with Detection Probability (GPD) and Green Function Penalty (GPF) constraints. The success ratio, defined as the proportion of iterations where each algorithm finds an acceptable solution to the maximum iterations across all algorithms, is analyzed in **Figure 9**. As the number of sensors increases, all algorithms achieve higher success rates due to the availability of more suitable sensors for selection. WLMTSS3 exhibits the highest success ratio because it enhances detection probability among active sensors, selecting those with higher remaining energy or lower consumption needs.

Figure 10 compares the WCSN lifetime achieved by the algorithms across varying numbers of sensors. The lifetime represents the number of iterations until algorithms find solutions meeting all constraints. WLMTSS3, employing joint dynamic threshold and sensor selection, achieves the highest lifetime, while WOLBSS shows the lowest due to its fixed threshold sensor selection approach. Algorithms using fixed thresholds reduce lifetime by over 50%, highlighting the advantage of dynamic threshold adjustment. Additionally, algorithms benefit significantly when instantaneous SNR information is available, improving performance by 50%. These findings underscore the effectiveness of dynamic threshold and sensor selection algorithms in WCSNs, particularly WLMTSS3, for prolonging network lifetime while enhancing energy efficiency and detection performance.

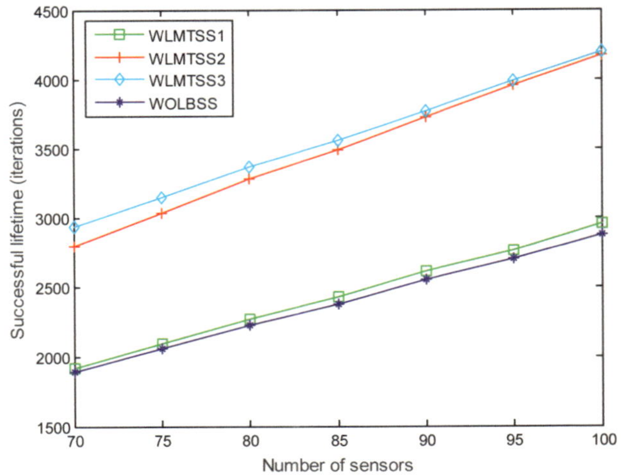

Figure 10.
The WCSN lifetime versus the number of sensors.

4. Conclusion

This chapter provides an in-depth exploration of Cooperative Spectrum Sensing (CSS) systems in wireless cooperative sensor networks (WCSNs), focusing on strategies to maximize network lifetime. Various methodologies are reviewed, each aimed at improving energy efficiency and prolonging network operation in scenarios like military and emergency environments where efficient spectrum usage is crucial. Efficient sensor selection using convex and game theory methods is highlighted as a key approach to enhance energy consumption and network longevity. Dynamic threshold level selection for energy detectors further optimizes network performance, with algorithms based on convex methods proving particularly effective in solving these optimization challenges.

Simulation results demonstrate significant improvements over existing methods, showing up to a 0.25 times increase in network lifetime and robustness against noise interference. Section 3 specifically delves into CSS lifetime maximization strategies, emphasizing the role of dynamic threshold optimization based on both instantaneous and average Signal-to-Noise Ratio (SNR). Comparative analyses indicate that algorithms like WLMTSS3 excel in extending WCSN lifetime, leveraging insights from instantaneous SNR for optimal dynamic threshold and sensor selection. These approaches not only enhance network lifetime but also manage complexity through linear algorithmic designs effectively. The presented algorithms were simulated assuming two notable applications IEEE802.15.4/Zigbee and IEEE802.11af.

In conclusion, the advancements presented underscore substantial progress in WCSN lifetime maximization. Future research may focus on adaptive algorithms that dynamically adjust to varying network conditions, further enhancing the robustness and efficiency of CSS systems in dynamic wireless environments.

Acknowledgements

My sincere appreciation goes to my academic supervisors, Professor G.A. Hodtani and Professor A. Ebrahimzadeh, whose guidance and expertise were instrumental in shaping the research direction of this book.

Special thanks to my husband Alireza, my daughter Nikoo, and my son Amir Ali for their patience and understanding during the long hours spent immersed in academic pursuit.

Lastly, I thank my students M.A. Karimi and M.J. Moosavi who helped edit the chapter.

Author details

Asma Bagheri
Department of Electrical Engineering, Esfarayen University of Technology, Esfarayen, North Khorasan, Iran

*Address all correspondence to: bagheri.asma84@yahoo.com

References

[1] Cichon K, Kliks A, Bogucka H. Energy-efficient cooperative spectrum sensing: A survey. IEEE Communications Surveys and Tutorials. 2016;**18**(3):1861-1886

[2] Joshi GP, Nam SY, Kim SW. Cognitive radio wireless sensor networks: Applications, challenges and research trends. Sensors. 2013;**13**(9): 11196-11228

[3] Ghasemi A, Sousa S. Spectrum sensing in cognitive radio networks: Requirements, challenges and design trade-offs. IEEE Communications Magazine. 2008;**46**(4):32-39

[4] Quan Z, Cui S, Sayed AH. Optimal linear cooperation for spectrum sensing in cognitive radio networks. IEEE Journal of Selected Topics in Signal Processing. 2008;**2**(1):28-40

[5] Bagheri A, Ebrahimzadeh A, Najimi M. Energy-efficient sensor selection for multi-channel cooperative spectrum sensing based on game theory. Journal of Ambient Intelligence and Humanized Computing. 2021;**12**(10):9363-9374

[6] Bagheri A, Ebrahimzadeh A, Najimi M. Game-theory-based lifetime maximization of multi-channel cooperative spectrum sensing in wireless sensor networks. Wireless Networks. 2020;**26**(6):4705-4721

[7] Maleki S, Chepuri SP, Geert L. Optimization of hard fusion based spectrum sensing for energy-constrained cognitive radio networks. Physical Communication. 2013;**9**:193-198

[8] Najimi M, Ebrahimzadeh A, Andargoli SH, Fallahi A. Lifetime maximization in cognitive sensor networks based on the node selection. IEEE Sensors Journal. 2014;**14**(7): 2376-2383

[9] Sklar B. Rayleigh fading channels in mobile digital communication systems part 1: Characterization. IEEE Communication Magzaine. 1997;**35**(7): 90-100

[10] Maleki S, Leus G, Chatzinotas S, Ottersten B. To AND or to OR: On energy-efficient distributed spectrum sensing with combined censoring and sleeping. IEEE Transactions on Wireless Communications. 2015;**14**(8):4508-4521

[11] Bagheri A, Ebrahimzadeh A. Statistical analysis of lifetime in wireless cognitive sensor network for multi-channel cooperative spectrum sensing. IEEE Sensors Journal. 2020;**21**(2):2412-2421

[12] Vien Q-T, Nguyen HX, Nallanathan A. Cooperative spectrum sensing with secondary user selection for cognitive radio networks over Nakagami-m fading channels. IET Communications. 2015;**10** (1):91-97

[13] Li P, Gua S, Cheng Z. Max-min lifetime optimization for cooperative communications in cognitive radio networks. IEEE Transactions on Parallel and Distributed Systems. 2014;**25**(6): 1533-1542

[14] Boyd S, Vanderberghe L. Convex Optimization. Cambridge, England: Cambridge University Press; 2004

[15] Asheralieva A, Quek T, Niyato D. An asymmetric evolutionary Bayesian coalition formation game for distributed resource sharing in a multi-cell device-to-device enabled cellular network. IEEE Transactions on Wireless Communications. 2018;**17**(6):3752-3767

[16] Song L, Li Y, Ding Z, Poor H. Resource management in non-orthogonal multiple access networks for 5G and beyond. IEEE Network. 2017;**31**(4):8-14

[17] Kim S. Game Theory Applications in Network Design. Pennsylvania, USA: IGI Global; 2014

[18] Moualeu JM, Ngatched TMN, Hamouda W, Takawira F. Energy-efficient cooperative spectrum sensing and transmission in multi-channel cognitive radio networks. In: IEEE International Conference on Communications (ICC). Sydney: IEEE; 2014

[19] Shapely LS. A value for n-person games. In: The Shapely Value. Princeton University Press; 1988. pp. 31-40

[20] Xu Y, Wang J, Wu Q, Anpalagan A, Yao Y. Opportunistic spectrum access in unknown dynamic environment: A game-theoretic stochastic learning solution. IEEE Transaction on Wireless Communications. 2012;**11**(4):1380-1390

[21] Lã QD, Chew YH, Soong B-H. Potential Game Theory: Applications in Radio Resource Allocation. New York City, USA: Springer International Publishing; 2016

[22] Bagheri A, Ebrahimzadeh A, Najimi M. Sensor selection for extending lifetime of multi-channel cooperative sensing in cognitive sensor networks. Physical Communication. 2017;**26**: 96-105

[23] Kaligineedi P, Bhargava V. Sensor allocation and quantization schemes for multi-band cognitive radio cooperative sensing system. IEEE Transaction on Wireless Communications. 2011;**10**(1): 284-293

[24] Flores A, Guerra R, Knightly E, Ecclesine P, Pandey S. IEEE 802.11 af: A standard for TV white space spectrum sharing. IEEE Communications Magazine. 2013;**51**(10):92-100

[25] Banerji S. Upcoming standards in wireless local area networks. Wireless and Mobile Technologies, arXiv preprint. 2013;**1**(1):arXiv:1307.7633

[26] Chiaravalloti S, Idzikowski F, Budzisz Ł. Power Consumption of WLAN Network Elements. Berlin: Technische Universit¨at Berlin; 2011

[27] Jaglan RR, Sarowa S, Mustafa R, Agrawal S, Kumar N. Comparative study of single-user spectrum sensing techniques in cognitive radio networks. Procedia Computer Science. 2015;**58**: 121-128

[28] Yu GC, Shao YB, Long H, Yue GX. Dynamic threshold based spectrum detection in cognitive radio systems. In: Networking and Mobile Computing. Beijing: IEEE; 2009

[29] Plata DMM, Reátiga ÁGA. Evaluation of energy detection for spectrum sensing based on the dynamic selection of detection-threshold. Procedia Engineering. 2012;**35**(1): 135-143

[30] Ebrahimzadeh A, Najimi M, Hosseini Andargoli SM, Fallahi A. Sensor selection and optimal energy detection threshold for efficient cooperative spectrum sensing. IEEE Transactions on Vehicular Technology. 2015;**64**(4): 1565-1577

Chapter 10

Ultra-Wideband Devices: Definition, Techniques, and Applications

Mahdi NoroozOliaei

Abstract

Nowadays, it is typical to use multitasking services more frequently. The Industrial, Scientific, and Medical (ISM) band is one of the several attempts to define the instruments according to these specifications and standards. Ultra-wideband technology enables secure ranging and precision sensing, creating a new utilization for wireless devices. In wireless technologies, ultra-wideband systems provide very high data rates as they can support high-speed services, especially in the long-term period and distance. Many applications in telecommunications, remote sensing, medical, space, etc., can be defined by bringing these types of fields into industrial operations. This chapter presents a practical view of the various techniques and applications of ultra-wideband devices.

Keywords: angle of arrival (AoA), equivalent isotropically radiated power (EIRP), power spectrum density (PSD), remote sensing, sensor, time of flight (ToF), ultra-wideband (UWB)

1. Introduction

In the early days of radio communication, Heinrich Hertz and Guglielmo Marconi's discoveries and innovations led to the development of pulse radio communication. Transmitting and receiving (transceiving) extremely narrow pulses with a duration of the order of nanoseconds, or equivalently in a wide spectrum of 500 MHz and beyond, defines ultra-wideband (UWB) concepts. This enables extremely precise time of flight measurements and UWB signals that are comparatively immune to interference.

It uses a significant amount of the radio spectrum to enable high-bandwidth, short-range communications at very low energy levels (low power consumption).

The UWB channel allocation according to the most recent IEEE standard introduced in 2020 is depicted in **Figure 1**. It encompasses the spectrums used in several telecommunication bands, including 5G and LTE [1].

Figure 2 also illustrates the Impulse Radio UWB Standard in IEEE [1].

Radar imaging using pulsed signals has traditionally made use of UWB. Orthogonal frequency division multiplexing (OFDM) is the foundation of modern

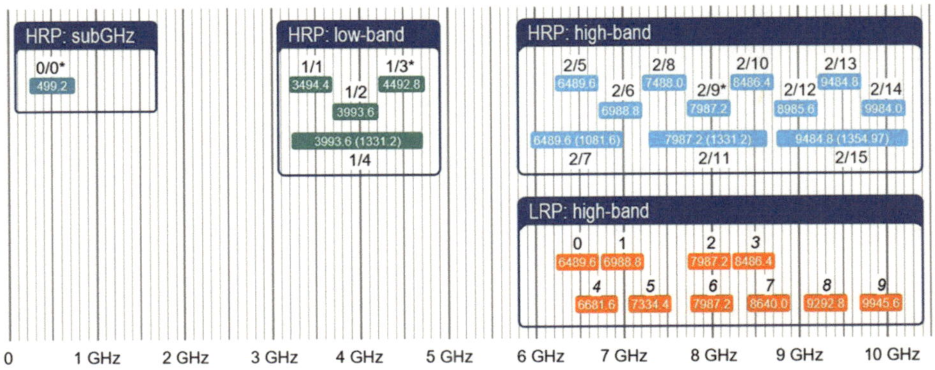

Figure 1.
IEEE 802.15.4z-based UWB channel allocation (March 2020) [1].

HRP UWB PHY High Rate Pulse repetition frequency			LRP UWB PHY Low Rate Pulse repetition frequency					
RDEV	ERDEV		RDEV			ERDEV		
base	BPRF	HPRF	base	extend	long-range	DF	enh. DF	DF w/ EPC
Modulation BPM-BPSK **Pulse rate:** 3.9 MHz 15.6 MHz 62.4 MHz	**Modulation** BPM-BPSK **Pulse rate:** 62.4 MHz	**Modulation** BPSK **Pulse rate:** 124.8 MHz 249.6 MHz	**Modulation** OOK **Pulse rate:** 1 MHz	**Modulation** OOK **Pulse rate:** 1 MHz	**Modulation** PPM **Pulse rate:** 2 MHz	**Modulation** PBFSK **Pulse rate:** 1 MHz 2 MHz 4 MHz	**Modulation** PBFSK **Pulse rate:** 1 MHz 2 MHz 4 MHz	**Modulation** PBFSK-PPM **Pulse rate:** 1 MHz 2 MHz

RDEV – Ranging device
ERDEV – Enhanced ranging device
BPM – Burst position modulation
BPRF – Base pulse repletion frequency
HPRF – Base pulse repletion frequency
PBFSK – Pulsed binary frequency shift keying
PPM – Pulse positioning modulation
EPC – Enhanced payload capacity
BPSK – Binary phase shift keying
DF – Dual frequency
OOK – On-off keying

Figure 2.
IEEE 802.15.4 standardization of impulse radio ultra-wideband (UWB) (Groups a, f, z) [1].

UWB systems for wireless connection communications. The ECMA-368 standard implements a multiband OFDM (MB-OFDM) structure, which is also used by the most significant UWB communications system, which is marketed by WiMedia® UWB Technology. High data rates with minimal power consumption are the goal of this standard. The USB Implementers Forum has chosen WiMedia® UWB technology for wireless USB, while the Bluetooth SIG, Inc. has chosen it for high-speed Bluetooth®.

In the license-exempt spectrum, ultra-wideband communication, as defined by IEEE 802.15.4 (groups a/f/z), allows for extremely accurate ranging and very low power communication. The technology is currently making its way into the consumer market, mostly because of its various uses in different industries mainly targeting mobile phones (smartphones with the UWB module) and automotive applications. Keyless entry, asset location, mobile data exchange, and secure payment are some examples of common use cases.

A new generation of ultra-wideband products is coming to the market. Wireless, ultra-precise UWB measurements are made possible by an ultra-wideband testing capability. The technology's operation is the question expressed here in the answers. Ultra-wideband is a radio technology that uses short signal pulses (**Figure 3**) with a wide bandwidth and a very low energy level for short-range communications.

DOI: http://dx.doi.org/10.5772/intechopen.1008294

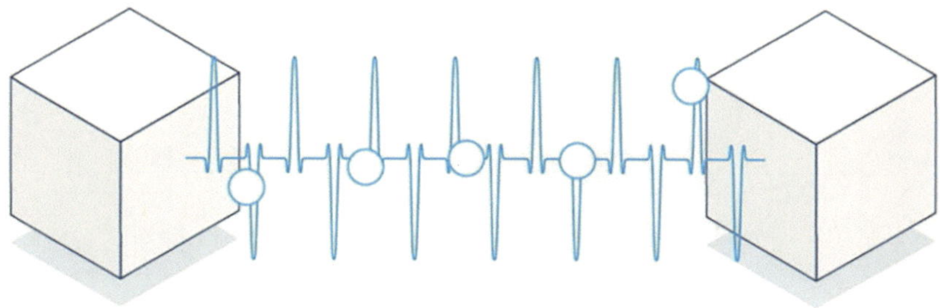

Figure 3.
Short signal pulses with a wide bandwidth [1].

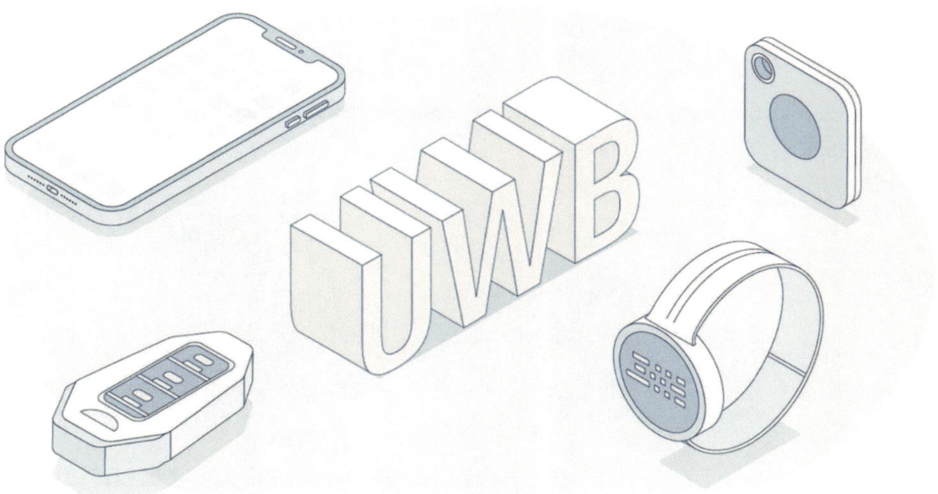

Figure 4.
New applications for wearables, cars, tags, cellphones, and other devices [1].

In 2002, ultra-wideband technology was initially employed, mostly for specialized industrial uses. More and more use cases were made possible by the gradual evolution of the standards over time, and the new z amendment included secure-ranging functionality while also enhancing it in other areas. As seen in **Figure 4**, this made it possible to create new applications for wearables, cellphones, tags, automobile access, and other devices.

The primary application of ultra-wideband technology is precise ranging for secure low-power communication.

Depending on the radio conditions, Bluetooth low energy (LE) ranging can provide a range precision of a few meters based on the strength of the received signal.

Figure 5 shows that UWB measures the current location with a variation of a few centimeters.

The precision of device-to-device communication is unmatched by UWB. With sharing and locating services, smartphone brands have recently begun using this accuracy. Recent tagging programs from Apple and Samsung allow users to pinpoint tiny UWB tag devices with remarkable precision. UWB's high bandwidth makes it helpful for data transfer over short distances and is already being used in certain smartphone designs.

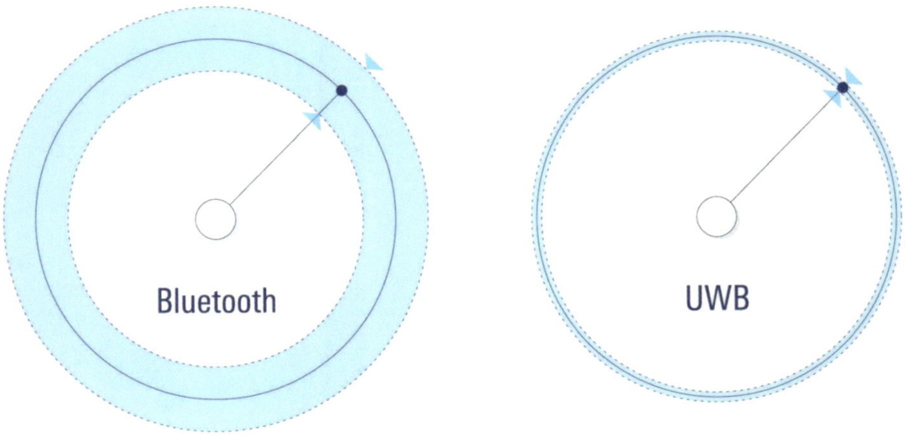

Figure 5.
(a) A few-meter bluetooth range precision and (b) a few-centimeter UWB variation [1].

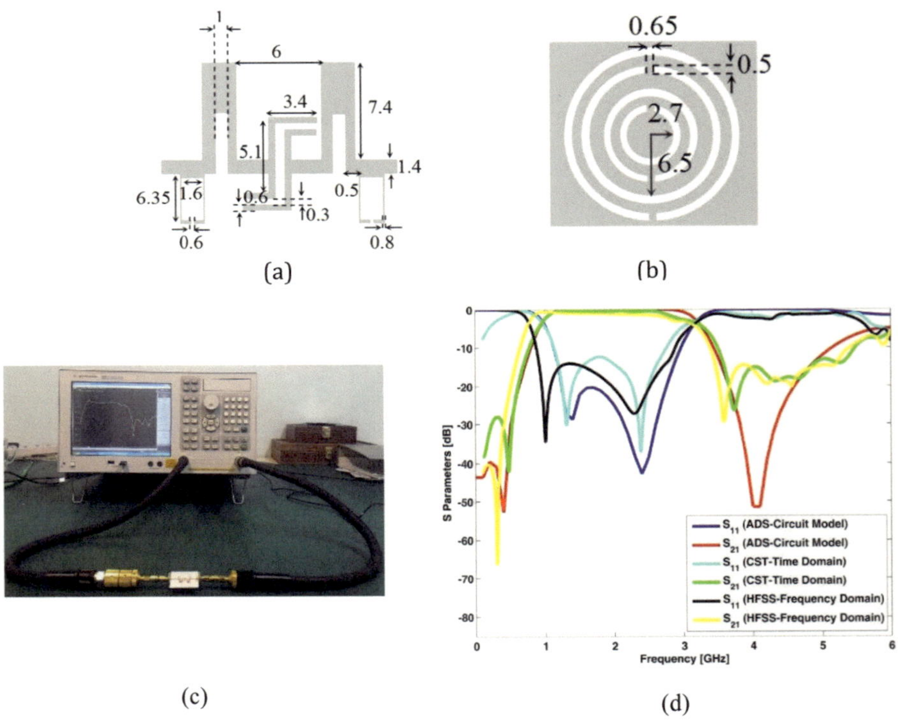

Figure 6.
ISM bandpass filter circuit usable for HRP-SubGHz UWB devices. (a) Top layer. (b) Bottom layer. (c) Measurement result. (d) Simulation results [2].

For instance, a bandpass filter circuit using in ISM band and HRP-SubGHz mentioned in **Figure 1** is being illustrated with the simulated and measured results as illustrated in **Figure 6**.

Another example of a UWB filter device is the following circuit for designing the concentric open-end ring resonators as depicted in **Figure 7**.

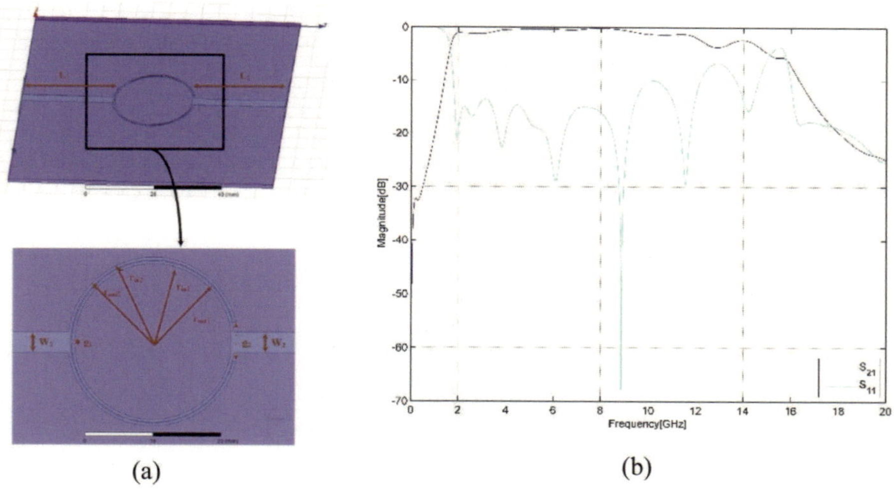

Figure 7.
UWB ISM bandpass filter circuit: (a) schematic of the simulation in the ANSYS HFSS software; (b) simulation results [3].

Time of flight (ToF) can be used to calculate the distance between two objects. As seen in **Figure 8**, it is the amount of time the signal must travel from the initiator to the responder.

Time measurement at the analog antenna is necessary for accurate ToF estimation. As a result, understanding the time difference between the digital time processor and the antenna is essential because this adjustment value may differ. UWB devices must be calibrated and validated. Therefore, to measure the ToF accurately, scientists and researchers must devise a clever approach. Ideally, UWB measurements should be conducted over the air. The key technological features of ultra-wideband (UWB) services are shown in **Figure 9**.

More precision in line-of-sight (LoS) and strong localization in non-line-of-sight (nLoS) situations are provided by UWB, which can also handle locations where many walls, people, and other obstructions would normally block these signals.

Furthermore, UWB sensors can identify if an object is stationary, approaching, or moving away.

Wireless communication that makes use of ultra-wideband frequency bandwidth is known as UWB. Its ability to facilitate high-precision positioning is its primary

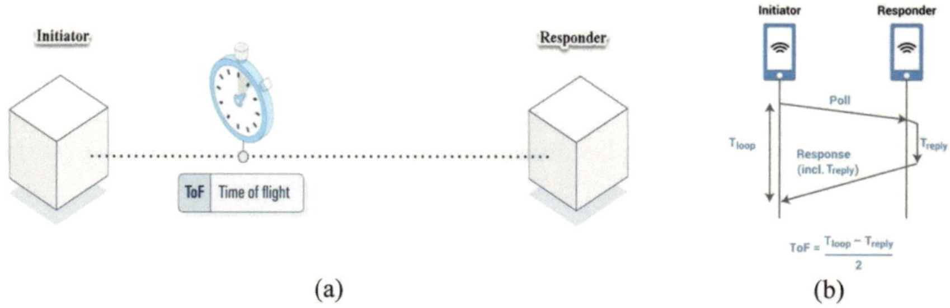

Figure 8.
(a), (b) The time-of-flight (ToF) calculation of the initiator and responder [1, 4].

Figure 9.
Technological features of ultra-wideband (UWB) [5].

characteristic. In recent years, UWB wireless communication has extended to consumer electronics. For instance, smartphones and luxury car smart keys are both integrated with UWB wireless communication to prevent loss and are employed for other uses. In the future, UWB wireless communication is anticipated to expand into numerous industries.

UWB technology, which involves the deliberate production and transmission of radio-frequency energy that disperses over a very wide frequency range, is used by UWB devices, which are a subset of short-range devices (SRD).

The benefits of the UWB standard include the ability to provide positioning capabilities to inside spaces and indoor environments, enable doors to open when approached with a UWB device, and provide centimeter-accurate location data. Additionally, UWB has very low power needs and secure device-to-device data transfers. UWB signals can easily share spectrum with other narrowband and wideband systems without interfering because of their extremely broad bandwidth and extremely low power density.

Time of flight (ToF) and angle of arrival (AoA) measurements, as well as device calibration techniques, are among the fundamental UWB test solutions that are developed by several well-known firms in partnership with industries for R&D, certification, and manufacturing.

DOI: http://dx.doi.org/10.5772/intechopen.1008294

Ultra-wideband technology's key characteristics are:

1. Accurate: Line-of-sight (LOS) or non-line-of-sight (NLOS) distance estimation down to less than 10 cm.

2. Reliable: It is stable against multipath effects due to the utilization of short ultra-wideband pulses.

3. Coexists: Functions outside of the crowded bands that Bluetooth and Wi-Fi use.

4. Low power: A transmitter with short airtime and low power helps prolong battery life.

5. Secure: It is more secure thanks to encryption and random number creation.

6. Real-time: Real-time location service is made possible by high refresh rates of up to 1000 times per second.

By integrating sensing and ranging, as enabled by UWB in the precise range and direction, i.e., time of flight (ToF) and angle of arrival (AoA), it is providing a new connected experience:

1. Asset finding

2. Navigation (especially indoor)

3. Track, trace, and warn (enterprise applications)

4. AR/VR anchor

5. Keyless access

6. Presence detection

7. Ticket validation

8. Vital sensing.

1.1 UWB technology's history and regulations

Since the 1960s, UWB technology has been developed in the United States (U.S.) first for military applications including radar technologies. Up until about 1994, the communication technology was a military secret. In the United States, the FCC began considering the civilian use of UWB in 1998 and approved it in 2002. Since then, there have been tremendous advancements in the research of UWB chipsets.

Since 2019, society has begun to genuinely recognize UWB technology. The advent of cell phones with UWB modules was the reason behind that transformation. As a result, UWB has received approval from several nations worldwide, even though it was previously unutilized in the civilian sector.

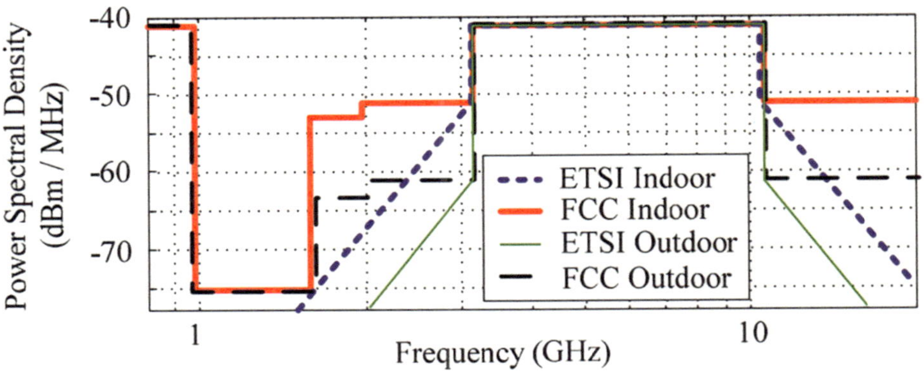

Figure 10.
Maximum power spectral density (PSD) levels for both indoor and outdoor environments as a function of frequency for FCC and European Telecommunications Standard Institute (ETSI) rules [6].

This is how UWB technology has been developed. In particular, the FCC has set aside 7.5 GHz from 3.1 to 10.6 GHz for UWB wireless technology in terms of frequency bandwidth requirements. However, the bandwidth allotted by nations and regions, such as those in Europe, Eurasia, East Asia, and Oceania, differs somewhat from that. Its usage is restricted to the 6.0–9.0 GHz spectrum, both indoors and outdoors (**Figure 10**). Ground penetrating radar (GPR) systems are designed to operate in a band below 1 GHz. GPR systems are unlikely to interfere with other systems because they aim their radar antennas toward the ground floor.

Although the IEEE 802.15.4a short-range wireless standardization, which was adopted in 2007 by the U.S. Institute of Electrical and Electronic Engineers (IEEE[1]), specifies preferential use for UWB, the majority of UWB modules use the specifications of channel number 9, which has a central frequency of 7987.2 MHz and a frequency bandwidth of 499.2 MHz.

1.2 A wideband technology for fine ranging

Two physical layers are defined by IEEE 802.15.4 standard. Low-rate pulse repetition frequencies (LRP UWB) ranging from 1 MHz to 4 MHz are covered by one layer. The other layer operates with high-rate pulse repetition frequencies (HRP UWB) between 4 and 250 MHz, and it is primarily utilized in modern applications. With a channel bandwidth of 499.2 MHz, the designated high-band channels in the 6–10 GHz range are used by the majority of applications.

FiRa—an acronym for "fine ranging"—is a group that works to expand the UWB ecosystem by guaranteeing device compatibility through certification and compliance initiatives. The three main UWB services that the FiRa Consortium focuses on are hands-free access control, device-to-device, and location-based services that leverage the most recent UWB-based secure-ranging technology as defined by IEEE 802.15.4z.

1 The U.S. Institute of Electrical and Electronic Engineers (IEEE) is the world's largest academic research organization in the electrical field. It is also a standardization organization in the same field.

1.3 Ensuring secure communication and accurate ranging

In addition to regulatory standards, the FiRa consortium, the UWB alliance, or the Car Connectivity Consortium's performance and interoperability requirements are crucial for UWB device testing.

In general, test solutions with the necessary dynamic range and the ability to create and analyze very wide signals are needed for the design, integration, and testing

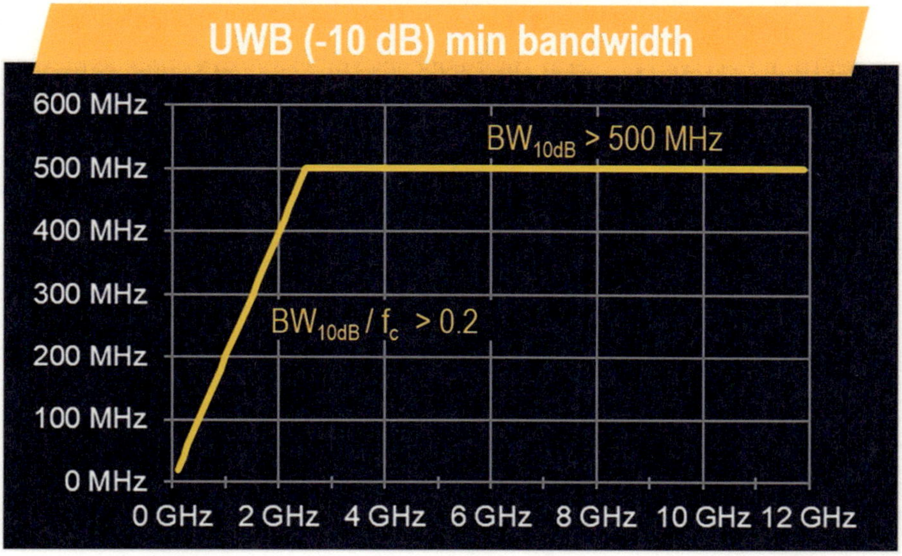

Figure 11.
Ultra-wideband (UWB) minimum bandwidth [1, 7].

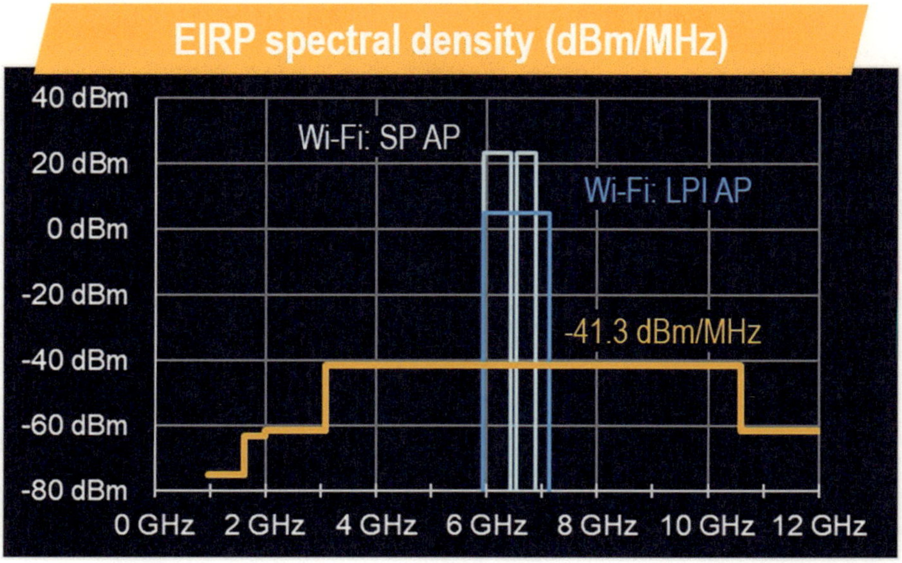

Figure 12.
Ultra-wideband (UWB) EIRP spectral density [1, 7].

of wideband radio devices at very low power. Verification and calibration of the time-of-flight measurement components are unique and necessitate a test system that permits extremely precise time measurements.

1.4 What is ultra-wideband about? The FCC definition from 2002

An ultra-wideband (UWB) transmitter is a purposeful radiator that, independent of the fractional bandwidth, has a UWB of at least 500 MHz or a fractional bandwidth of more than 0.20 at any given time. Accordingly, **Figure 11** illustrates this definition.

Equivalent isotropically radiated power (EIRP) is the maximum signal strength measured from the UWB device at any frequency and in any direction, expressed in dBm. According to the FCC regulation and **Figure 12**, an ultra-wideband device's radiated emissions between 3.1 GHz and 10.6 GHz cannot surpass the average limit of −41.3 dBm/MHz.

2. Features of UWB wireless communication

An example of a USB wireless communication standard is the IEEE 802.15.4z standard. One technique makes use of impulse radio, which employs short-duration pulse messages. Here are a few of its primary attributes:

- High precision ranging and positioning
- High security (secure ranging)
- Low interference with other communications
- Low power consumption wideband.

In UWB wireless communications, a pulse lasting around 2 ns is transmitted as data, as seen in **Figure 13**. One advantage of these short-duration pulses is their excellent resolution in terms of positioning and range.

Notably, the power spectrum in the frequency domain fills a wideband when the pulse duration in the time domain is narrow.

UWB wireless communication's transmission power (power spectrum density or PSD) in that frequency range is qualitatively depicted in **Figure 14**. For instance,

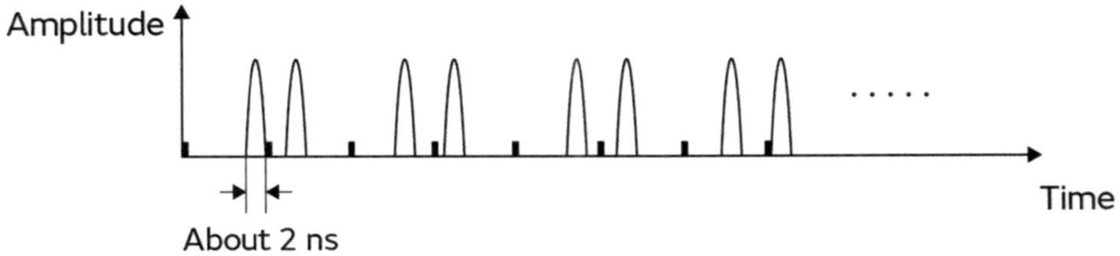

Figure 13.
Ultra-wideband (UWB) for showing its short duration of pulses [8].

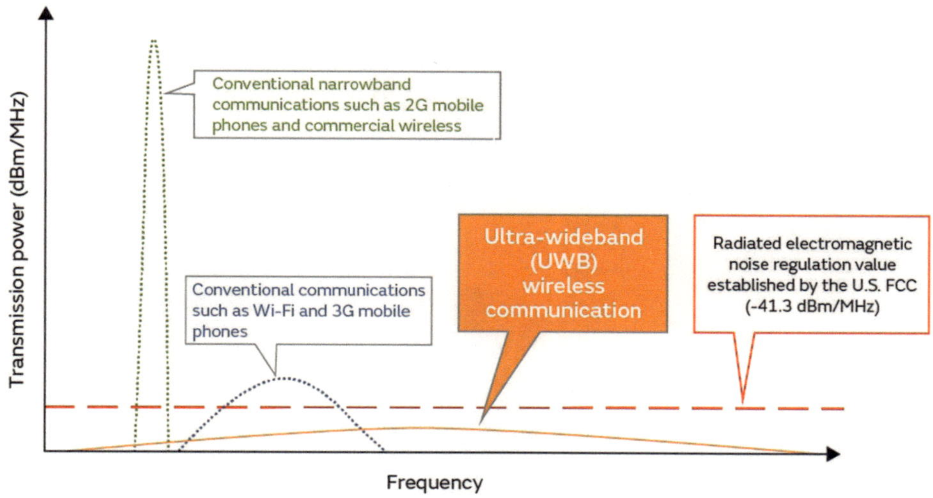

Figure 14.
Example showing a qualitative comparison of the UWB wireless communication method's power spectrum density bandwidth with that of other communication methods [8].

when comparing UWB wireless communication to the frequency bandwidth utilized by Wi-Fi or 2G–3G mobile phones, the former is significantly more extensive.

UWB wireless communication has a lower peak transmission power than other communication methods and a lower noise level regulation value than digital equipment that emits radio waves. The U.S. Federal Communication Commission (FCC)[2] established a radiated electromagnetic noise regulation value of −41.3 dBm/MHz[3] (75 nW/MHz).

Because UWB wireless communication uses a wide frequency band and transmits at a power level lower than noise, it has the advantage of having less interference with other communications. Furthermore, it has excellent security and allows for communication with low power consumption because the communication itself is unknown to third parties.

UWB will become ubiquitous, make surroundings aware, and increase wireless interaction between human beings and objects in people's lives.

Nowadays, UWB is utilized in a wide range of devices, including laptops, cell phones, baby phones, sleep monitors, lights, and air conditioners for smart home use.

Soon, everyone will be able to use the technology in cars to make sure that a child or pet is not inadvertently left behind, and UWB will be able to verify that a convict is alive and breathing in prison cells. The laptop can even notify the person if they are about to get sick, and smart speakers will be able to route the sound to the appropriate areas of the room.

In a variety of fields, such as public safety, building, engineering, science, medicine, consumer applications, information technology, multimedia entertainment, and

[2] The U.S. Federal Communication Commission (FCC) is a government agency that is involved in managing and regulating all communications, whether wireless or wired, in the United States.

[3] dBm/MHz refers to the power level per 1 MHz frequency width (power spectrum density).

Figure 15.
Ultra-wideband (UWB) applications [5].

transportation, UWB applications can benefit the general public, consumers, businesses, and industries (**Figure 15**).

3. UWB in automotive radar, level probing radar, etc.

UWB connection is already being included in the designs of major automakers like BMW. Physical or digital car keys that communicate with a vehicle based on proximity can incorporate the accurate distance measurement provided by UWB. Without ever taking the key or smartphone out of their pocket, a user can approach their car and unlock it. Because UWB's radio signals are independent of other wireless connections and cannot be intercepted or transmitted, they also provide an additional degree of security. Compared to other digital keys that employ other protocols, this makes it far more secure.

24 and 79 GHz are the two frequency ranges that are most frequently used in the automotive industry; the latter is best suited to replace the former.

3.1 24 GHz UWB automotive short-range radar

Nowadays, the main factor contributing to the ease of living in developing nations is the automation of essential industries. Automotive short-range radar (SRR) systems, which are based on telecommunication network applications, are one of these businesses.

It has been suggested to use asteroid unit cells to enhance the impedance and radiation bandwidth of an amplitude tapered grid array antenna (GAA) for ultra-wideband vehicle radar sensors, with improvement factors of 25 and 10.4%, respectively [9].

Any object inside a designated area can have its movement and speed detected by the 24 GHz radar. As a result, security and anti-theft systems can make extensive use of it.

Additionally, this frequency solution makes sure that the advanced driver assistance system (ADAS) is stable and that it calculates the distance between a car and an obstruction. It is notable that every vehicle or drone must have energy efficiency and distribution as key components.

The following categories apply to applications of mid-range radars (MRRs) to short-range radars (SRRs) [10, 11]:

1. Lane change assistance (LCA)
2. Blind spot detection (BSD)
3. Collision mitigation (CM)
4. Parking aid (PA)
5. Rear cross traffic alert (RCTA)
6. Traffic sign recognition (TSR).

Point-to-point transmission, satellite communication, and vehicle radar systems are just a few of the potential uses.

3.2 79 GHz UWB automotive short-range radar: Spectrum allocation and technology trends

Numerous features of automotive radar improve the facilities for driver's convenience and safety. The development of systems that enhance driver's ability to identify objects in poor optical visibility or those concealed in their blind spots when parking or changing lanes is made possible by the precise measurement of the distance and relative speed of objects in various directions, such as in front of, behind, or on the sidewalls of an automobile. This radar technology has demonstrated its suitability for use in automobiles for many years. An overview of uses for short-range radar (SRR) is presented in **Figure 16**.

The advantages of radar over lidar, video (with image processing), or its optical counterpart are as follows:

- Direct distance and speed measurement
- Robust against weather influences and pollution
- Unaffected by light
- Measuring both moving and stationary items on the road and nearby
- Invisible integration behind electromagnetically transparent materials for instance bumpers.

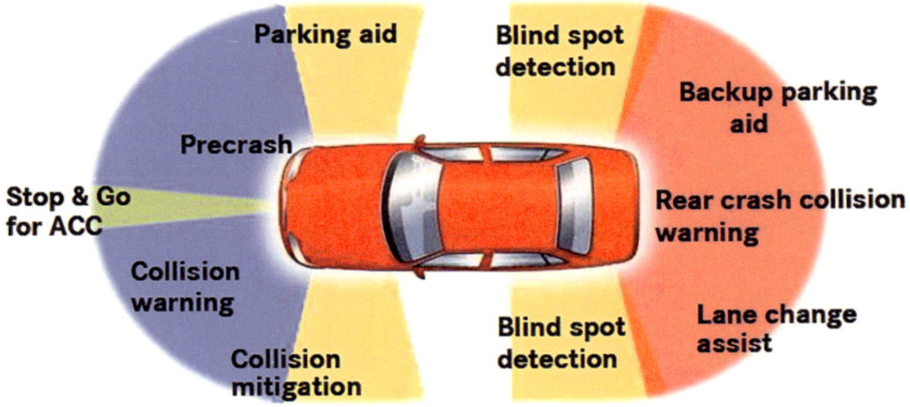

Figure 16.
Applications for safety and support using short-range radar (SRR) [5].

Notably, 77 GHz radar is used by numerous automakers for automated cruise control (ACC). It has even been employed recently for collision or pre-crash mitigation. In 2005, the Mercedes-Benz S-class was equipped with a UWB short-range radar that operates at 24 GHz. For both new and existing driver assistance and safety systems, it is a crucial technological component. A UWB bandwidth must have a high enough range resolution in the centimeter range scale to identify smaller objects and vulnerable road users, such as children and motorcycle riders, to foresee and mitigate potential collisions.

In comparison with the frequency range of around 24 GHz, the frequency range of about 77–81 GHz offers the following various benefits:

1. One common technology platform that is available for both long-range radar (LRR) and short-range radar (SRR) (frequency range 76–81 GHz)

2. Reduced weight and size

3. A higher doppler sensitivity

4. A higher angular resolution with moderate antenna aperture dimensions

Automotive short-range radar is a crucial piece of technology for active safety and comfort features in cars both now and in the future. For time-sensitive safety and accident mitigations like pre-crash, the UWB approach's real-time high-range resolution is especially significant and intriguing. According to the frequency restrictions, UWB automobile short-range radar (SRR) must switch from the 24 GHz band to the 79 GHz band. With the same technology platform for long-range radar (LRR) and UWB SRR, the 79 GHz frequency range provides the same application as the 24 GHz range. Additionally, frequency-dependent metrics like velocity and angular resolution are greatly enhanced.

For engineering structures to be safe and serviceable, distance measuring is important. Ultra-wideband (UWB) sensors have recently provided a different real-time remote sensing option for distance measurement. Although UWB sensors are compact, inexpensive, energy-efficient, and highly weather-resistant, their ranging accuracy is still quite low.

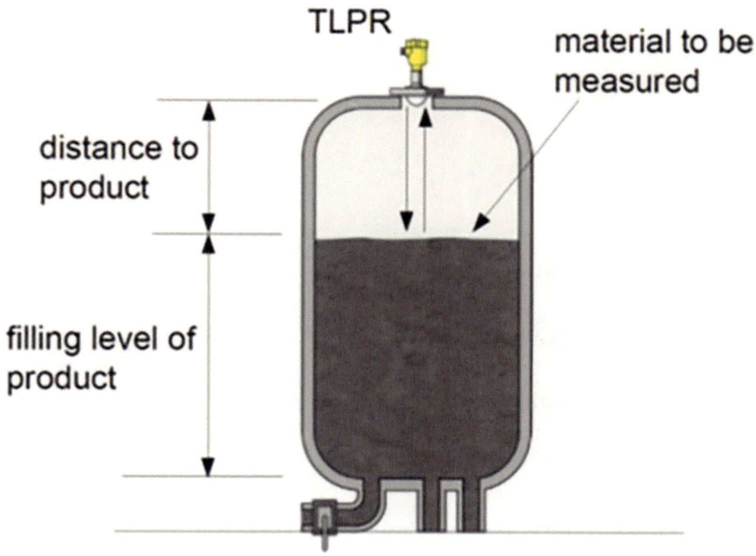

Figure 17.
Solids and liquids within a metallic tank can be detected using a tank level probing radar (TLPR) [12].

3.3 Tank level probing radar (TLPR)

Radio-frequency (RF), frequency-modulated continuous wave (FMCW), or comparable wideband techniques are the foundation of tank level probing radar (TLPR) sensors. TLPRs are only permitted to be mounted and radiated inside metallic tanks or containers made of similar attenuating material, such as concrete. Other than that, these sensors are nearly identical to level probing radars (LPRs). The time-of-flight method with an FMCW-modulated radar signal is the same measurement approach as the LPRs.

In many sectors, TLPRs are used similarly to LPRs to measure the distance to the surface of various materials and substances (nearly liquids and solids) inside confined and closed containers. This allows one to infer the amount of filling product indirectly. **Figure 17** provides an overview of measurement and how tank level probing radars are set up.

4. UWB in medical technologies

For short-distance, high-speed applications, UWB is a suitable technology to use. Because they operate at the baseband, the noise-like feature relies on ultra-short waveforms and does not require IF processing.

UWB has multiple uses in medical technologies, including:

1. Patient motion monitoring

2. Vital signs monitoring of the human body

3. Monitoring of medicine storage

4. Cardiology imaging
5. Pneumology Imaging
6. Obstetrics Imaging
7. Ear-nose-throat Imaging
8. Underwater medicine measurement
9. Space medicine measurement
10. Sports medicine measurement
11. Military medicine
12. Rubble rescue radar for emergency medicine
13. Retinal, eye implants, and sensors [13, 14]
14. Tumor detection
15. Cancer detection.

4.1 Ultra-wideband (UWB) sensor with low complexity for wirelessly detecting and tracking human body movements

These kinds of gadgets are an affordable option that offers strong performance against obstructions and interferences together with millimeter range precision.

A wireless sensor scheme has been developed based on the ultra-wideband (UWB) technology that offers a highly accurate mechanism for the detection and measurement of human body movements (**Figure 18**).

The sensor is especially well-suited for identifying minute motions of the human body, such as respiratory chest movements because it can deliver precision in the millimeter range. Features of UWB signals include low energy consumption, minimal interference, low electromagnetic radiation, and the capacity to pass through concrete and other impediments.

These attributes open up a wide range of possible uses, such as interior tracking and localization, security and surveillance, medical instruments, and search and rescue, to mention a few.

Even in extremely noisy and non-isolated situations, the method allows for reliable performance. The development of the low-cost portable device and a reduction in hardware complexity are the results of the innovative subsampling approach.

These kinds of sensors have the following advantages:

1. Offering a low-cost way to wirelessly track respiratory motions
2. High degree of precision in tracking and detection within millimeters
3. Capable of operating in highly noisy environments

Figure 18.
Human body movement detection and tracking via wireless [5].

4. Resilience in the face of changing background clutter and obstacles

5. Low-power and energy-efficient radiation that is appropriate for portable electronics and medical applications

Additionally, these kinds of sensors have some uses:

1. Medical and fitness (health monitoring, diagnosis of respiratory disorders)

2. Surveillance (through wall monitoring)

3. Search and rescue (survivors under rubble)

4. Indoor tracking and localization (smart homes).

The use of UWB technology allows for the detection and sensing of barriers and obstacles through the use of low-power, low-electromagnetic radiation (EMR) pulses. The novel subsampling scheme enables us to overcome the Nyquist sampling constraint, allowing us to achieve millimeter range precision using UWB signals while sampling much below the Nyquist rate (**Figure 19**). For example, typical approaches require a sample frequency of 80 GHz to achieve a 2 mm precision. With this sensor, just 2.1 GHz is needed to get the same level of accuracy! Significant cost reductions and the potential for the creation of a portable, affordable commercial device result from the decrease in hardware complexity.

Additionally, the proposed processing algorithms enable us to effectively deal with and eliminate background clutter, noise, and interference, even in harsh and non-isolated environments. The end product is a very reliable and accurate sensor that offers an affordable way to wirelessly detect and track motions of the human body.

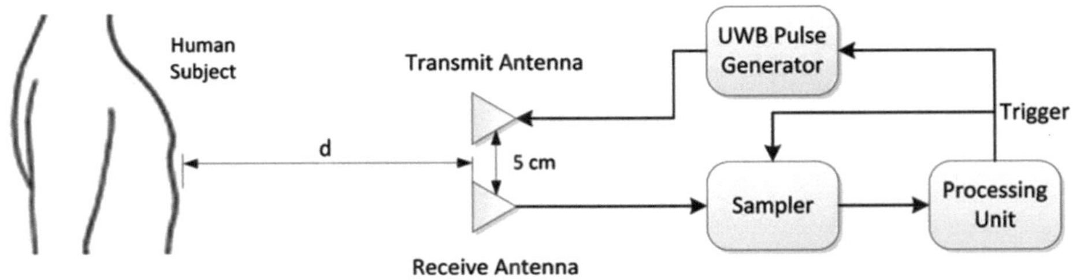

Figure 19.
Operation of the UWB sensor [15].

5. UWB in structural and civil engineering

When building and maintaining large-scale engineering structures like bridges that are subjected to time-varying loads, distance measuring is essential. The loads include temperature impacts, building loads, and operating loads like cars and pedestrians. Cracks and buckling are examples of the serious damage that can result from structural deformations.

According to the Load and Resistance Factor Design (LRFD) Bridge Design Specifications of the American Association of State Highways and Transportation Officials (AASHTO), the maximum deformation of highway bridges is restricted to 1/800 of the span length. Other specifications have similar requirements.

For example, highway bridges and high-speed railway bridges have maximum deflections of 1/600 and 1/1600 of the span length, respectively. Bridge deformation measurements reduce the likelihood of disastrous outcomes and assist in detecting anomalies early.

Regarding distance measuring, UWB sensors have been used for robot placement and navigation and are compact, inexpensive, energy-efficient, and highly weather-resistant.

Since UWB signals are acquired in the time domain, UWB technology has emerged as a new method of measuring distance. To find the distance between two UWB sensors, one uses the time of flight (TOF) of UWB signals.

Nevertheless, there are difficulties in using UWB sensors to monitor structural deformations because doing so typically calls for both high frequency and high precision. Millimeter-scale accuracy has been attained by current UWB-ranging techniques.

6. UWB in remote sensing

Gathering data about an object or phenomenon from a distance without making direct touch or physical contact is known as remote sensing.

Sensors in space or the air are used to collect data.

6.1 Microwave remote sensing

There are two forms of microwave remote sensing: passive and active. The radiation coming from the particular objects being observed is collected by the passive sensors. These radiations are either the energy that the items emit or the energy that they reflect when the sun's energies strike them. Instruments designated by radiometers gather the signal of interest and the radiation that is released.

Signals are transmitted by the active instruments, which then measure the scattered or reflected radiation from a lighted and illuminated area.

Electromagnetic radiation with frequencies ranging from less than 45 MHz to a terahertz or higher is used to measure microwave remote sensing.

For extremely sensitive low-frequency measurements, the soil moisture estimates are the most accurate. Therefore, the 1.4 GHz operating frequency is suitable for the sensors now in use in this industry.

To accurately estimate a physical parameter, some measurements must be made across a range of frequencies. For instance, in the case of water vapor absorption, observations will be made at two frequencies that surround the peak, which is around 18 and 23 GHz, and which measure the amount of water in the atmosphere.

To facilitate communication between the remote sensing community and the field of radio-frequency regulation, the IEEE Geoscience and Remote Sensing Society (GRSS) established the Frequency Allocations in Remote Sensing (FARS) Technical Committee in 2000.

6.2 Unmanned aerial vehicle (UAV)

Uses of unmanned aerial vehicle (UAV) radar include remote sensing. It has many possibilities and has been used more and more in recent years. Not only are UAVs and drones being used for military purposes, but they are also being used for precision agriculture, 3D mapping, aerial inspection, search and rescue, indoor industrial deployment, and even capturing and intercepting other UAVs. Low-power UWB

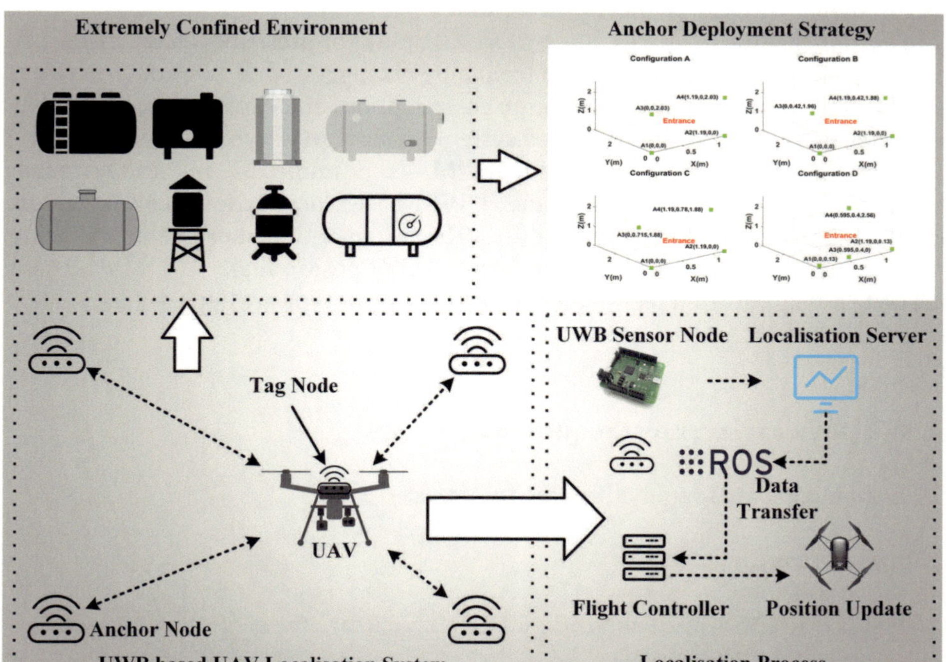

Figure 20.
UAV localization with UWB [16].

Figure 21.
UAV prototype in a lab experiment setting [16].

modules are lightweight, compact, and capable of centimeter-level range precision, making them suitable for UAV localization.

Ultra-wideband (UWB) transceivers with two-way time-of-flight (TW-ToF) measurements are used by UAVs to determine their range.

In extremely limited spaces, like tiny storage tanks or pressure vessels, ultra-wideband (UWB)-based localization technology is suggested. When constructing the UAV system, the weight, high energy consumption, and expense of comparable technologies, such as the LiDAR-based system, make it an unavoidable issue. This contradicts the use of micro-aerial vehicles (MAVs) or mini-UAVs in these extremely constrained spaces. The ultra-wideband (UWB)-based localization technology is used in severely constrained locations to address the aforementioned problems and achieve low energy consumption, low cost, and precise UAV positioning.

The UWB application techniques on the UAV prototype are depicted in **Figures 20** and **21**.

7. UWB physical layer test requirements

The following are roughly three testing areas:

7.1 Standard conformance

The IEEE 802.15.4-2020 and IEEE 802.15.4z-2020 are the subjects of the action.

1. Operating frequency bands

2. Channel assignments

3. Baseband impulse response

4. Transmit power spectral densities (PSD) mask

5. Chip rate clock and chip

6. Carrier alignment

7.2 Regulatory compliance

This is being carried out by FCC parts 15ʃ15.519, 15ʃ15.517/ETSI EN301 489-33, EN 302065, EN 303883:

1. Operating bandwidth

2. Mean power spectral density

3. Maximum value of peak power

4. Other emissions

5. Receiver spurious emissions

6. Detect and avoid (DAA)

7. Low duty cycle (LDC)

7.3 Interoperability certification

It is carried out following the CCC Consortium, UWB PHY Conformance, and FiRa Consortium:

1. Packet format

2. Power spectral density mask

3. Frequency tolerance, and timing

4. Baseband impulse response

5. NRMSE (**Figure 22**)

6. Packet reception sensitivity

7. Dirty packet tests

8. First path dynamic range

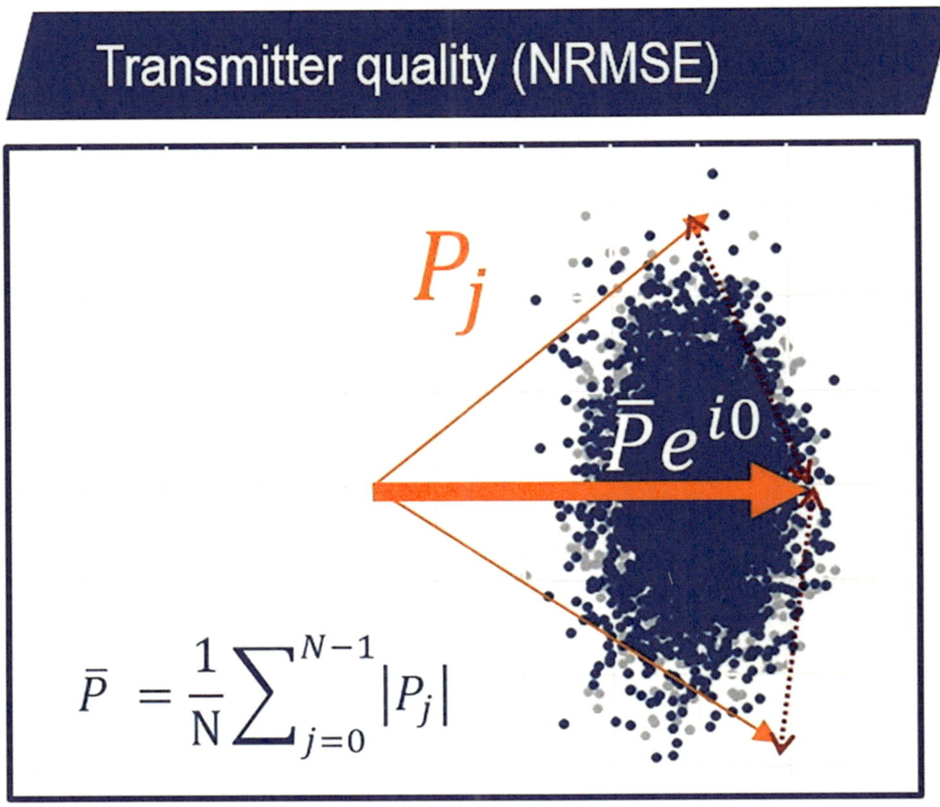

Figure 22.
Measurement of transmitter quality (NRMSE) [1].

8. Specific UWB measurements (IEEE, FiRa, etc.)

A few particular UWB metrics can be cited as examples:

1. Transmit power spectrum density (**Figure 23**)

The transmitted spectrum must be smaller than −10 dB relative to the maximum spectral density for.

$$0.65 / T_P < |f - f_c| < 0.8 / T_P$$

And −18 dB for $|f - f_c| > 0.8/T_P$

2. Impulse response (**Figure 24**)

The time unit is pulse duration (T_P), and the time domain mask should confine the pulse form by scaling the pulse's peak magnitude to a value of 1.

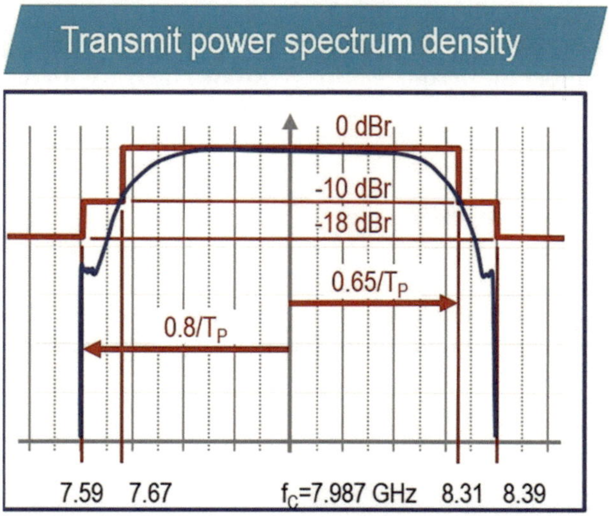

Figure 23.
Measurement of transmitted power spectrum density [1].

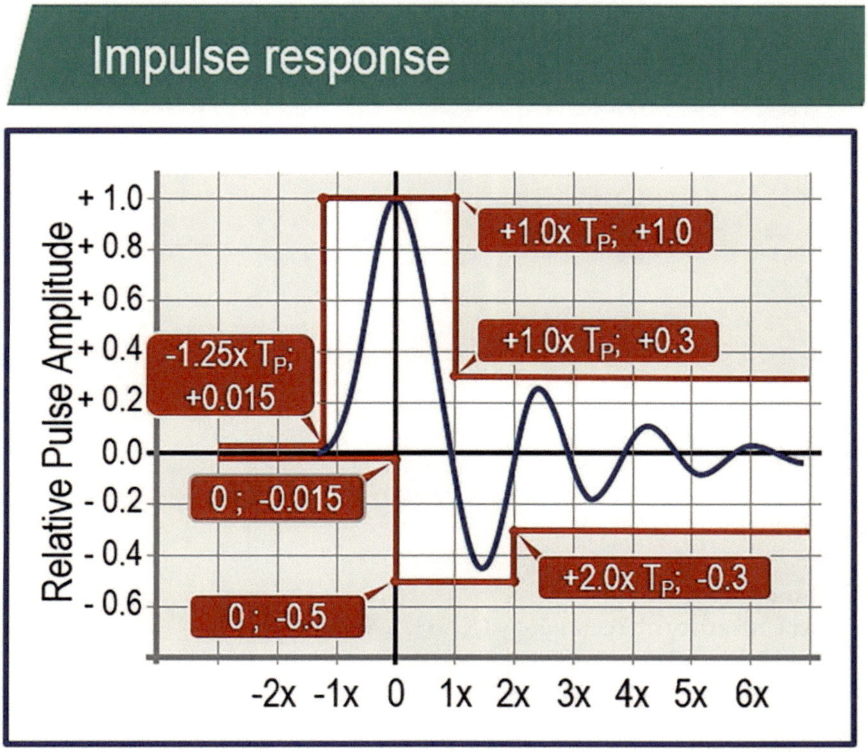

Figure 24.
Impulse response measurement [1].

3. Transmitter quality (**Figure 22**)

The normalized root mean square error (NRMSE) metric with the mean pulse amplitude $\bar{P}$ should be used to measure the quality of the transmitted signal.

$$NRMSE = \sqrt{\frac{1}{N}\sum_{j=0}^{n-1}\frac{\left|P_j - \bar{P}e^{i0}\right|^2}{\overline{P^2}}} \tag{1}$$

Lastly, it should be mentioned that, by 2030, the worldwide UWB market is expected to grow to over 1.8 billion units. UWB is a rapidly expanding technology [17].

9. Conclusions

This chapter introduced several industrial uses for ultra-wideband technology. These uses range from medicine to radar-related concepts. It was said that the exact and accurate range of this technology makes it frequently employed. The majority of measurements and standards were covered, with providing some examples. In light of this remarkable advancement in the industry, it will be a promising technology for people and customers in the future. From 2019 to the present, smartphones were also created using UWB modules.

Author details

Mahdi NoroozOliaei
K. N. Toosi University of Technology (KNTU), Tehran, Iran

*Address all correspondence to: mahdi.norooz@email.kntu.ac.ir

References

[1] Rohde & Schwarz GmbH & Co. KG, Ultra-Wideband (UWB) Technology and Testing. 1933. Available from: https://www.rohde-schwarz.com/home_48230.html [Accessed: January 01, 1983]

[2] Oliaei M, Tayarani M, Karami M. Compact microstrip Bandpass filter improved by DMS and ring resonator. Progress In Electromagnetics Research Letters. 2014;**45**:7-12. DOI: 10.2528/PIERL13121302

[3] Karami M, Sadeghzadeh RA, Oliaei M. Concentric open end rings resonator filter. In: The 36th Progress in Electromagnetics Research Symposium (PIERS)-PIERS Proceedings, Prague, Czech Republic. 6-9 July 2015. pp. 2646-2650

[4] FiRa Consortium. Why Ultra-Wideband Matters. 2019. Available from: https://www.firaconsortium.org/resource-hub/blog/why-ultrawideband-matters [Accessed: September 11, 2023]

[5] Jorge F, Soldo Y, Rosello J, Dreis M. Earth Observation Sensors in Purely Passive Bands and UWB Technologies: A Risk Analysis for Space-Borne Passive Remote Sensing Applications within the CEPT. Reading, United Kingdom: The European Space Agency; 14-18 February 2022

[6] Ramos A, Lazaro A, Girbau D, Villarino R. RFID and Wireless Sensors Using Ultra-Wideband Technology. UK: ISTE Press—Elsevier; 2016. DOI: 10.1016/C2015-0-01232-1

[7] Glover IA, Atkinson R. In: Uttamchandani D, editor. Handbook of MEMS for Wireless and Mobile Applications – Chapter 2: Overview of Wireless Techniques for Use with MEMS. 1st ed. UK: Woodhead Publishing Limited; 2013 [Hardback ISBN: 9780857092717, eBook ISBN: 9780857098610]

[8] Murata Manufacturing Co., Ltd. (Innovator in Electronics). What is Ultra-Wideband (UWB) Wireless Communication? 1944. Available from: https://article.murata.com/en-us/article/what-is-uwb-wireless-communication [Accessed: September 27, 2023]

[9] Alsath MGN, Lawrance L, Kanagasabai M. Bandwidth enhanced grid Array antenna for UWB automotive radar sensors. IEEE Transactions on Antennas and Propagation. 2015;**63**(11):5215-5219. DOI: 10.1109/TAP.2015.2478143

[10] Mohan Rao U, Verma AK, Darimireddy NK, Fofana I, Park C-W, Vedik B. Perspectives of convertors and communication aspects in automated vehicles, part 1: Convertors and condition monitoring. Energies. 2021;**14**(7):1795. DOI: 10.3390/en14061656

[11] Darimireddy NK, Mohan Rao U, Chan-Wang Park I, Fofana MS, Verma AK. Perspectives of convertors and communication aspects in automated vehicles, part 2: Printed antennas and sensors for automotive radars. Energies. 2021;**14**(6):1656. DOI: 10.3390/en14061656

[12] ECC Report 334, "UWB radiodetermination applications in the frequency range 116-260 GHz", Approved 28 January 2022, Amended 3 February 2023

[13] NoroozOliaei M, Riazi Esfahani H, Abrishamian MS. Graphene coated dielectric resonator antenna for modeling the photoreceptors at visible

spectrum. Heliyon. 2022;**8**(6):e09611. DOI: 10.1016/j.heliyon.2022.e09611

[14] NoroozOliaei M, Riazi Esfahani H, Abrishamian MS. Modeling of dielectric resonator antenna array for retina photoreceptors. Heliyon. 2023;**9**(3):e13794. DOI: 10.1016/j.heliyon.2023.e13794

[15] King Abdullah University of Science and Technology (KAUST) Innovation, License Technologies, Electrical Electronics Info & Communications [Internet]. 2009. Available from: https://innovation.kaust.edu.sa [Accessed: 2020]

[16] Yang B, Yang E, Yu L, Loeliger A. High-precision UWB-based localisation for UAV in extremely confined environments. IEEE Sensors Journal. 2022;**22**(1):1020-1029. DOI: 10.1109/JSEN.2021.3130724

[17] Sabban A, editor. Innovations in UltraWideband Technologies. London, UK: IntechOpen; 2021